Introductory

NUCLEAR REACTOR STATICS

Introductory

NUCLEAR REACTOR STATICS

Karl O. Ott   Winfred A. Bezella

Revised Edition

AMERICAN NUCLEAR SOCIETY
La Grange Park, Illinois, USA

Library of Congress Cataloging in Publication Data

Ott, Karl O. (Karl Otto), 1925–
Introductory nuclear reactor statics.

Includes bibliographies and index.
1. Nuclear reactors. I. Bezella, Winfred A.
II. Title.
QC786.5.087 1989 621.48'31 88-35131
ISBN 0-89448-033-2

ISBN: 0-89448-033-2
Library of Congress Catalog Card Number: 88-35131
ANS Order No. 350013

555 North Kensington Avenue
La Grange Park, Illinois 60525

Printed in the United States of America

Contents

PREFACE

Traditionally, nuclear fission in critical reactors or in sub- or supercritical systems has been viewed as a "chain reaction." However, in practice, the applied theoretical and computational treatment of the neutron populations in these systems has evolved along a different, more mathematical route, without explicit use of the chain reaction concept. This mathematical approach is fully reflected in advanced textbooks, but only to a lesser extent in introductory texts.

The selection of the material and the structure and style of its presentation in this textbook are guided by the desire to provide—on an introductory level—a good understanding of the way nuclear reactors are described in the mathematical theory, the basis for practical calculations. The authors believe that such an understanding is important for performing typical nuclear engineering tasks that include defining problems, selecting solution approaches, evaluating the results, and judging possible deficiencies in approaches, problems as they appear in reactor operation and development, or in the further refinement of computational methods.

The structure of the presentation results from the notion that after the introduction of the necessary descriptive concepts (for example, the various neutron fluxes in Chapter 1) and the derivation of the basic neutron balance equations (beginning of Chapter 2), one is confronted with a complicated multidimensional mathematical problem. Therefore, the main body of the text is structured along the lines of the typical approach to the solution of such problems: namely, investigating separately the dependencies along individual dimensions, thereby developing suitable approximations that finally allow an approximate treatment of the full problem.

Prior to entering in this scheme, the general features of the solutions of reactor statics problems are derived and discussed extensively in Chapter 2. This is probably done more explicitly and in greater detail than in most other textbooks. These general features are independent of the particulars of subsequent approximations; they include the introduction of reactor criticality, general features of static solutions for sub- and supercritical systems, of homogeneous versus inhomogeneous problems, and flux separation and bucklings (material and geometric) including the respective eigenvalue problems. It is felt that this chapter provides the needed foundation for the subsequent development of reactor theory.

Next the dependencies along the individual dimensions are investigated separately. Chapter 3 is devoted to the treatment of the pure space dependence. Chapters 4 and 5 deal with the energy dependencies, i.e., with the neutron spectrum. In Chapter 4, the spectrum resulting from interactions with cross sections with smooth energy dependencies is investigated and in Chapter 5, the resonance cross sections are included in the description. Chapter 6 contains a brief survey of the treatment approaches of angular dependencies. The presentation is limited to a brief survey, since transport theory is clearly beyond the scope of this text. Only the derivations of the one-group P_1 equations, the diffusion equation, and the transport correction are presented in detail. Chapter 7 then contains the multigroup diffusion theory in which information from the investigation of the individual dimensions is combined to obtain an approximate solution of the entire problem.

Some homework problems are given after each chapter. Also included are detailed review questions that not only help to guide the student through a thorough review of the material, they also point out which aspects, in the opinions of the authors, should be emphasized in a review.

Although the physically motivated and intuitively appealing concept of the "chain reaction" was not used as the basis of this text, some of the historical chain-reaction-based concepts are presented. This is done not only to provide some historical perspective but also because these concepts (for example, the components of the four-factor formula) frequently appear in a more qualitative manner in the literature.

The entire approach of presenting this material is more mathematical than textbooks normally written for the senior undergraduate and beginning graduate student. However, the mathematics is kept simple so that a senior nuclear engineering student with an average mathematical background can handle it. Some specific mathematics is presented as needed and partially summarized in the appendices. In addition, a basic knowledge of physics is important. A practical prerequisite is the introductory nuclear engineering material, especially reactor cores, etc., as the theory is concerned with the neutron behavior in these systems. Some graduate students with majors in mathematics or physics have taken the course successfully without formally fulfilling this prerequisite.

The content and order of the material in this text is consistent with a full three-hour course in reactor statics. Therefore, other areas had to be deleted, such as interaction of alpha, beta, and gamma rays with matter (hardly any specific knowledge is needed here); nuclear reactions (only the key aspects are presented as needed); nuclear fuel cycle (special texts should be consulted); and reactor kinetics or dynamics (which is the topic of a subsequent course).

The material of this reactor statics text has been taught by one of the authors (K.O.O.) and also by Dr. Owen Gailar at Purdue University for many years. It has been developed into its present form through interaction with several colleagues and with many students, considering their reactions and suggestions. Students who have contributed in various ways include: Bob Borg, Bob Burns, Charles Fraime, Nelson Hanan, Ken Koch, Fred Krauss, Larry Luck, Pat McDaniel, Don Malloy, and Paul Maudlin. Their input and assistance is gratefully acknowledged, as well as Dr. Gailar's beneficial suggestions.

The authors also wish to thank Prof. R. L. Murray for his thoughtful and detailed review of the manuscript and his many valuable suggestions that helped to considerably improve the manuscript. Special thanks are due to Prof. P. S. Lykoudis, head of the School of Nuclear Engineering at Purdue University, for his encouragement and support. Also the help of the Computer Science Center of Purdue University in running the programs for many of the figures in this text is gratefully acknowledged. We thank M. H. Feldman of the Computer-Aided Design and Graphics Laboratory at Purdue University for the computer graphics of the jacket design that he calculated with a program developed during research for his MS thesis; the design is based on a flux reconstruction by K. R. Koch, using a two-group calculation by O. H. Gailar, also of Purdue University. In addition, the authors have incurred a debt to others who helped in writing this book; to the ANS publishing staff and in particular the competence of Lorretta Palagi in editing a difficult manuscript; and also to the secretaries, Mmes. Georgia Ehrman and Linda Dallinger, who patiently typed the text in its several versions.

Finally, heartfelt thanks are due the authors' wives, who contributed indirectly in every phase of this book's preparation by providing their time and support.

The second edition has been improved by numerous revisions throughout the text. These include an early introduction of the multigroup equations, still deferring the group constant definition and generation to Chapter 7, after the treatment of the neutron weighting spectra.

Four smaller sections have been added: The flux separation and the resulting concepts have been demonstrated by an explicit treatment on the two-group level, yielding the material buckling, B_m^2, the higher B^2 eigenvalue, and the associated two-group eigenspectra (Sec. 2-7). The application of these concepts to an analytical solution of the two-group diffusion for one-dimensional slab geometry is given in Sec. 3-6. We gratefully acknowledge the help of Nasir Mirza, who provided all the numerical and graphical information for these two sections.

At the end of Chapter 7, a short section is devoted to the current most widely used computer codes for flux calculations, applying multigroup constants for the energy and finite differences for the spatial representation. This listing includes the two major group constant generation codes applied in the United States as well as the more widely used diffusion and transport (S_n) theory programs. We greatly appreciate the help of Joel Rhodes for providing the information for this subsection.

Furthermore, a section on nodal methods has been added. In the writing of this section, we benefited greatly from Dr. Tom Downar's experience and are grateful for his advice. We also wish to acknowledge the help of Chi-Wen Tsaoi and Chuck Wemple in the proofing of this revision.

Karl O. Ott
Purdue University

Winfred A. Bezella
Argonne National Laboratory

January 1989

One

BASIC NEUTRON PHYSICS CONCEPTS

1-1 Introduction

A nuclear reactor is an assembly in which nuclear energy is generated by the self-sustaining chain reaction between neutrons and fissionable or fissile nuclei. One can distinguish between "fissionable" and "fissile" nuclei: "Fissionable" is the general term indicating that a nucleus can be fissioned, while "fissile" nuclei are those that can be fissioned by neutrons of all energies, especially by "thermal" neutrons. The major fissile nuclei that are fissioned in uranium-fueled thermal reactors are ^{235}U and ^{239}Pu. Nuclei that can be fissioned only by neutrons with an energy above a certain threshold value are fissionable, but not fissile; examples are ^{238}U and ^{232}Th. It should be noted that although ^{238}U and ^{232}Th are not fissile, they can, upon absorption of a neutron and several subsequent beta particle decays, be transmuted into other fissile species; in this case, ^{239}Pu and ^{233}U, respectively. Nuclides that can produce fissile nuclei are called "fertile" with the transformation process generally referred to as—depending on the fissile production efficiency—either "conversion" or "breeding."

The efficient and safe operation of a nuclear chain reaction represents the major challenge to the nuclear engineer. An understanding of the physical processes of a nuclear reactor is essential to this task. To provide this understanding of the nuclear physical processes and to develop computational methods for the prediction of these processes, "nuclear reactor theory" or the theory of "reactor physics" is studied.

For the prediction of the performance of a nuclear reactor, the behavior of the neutrons and their interactions with the nuclei of the reactor materials are of primary importance. The probability of these interactions is expressed in terms of "cross sections." The needed cross sections and other relevant nuclear data are largely obtained experimentally, with nuclear theory essentially playing a supporting and completing role. Basic to the theoretical description are the theories and laws of nuclear physics relating to interactions between neutrons and nuclei. For detailed discussions and descriptions of these processes, nuclear physics texts should be consulted, for example, see Refs. 1, 2, and 3. The nuclear physics input into reactor theory is described here only as far as it is needed for development of the theory. In general, nuclear theory will be discussed in this text where and when it is required for the understanding of reactor physics concepts and methods. The basic concepts, such as cross sections and some other important data, are presented in this chapter.

Before presenting the basic reactor physics concepts, a digression is in order: In the next section, the general relationship between reactor physics, design, and reactor engineering is discussed.

1-2 Engineering Requirements of Reactor Physics

The reactor design engineer or physicist is called upon to provide the essential physics information required in reactor design. Several characteristics of the reactor must be established to ensure that the reactor can be started up, operated at a steady state over its entire operating lifetime, and shut down as needed in a safe and economic manner. Some of the parameters and phenomena, which must be predicted with sufficient accuracy, are listed in order to demonstrate the variety of information to be provided by the nuclear engineer:

1. the amount of fuel (critical mass) or size of assembly (critical size) required for steady-state reactor operation
2. the spatial distribution of the power generation throughout the operational period
3. the type, the amounts, and the positioning of materials within the core as required for control of the nuclear reactor at all times
4. the effects of temperature, thermal expansions, and other density changes on the reactor performance
5. the neutron reaction rates leading to fuel "burnup" and "buildup" of fissile material through conversion
6. the production and further transformation of fission products as a function of core lifetime
7. the economic performance, which depends strongly on prediction of an optimum fuel reloading to achieve an optimum power distribution
8. the time-dependent physics parameters required for effective reactor control and safe shutdown

9. the time-dependent behavior of the spatial power distribution during off-normal and accident situations in order to ensure the safety of the reactor operating personnel and the general public
10. the reaction rates leading to material damage (e.g., fuel and steel swelling) and the effect of radiation on nonnuclear phenomena (e.g., radiation-enhanced creep)
11. the radiation levels and activity buildup (e.g., in coolant)—information that is necessary for the design of radiation shields and for waste management control.

The relative importance of the prediction of these and other neutron reactor parameters varies, of course, with the particular type of reactor system: High accuracy is required for commercial power reactors; less accuracy for most experimental reactors.

Nearly all nuclear parameters and phenomena of interest in reactor design, operation, and control are based on neutron reactions with nuclei in the reactor. This becomes evident by inspecting the above list of parameters and phenomena, which are briefly discussed as follows.

The *critical mass* of fuel depends on the balance of neutron fission reactions with absorption reactions and neutron leakage. The leakage is strongly influenced by the neutron scattering reactions.

The *power distribution* is directly related to the fission rate. Most of the energy is deposited in close proximity to the fission event by slowing down of the two fission products. Part of the energy ($\simeq 15\%$) is deposited in a larger volume since it involves migration of neutrons and gamma quanta over some distance from the originating fission event.

Reactor control is primarily concerned with the variation of neutron capture in special devices (control rods); the variation is achieved, for example, by changing the depth of insertion of a control rod. To achieve this control in a minimal space, special materials with high neutron capture cross sections are chosen as control materials (e.g., cadmium, hafnium, boron, and ^{10}B). High concentrations of strong absorbers represent sudden and strong "sinks" for neutrons; they cause strong perturbations in the neutron field. The theoretical treatment of concentrated strong sinks requires the development and application of special methods.

The fission rate is the basis for power production. From the power production rate at various points in the reactor and from its thermodynamic and fluid dynamic characteristics, one can calculate the temperatures in the fuel, cladding, and coolant together with the coolant velocities. This combined treatment is a special subject often called "reactor thermohydraulics." The resulting temperatures can have an impact on the neutron reactions (e.g., in the fuel), or they can, through density changes, impact on all reaction rate densities. This is the basis of what is called "feedback," which plays a predominant role in the control and stability of the reactor. Also, control materials can suffer long-term changes. The long-term variation of control materials is sometimes employed (as "burnable poison") in pressurized water reactors to meet the desired long-term operational goal (fuel burnup guarantee).

The generation rate of *fission products* (FPs) is obviously equal to the fission rate. The further transformation of the FPs is determined by the interplay of the nuclear transmutations through beta decay and neutron capture. Part of the FPs have high neutron cross sections; they are therefore called neutron "poisons." In addition, FPs, especially the gaseous (krypton, xenon, and bromine) and the volatile ones (e.g., iodine and cesium), migrate within the fuel; hence their spatial distribution depends on time. Fission product decay and migration are examples of phenomena not directly related to neutron reactions.

Long-term effects obviously depend on the intensity of the neutron population, which is different at different points in the reactor. Therefore, the reactor composition has a long-term variation and the fuel reloading scheme must be based on the anticipation of these variations and of their impact on the nuclear chain reaction and the spatial power distribution.

Steady-state operation is controlled by several properly positioned rods. If steady-state operation were to be disturbed by some kind of an incident, often related to some gradual or sudden failure (e.g., coolant flow restriction), the "shutdown" or "scram" system will quickly reduce the power by inserting the control rods into the fueled core region, which shuts the reactor down. It is possible, however, that the power could increase before the shutdown becomes effective. This leads to a power "transient," which is influenced by a number of *time-dependent physics parameters*.

The possible power transients must be carefully investigated to assure effective shutdown. Every technical system has a certain probability of failing to respond properly and although the probability of these events is very small in the case of nuclear reactors, a power transient or other type of accident in the core can lead to major damage and possibly to the release of radioactivity. Therefore, such possibilities must be investigated. The description of transient behavior of the neutron flux is the subject of reactor dynamics, which is also part of nuclear reactor theory.

Neutrons emerging from fission have fairly high energies (mostly between 0.5 and 3 MeV). The neutron energy, E, is subsequently reduced; the neutrons are "slowed down." In the largest part of the energy range (for $E \gtrsim 30$ eV), neutrons are so energetic that some nuclei that suffer a collision with a neutron may be knocked out of their position, leading to material damage. These "dislocations" can perturb the crystal lattice structure of materials (e.g., in stainless steel) or separate a molecule (e.g., by knocking out the oxygen nucleus in an H_20 molecule). The original damage is generally reduced by "healing" of dislocations and recombination of atoms to molecules.

Various radiation intensities resulting from buildup of radioactivity must be calculated: The coolant in the core becomes radioactive; fuel elements are unloaded and transported; fuel is reprocessed and refabricated into new fuel elements. Thus, radiation of various levels is emitted during the entire fuel cycle. The residue of the reprocessed fuel contains FPs and some traces of fuel. This residue, often called "radioactive waste," must be stored or disposed of properly.

Before leaving the general discussion of engineering requirements (and thus of the purpose and objectives of nuclear reactor theory) in order to embark on the basic reactor physics concepts, it appears appropriate to say a few words about the relationship, or interrelationship, between the physics and engineering aspects of reactor design. As seen from the preceding list of parameters, which transcends merely the physics aspects, others are based on a combination of nuclear physics, thermohydraulics, and various design features. To calculate many of these parameters of interest, the wide spectrum of engineering disciplines must be brought into play. An interplay between nuclear physics analysis and thermohydraulic analysis is required, for example, in order to obtain temperature distributions. Other examples of close interaction and feedback between the reactor physicist and mechanical engineer, metallurgist, control engineer, chemical engineer, and health physicist can be seen from a close examination of the engineering requirements of reactor physics presented above. It therefore must be kept in mind that the physics of reactor theory relies on, is supported by, and interacts with a wide range of engineering disciplines. The engineering aspects of reactors are not treated in this book; the concentration is almost exclusively on the physics of reactor theory. For a discussion of the engineering aspects of nuclear reactor design, several excellent texts and introductory books are available, such as Refs. 4, 5, and 6.

1-3 The Fission Process

The central physical process underlying the device called a "nuclear reactor" is the fission reaction, in which a neutron splits a fissionable nucleus into two parts yielding additional neutrons. Liberation of energy is also a direct result of this fission process and this is of course the primary motivation for building nuclear reactors.[a] The two parts into which a nucleus is split in a fission reaction, the so-called "fission products," are radioactive. This process can be considered to consist of two steps. In the first, a heavy isotope X of mass A_0 and charge Z_0 absorbs a neutron n:

$$n + {}^{A_0}_{Z_0}X \longrightarrow {}^{A_0+1}_{Z_0}X^* \text{ (excited nucleus)} \quad , \tag{1.1}$$

forming an "excited" nucleus ${}^{A_0+1}_{Z_0}X^*$, which may subsequently split into two isotopes (FPs), and release energy and ν additional neutrons:

$${}^{A_0+1}_{Z_0}X^* \longrightarrow {}^{A_1}_{Z_1}X + {}^{A_2}_{Z_2}X + \nu n + \text{energy} \quad , \tag{1.2}$$

where

$$A_0 = (\nu - 1) + A_1 + A_2$$

and

$$Z_0 = Z_1 + Z_2 \quad .$$

Other decay schemes of ${}^{A+1}_{Z}X^*$ consist of neutron reemission or of gamma emission; these are the neutron scattering and capture reactions, respectively. The fission process described by Eq. (1.2) is of primary interest. The average number of neutrons *per fission*, ν, ranges between a value of 2 and 3 for most commonly used fissionable isotopes, although up to 6 neutrons can be released during any individual fissioning event. The precise value of ν is obtained experimentally; it depends on the specific fissioning isotope. The value of ν depends also on the energy of the fission-initiating neutron; it increases by 10 to 30%, largely linearly, over an energy range of several MeV.

An important parameter, proportional to ν, is the average number of neutrons emitted per neutron absorbed; it is generally denoted by η. Since the parameter η incorporates or accounts for the competition of the fission process with other events in describing the decay of isotope ${}^{A_0+1}_{Z_0}X^*$ in Eq. (1.1), it is a function of neutron energy. Table 1-I shows

---

[a]Other purposes include isotope production (both fissionable isotopes as well as medicinal isotopes), propulsion, direct electrical power generation, testing of materials, safety testing, as well as general research.

TABLE 1-I
Neutrons Emitted from Fissile and Fissionable Isotopes*

Isotope	Slow Neutrons (E = 0.025 eV)		Fast Neutrons (E = 2 MeV)	
	ν	η	ν	η
^{235}U	2.43	2.08	2.6	2.4
^{239}Pu	2.89	2.03	3.1	2.9
^{233}U	2.50	2.31	2.6	2.5
^{238}U	0	0	2.6	2.4
^{232}Th	0	0	2.2	~1.4

*Source: Evaluated Nuclear Data File ENDF-IV.

typical values of ν and η for several common fissionable isotopes for two initiating neutron energy values.

An essential feature of the fission process illustrated by Eqs. (1.1) and (1.2) is the release of energy, which is ~200 MeV per event. The two generally unequal fission fragments, $^{A_1}_{Z_1}X$ and $^{A_2}_{Z_2}X$, possess energy in the form of kinetic energy as well as in the form of radioactivity. Since the fragments are relatively massive, their range of travel is short and they transfer their kinetic energy into heat within the fuel rods whence they were born. Both beta and gamma-ray emissions are due to radioactive decay of FPs. An approximate breakdown of the energy release is shown in Table 1-II. Note that the 11 MeV of neutrino energy is unavailable because of its small reaction cross sections.

TABLE 1-II
Typical Energy Release From Fission

Particles	Energy/Event (MeV)
FPs (kinetic energy)	167
Fission neutrons	5
Neutrinos (escape from reactor)	11
Prompt gamma rays	7
Beta particles from FP decay	5
Gamma rays from FP decay	6
TOTAL (from fission)	201
Gamma energy from capture of ~1.5 neutrons	~10
TOTAL	211

Also of importance in the fission process is the energy of the neutrons, both of the initiating neutron as well as of the emitted fission neutrons. The fission reaction probability (as we will see) is quantitatively described by a physical parameter called a "cross section," which depends strongly on the neutron energy. The slower the neutrons the better they are at causing fission in fissile nuclei. Therefore, in the present types of power reactors, the neutron energy is reduced as much as possible, i.e., until it becomes comparable to the thermal energy of the nuclei in the core. One then has so-called "thermal neutrons." Fission in nonfissile but fissionable isotopes (e.g., ^{238}U) has an even stronger energy dependence through the existence of a fission threshold. Figure 1-1 illustrates the fission cross sections for ^{235}U and ^{238}U. The peaks in the cross sections (only a very few are indicated) result from "resonance" effects, which are discussed in Chapter 5. Table 1-III gives typical fission and capture cross sections for thermal neutrons.

TABLE 1-III

Typical Thermal Neutron Fission and Capture Cross Sections for Some Reactor Materials ($E = 0.025$ eV)*

	$\sigma_f(b)$	$\sigma_c(b)$
^{232}Th	0.00004	7.40
^{233}U	531	47.1
^{234}U	< 0.65	100
^{235}U	582	98.6
^{236}U	0	5.2
^{238}U	0	2.70
^{238}Pu	16.5	547
^{239}Pu	742	269
^{240}Pu	0	290
^{241}Pu	1009	368
^{242}Pu	< 0.2	18.5
Hydrogen	0	0.33
Deuterium	0	0.00053
Helium	0	0
Oxygen	0	0.00027
Carbon	0	0.0034
Iron	0	2.55
Nickel	0	4.43
Cadmium	0	2450
Boron	0	759
^{10}B	0	3837

*Source: S. F. Mughabghab and D. I. Garber, "Resonance Parameters," Vol. 1 of *Neutron Cross Sections*, 3rd ed., BNL 325, Brookhaven National Lab. (June 1973).

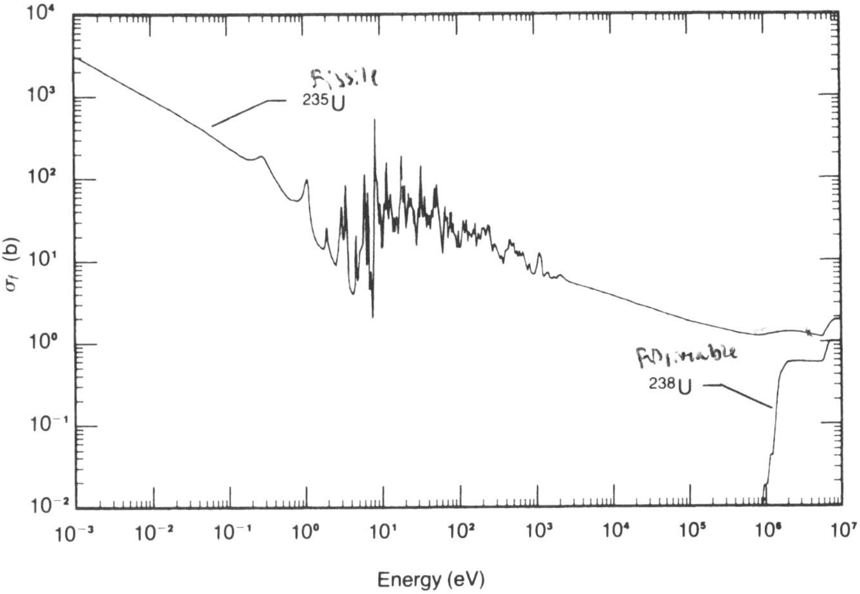

Fig. 1-1. Fission cross sections of ^{235}U and ^{238}U as a function of the neutron energy in electron volts. [Source: S. F. Mughabghab and D. I. Garber, "Resonance Parameters," Vol. 1 of *Neutron Cross Sections*, 3rd ed., BNL 325, Brookhaven National Lab. (June 1973).

The energy distribution of neutrons, $\chi(E)$, emitted during the fissioning event is illustrated in Fig. 1-2 for ^{235}U. Fission neutrons are emitted in the megaelectron volt range (mean energy of ~2 MeV).

Fissions in thermal reactors are primarily induced by thermal neutrons. This dichotomy in energy, between the production at high energies and the fissioning at thermal energies, requires that light mass materials called "moderators" be employed to slow the fast neutrons down to thermal energies. This is not to say that reactors cannot operate with predominantly high-energy neutrons. In fact, in reactors of this type, called "fast" reactors, one goes to great lengths to prevent neutron moderation. Commercial reactors employ light water, heavy water, or graphite as moderators—the pressurized water reactor and the boiling water reactor (BWR) as well as gas-cooled reactors. A typical gas-cooled reactor is the high-temperature gas-cooled reactor, which employs graphite as the moderator with helium serving as the coolant. The liquid-metal fast breeder reactor, using liquid-metal sodium as the heat-removing medium, is still in the developmental stage.

For further details about the fission process as it relates to nuclear

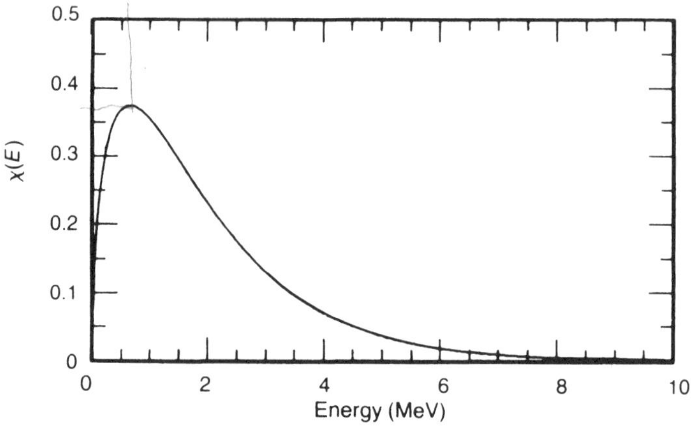

(a)

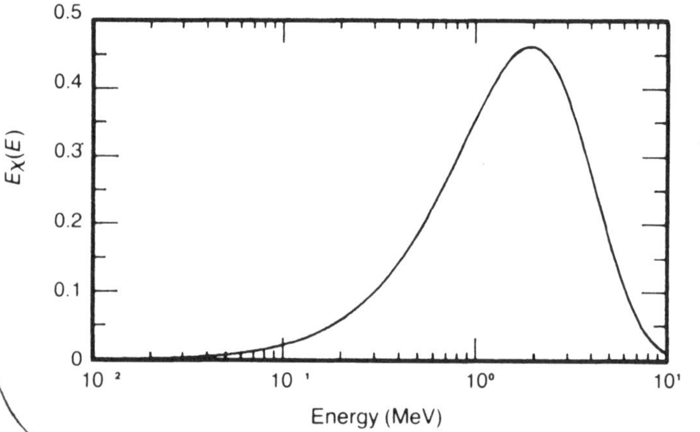

(b)

Fig. 1-2. Energy distribution of fission neutrons emerging from ^{235}U after fission by thermal neutrons: (a) $\chi(E)$ as a function of E on a linear scale and (b) $E\chi(E)$ as a function of log E. Note that $E\chi(E)$ needs to be plotted so that the ordinate times $d$log E represents the contribution to the integral of χ.

reactor theory, several older texts are available that could be consulted.[7–9]

1-4 Fluxes, Currents, and Sources

The basic conceptual quantity used to describe the average behavior of a neutron population in a reactor or in any other neutronic system is the neutron "flux." One must distinguish between various kinds of neutron fluxes, depending on the degree of detail needed in the description of a particular problem. In this book, the Greek symbol ϕ is used to denote all types of neutron fluxes; the respective variables are given in parentheses. The reader is cautioned, however, that the literature occasionally contains other symbols for the various types of neutron fluxes (for example, see Table 1.1 of Ref. 8).

The most general description of a neutron population includes a description of its *stochastic variation* about an average value by using probabilities to describe the deviations from the average value. The description as a statistical phenomenon is only required for the analysis of special experiments, and not for normal design, fuel cycle, or safety calculations. It is therefore not discussed further in this text.

The most detailed description of the average neutron flux contains all the seven variables that the average flux can depend on, including the two angular variables[b]; this is the *time-dependent angular flux*, $\phi(\boldsymbol{r},E,\boldsymbol{\Omega},t)$. The corresponding *time-independent angular flux* is denoted by $\phi(\boldsymbol{r},E,\boldsymbol{\Omega})$. The dimension[c] of the angular flux is

$$\text{dimension of } \phi(\boldsymbol{r},E,\boldsymbol{\Omega}) = \left[\frac{\text{neutrons}}{\text{cm}^2\text{s } U(E)\ U(\Omega)}\right] \quad , \tag{1.3}$$

with $U(E)$ and $U(\Omega)$ being the units of energy and solid angle, respectively. The verbal definition of the angular flux is formulated as:

> $\phi(\boldsymbol{r},E,\boldsymbol{\Omega})\, dE\, d\Omega$ is the number of neutrons passing at $\boldsymbol{r}$ through an area of 1 cm^2, perpendicular to $\boldsymbol{\Omega}$, per second with an energy in the interval dE at E and a direction in the interval $d\Omega$ at $\boldsymbol{\Omega}$.

Figure 1-3 illustrates the concept of the angular flux.

The neutron flux can be separated into or is composed of two more basic quantities, the neutron density, $n(\boldsymbol{r},E,\boldsymbol{\Omega})$, and neutron velocity, v:

$$\phi(\boldsymbol{r},E,\boldsymbol{\Omega}) = vn(\boldsymbol{r},E,\boldsymbol{\Omega}) \left[\frac{\text{cm}}{\text{s}}\right]\left[\frac{\text{neutrons}}{\text{cm}^3\ U(E)\ U(\Omega)}\right] \quad . \tag{1.4}$$

Note that v is the scalar neutron velocity and not the velocity vector, $\boldsymbol{v}$, which is given by:

$$\boldsymbol{v} = v\boldsymbol{\Omega} \quad . \tag{1.5}$$

---

[b]See Eq. (1.12) for a description of the angular direction $\boldsymbol{\Omega}$.

[c]The statement of the dimensions will often be augmented by additional information, which may be called "pseudo-dimensions"; for example, "neutrons" in the above dimensions.

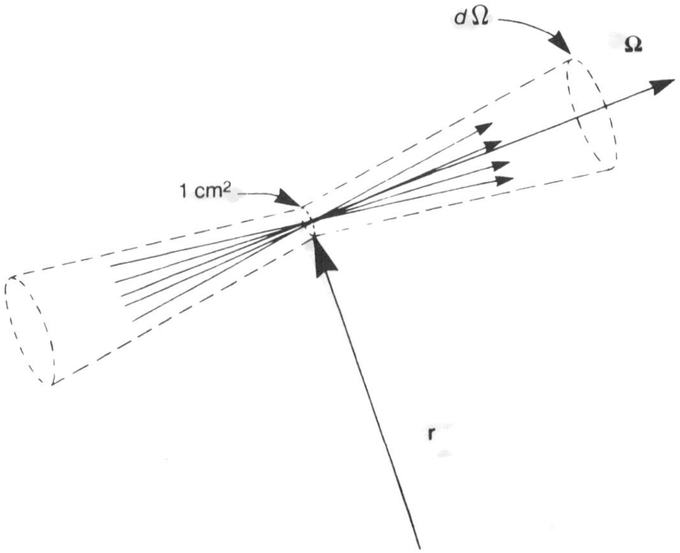

Fig. 1-3. Illustration of the angular flux, $\phi(\boldsymbol{r}, E, \boldsymbol{\Omega})$, at $\boldsymbol{r}$ in direction $\boldsymbol{\Omega}$ in $d\Omega$.

In some older texts, the neutron density is multiplied with the velocity vector and the resulting quantity is called "vector flux."[10] This concept has since been abandoned and is mentioned only to emphasize that the angular flux is a scalar and not a vector. Abandoning the concept of a vector flux allows one to also drop the term "scalar flux" since this distinction is no longer needed.

There are cases and features for which the detailed neutron angular dependence is unimportant. In particular, the reaction rates result from the combined effect of neutrons coming from all directions. For all these applications, one integrates the angular flux over all directions and obtains the most important concept of reactor physics, the *neutron flux:*

$$\phi(\boldsymbol{r},E) = \int_{\Omega} \phi(\boldsymbol{r},E,\boldsymbol{\Omega})\, d\Omega \quad , \tag{1.6}$$

with $d\Omega = d\alpha \,\sin\vartheta \, d\vartheta$; see Eq. (1.12) for the coordinate representation of $\boldsymbol{\Omega}$.

The dimensions of the neutron flux, sometimes merely referred to as "flux," follow directly from Eqs. (1.3) and (1.6):

$$\text{dimensions of } \phi(\boldsymbol{r},E) = \left[\frac{\text{neutrons}}{\text{cm}^2\text{s}\; U(E)}\right] \quad . \tag{1.7}$$

Note that $\phi(\boldsymbol{r},E)$ is the *integrated and not the average* angular flux. The fact that no angular variable appears in the argument of the flux does not mean that the angular distribution of the neutrons is isotropic; it only means that the angular dependence has been eliminated through integration. In the special case of an isotropic angular flux, one has:

$$\phi(\boldsymbol{r},E,\boldsymbol{\Omega}) = \frac{1}{4\pi}\phi(\boldsymbol{r},E) \quad . \tag{1.8}$$

The neutron flux may be physically understood as the number of neutrons of energy E, in the interval dE, that penetrates a sphere of a 1-cm^2 cross section, located at $\boldsymbol{r}$, per second. Figure 1-4 illustrates the definition of the neutron flux as the integrated angular flux. Note that the angular flux for neutrons moving to the right or left of Fig. 1-4 is different so as to indicate a dependency of $\phi(\boldsymbol{r},E,\boldsymbol{\Omega})$ on $\boldsymbol{\Omega}$.

The neutron flux is the key concept for the calculation of the reaction rates between neutrons and nuclei; this is the fundamental mission

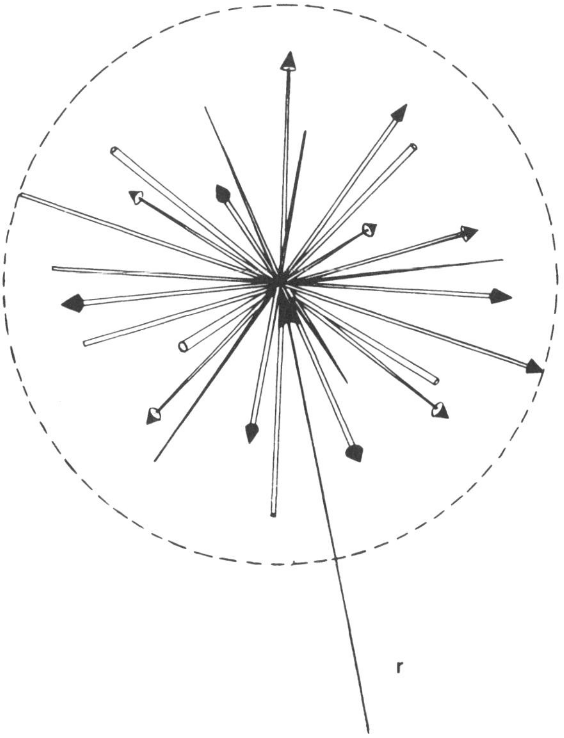

Fig. 1-4. Illustration of the flux, $\phi(\boldsymbol{r},E)$, at $\boldsymbol{r}$.

in reactor analysis. An interpretation of the flux, different from the "integral" interpretation of Eq. (1.6), is the interpretation as "total track length" per unit volume and per unit time. This interpretation is often useful in the context of reaction rates.

In devising approaches for the practical calculation of the neutron flux, one often does not need to consider the detailed energy dependence. One then integrates with respect to energy and indicates the range of integration by a descriptive adjective;

$$\int_0^\infty \phi(\mathbf{r},E)\,dE = \phi(\mathbf{r}) = \text{“total” neutron flux} \left[\frac{\text{neutrons}}{\text{cm}^2\text{s}}\right] ,$$

$$\int_0^{E_{th}} \phi(\mathbf{r},E)\,dE = \phi_{th}(\mathbf{r}) = \text{“thermal” neutron flux} \left[\frac{\text{neutrons}}{\text{cm}^2\text{s}}\right] ,$$

$$\int_{E_{th}}^\infty \phi(\mathbf{r},E)\,dE = \phi_f(\mathbf{r}) = \text{“fast” neutron flux} \left[\frac{\text{neutrons}}{\text{cm}^2\text{s}}\right] ,$$

and

$$\int_{E_g}^{E_{g-1}} \phi(\mathbf{r},E)\,dE = \phi_g(\mathbf{r}) = \text{“group g” neutron flux} \left[\frac{\text{neutrons}}{\text{cm}^2\text{s}}\right] , \quad (1.9)$$

where E_{th} denotes the upper end of the thermal neutron range; E_g and E_{g-1} give the lower and upper limits of "group g."

There are changes in the neutron population not only through reactions but also through the motion of the neutrons themselves. The "net motion" of neutrons is described by the *neutron current*, which is defined as the integral of $\mathbf{\Omega}$ times the angular flux:

$$\mathbf{J}(\mathbf{r},E) = \int_\Omega \mathbf{\Omega}\phi(\mathbf{r},E,\mathbf{\Omega})\,d\Omega = \text{neutron current} \left[\frac{\text{neutrons}}{\text{cm}^2\text{s}\,U(E)}\right] . \quad (1.10)$$

The fact that $\mathbf{J}$ describes the "net" motion can be illustrated by rewriting Eq. (1.10) as the integral of the difference between the angular fluxes of neutrons in the $\mathbf{\Omega}$ and the $-\mathbf{\Omega}$ directions, extended only over half the solid angle:

$$\mathbf{J}(\mathbf{r},E) = \int_{h(\Omega)} \mathbf{\Omega}[\phi(\mathbf{r},E,\mathbf{\Omega}) - \phi(\mathbf{r},E,-\mathbf{\Omega})]\,d\Omega . \quad (1.11)$$

Here, $h(\Omega)$ denotes half of the solid angular space. Thus, only the differences of the angular fluxes of opposite directions contribute to the current, and $\mathbf{J}$ is the integral of these differences. This indicates the nature of $\boldsymbol{J}$ as the basic concept for the description of the net motion.

The current $\boldsymbol{J}$ is a vector pointing in the direction of the net (neutron) motion. Thus, Eq. (1.10) actually consists of three equations, one for each of the components of the vector $\boldsymbol{J}$. The vector character of the right side is introduced by $\boldsymbol{\Omega}$; the three components, Ω_x, Ω_y, and Ω_z, are given for a standard spherical coordinate system (see Appendix A):

$$\boldsymbol{\Omega} = \begin{pmatrix} \Omega_x \\ \Omega_y \\ \Omega_z \end{pmatrix} = \begin{pmatrix} \sin\vartheta \cos\alpha \\ \sin\vartheta \sin\alpha \\ \cos\vartheta \end{pmatrix} \quad , \tag{1.12}$$

with ϑ being the angular distance from the z axis, and α the azimuthal angle. Inserting Eq. (1.12) into Eq. (1.10) gives the three components of the current:

$$\boldsymbol{J} = \begin{pmatrix} J_x \\ J_y \\ J_z \end{pmatrix} = \int_{\Omega} \begin{pmatrix} \Omega_x \\ \Omega_y \\ \Omega_z \end{pmatrix} \phi(\boldsymbol{r},E,\boldsymbol{\Omega})\, d\Omega \quad . \tag{1.13}$$

In physics, the quantity defined by Eq. (1.13) is normally called the "current density," and is usually denoted by $\boldsymbol{j}$. The "current," then, describes the net motion through a given area, $\boldsymbol{S}$, different from the unit area. It is calculated by integration over the current density:

$$\text{current through } S = \int_S \boldsymbol{j}\cdot d\boldsymbol{S} \quad . \tag{1.14}$$

The integration of Eq. (1.14) contains the "dot product" between $\boldsymbol{j}$ and the unit vector perpendicular to the surface element $d\boldsymbol{S}$. This is needed to account for the different contributions of $\boldsymbol{j}$ depending on its relative direction to $d\boldsymbol{S}$. Apparently the current through $d\boldsymbol{S}$ is largest when $d\boldsymbol{S}$ is perpendicular to $\boldsymbol{j}$; it decreases when $d\boldsymbol{S}$ becomes more and more parallel to $\boldsymbol{j}$. This decrease is described by the dot product. These concepts can be used in the same way for neutrons. However, the area integral is not normally employed in reactor physics. Hence, one replaces the term "current density" by the simpler term "current" even though $\boldsymbol{J}$ in Eq. (1.13) corresponds to the current density in physics.

The detailed energy dependence of the current can be reduced in the same way as for the flux. The same adjectives are applied to denote the energy integrated currents, e.g.:

$$\int_0^{\infty} \boldsymbol{J}(\boldsymbol{r},E)\, dE = \boldsymbol{J}(\boldsymbol{r}) = \text{"total" neutron current} \left[\frac{\text{neutrons}}{\text{cm}^2\text{s}}\right] \quad , \tag{1.15}$$

$$\int_{E_g}^{E_{g-1}} \boldsymbol{J}(\boldsymbol{r},E)\, dE = \boldsymbol{J}_g(\boldsymbol{r}) = \text{“group } g\text{” neutron current} \left[\frac{\text{neutrons}}{\text{cm}^2\text{s}}\right] \quad ,$$

and similarly for the thermal and fast neutron currents.

Note that neutron fluxes are defined as certain integrals of the angular flux. Since the angular flux is not needed for reaction rate calculations, one devises methods for the calculation of neutron flux and current without using the angular flux as a computational basis (see Chapter 2). But these methods yield only approximate fluxes—for the very reason that they avoid using the angular flux explicitly. However, in many cases, the directly calculated fluxes are sufficiently accurate and no appreciable error is introduced in their use.

Three types of neutron sources are of importance in reactor problems:

1. independent sources (i.e., sources with an intensity independent of the neutron flux; independent sources are often located outside of the reactor and are therefore frequently called “external” sources)
2. flux-dependent sources [the major one is the fission neutron source; the two minor ones result from $(n,2n)$ and from (γ,n) reactions]
3. delayed neutron sources (they depend on the flux history through the production of “precursors” at earlier times and their subsequent radioactive decay, followed by an instantaneous release of a neutron).

The following notation is introduced to denote the source densities (a time dependency is omitted as for fluxes and currents above):

$$S(\boldsymbol{r},E,\boldsymbol{\Omega})\, dE\, d\Omega = \text{independent source} \left[\frac{\text{neutrons}}{\text{cm}^3\text{s}}\right] \text{in } dE\, d\Omega \quad ,$$

$$S_f(\boldsymbol{r},E)\, dE = \text{fission neutron source} \left[\frac{\text{neutrons}}{\text{cm}^3\text{s}}\right] \text{in } dE \quad ,$$

and

$$S_d(\boldsymbol{r},E)\, dE = \text{delayed neutron source} \left[\frac{\text{neutrons}}{\text{cm}^3\text{s}}\right] \text{in } dE \quad . \qquad (1.16)$$

The angular and energy dependencies may be reduced through integration in the same way as for the flux and current. Fission and delayed neutron sources are isotropic.

1-5 Microscopic Cross Sections and Other Nuclear Data

Microscopic cross sections represent a measure of the probability of reactions between reaction centers (e.g., nuclei) and a beam of particles (e.g., neutrons) passing through these reaction centers. Therefore, they depend on the properties of two particles, i.e., nuclei and neutrons.

For our application, microscopic cross sections, σ, are expressed in a form that assumes the target nucleus to be at rest in the laboratory system. If the thermal motion of the target nucleus is important, which is the case in the thermal range and within resonances, one defines the cross sections between the neutron beam and target nuclei in thermal motion. Then only the "target" is at rest in the laboratory system and not the individual nuclei (see Secs. 4-7 and 5-7).

Cross sections depend only on the *relative* energy between nucleus and neutron. For practical use, the cross sections are expressed as a function of the *energy* of the neutron in the laboratory system of coordinates. The cross sections per nucleus are called "microscopic cross sections" and are measured in barns. A barn is equal to 10^{-24} cm^2 or 10^{-28} m^2.

The total cross section, $\sigma_t(E)$, is composed of several partial[d] cross sections:

$$\sigma_t(E) = \begin{cases} \sigma_s(E), \text{ elastic scattering} \\ \left.\begin{aligned} &+\sigma_c(E), \text{ capture} \\ &+\sigma_f(E), \text{ fission} \end{aligned}\right\} = \sigma_a(E), \text{ absorption} \\ +\sigma_{in}(E), \text{ inelastic scattering} \\ +\sigma_{n,2n}(E), \ (n,2n) \text{ reactions} \end{cases} \quad . \tag{1.17}$$

The elastic and inelastic scattering cross sections are often combined to form the "scattering" cross section, and their adjectives are dropped:

$$\sigma_S(E) = \sigma_s(E) + \sigma_{in}(E) \quad . \tag{1.18}$$

If the inelastic cross section is negligible or zero, then $\sigma_S(E)$contains only the elastic cross section and $\sigma_s(E) = \sigma_S(E)$.

Figure 1-5 illustrates some of the concepts discussed in this section. The upper two diagrams illustrate the angular flux, i.e., a neutron beam impinging vertically on a square centimeter, seen once from the side and once from the front. The middle diagram shows a microscopic cross

---

[d]The partial cross sections may be zero for some energy ranges, such as $\sigma_f(^{238}U)$ for a neutron energy below the ^{238}U fission threshold. Also inelastic scattering and $(n,2n)$ reactions are threshold reactions.

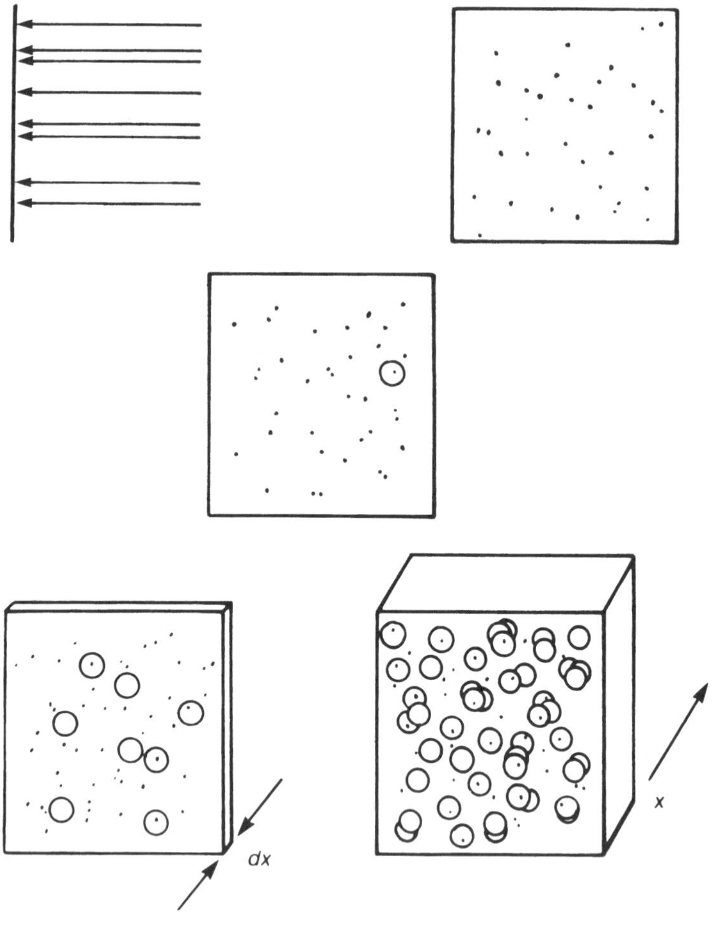

Fig. 1-5. Illustration of the angular flux and the micro- and macroscopic cross sections.

section as a circle (actually cross sections do not have a particular shape). The number of neutrons in a flux beam that "collides" with the cross section represents the microscopic reaction rate of the flux beam in terms of numbers of reactions per second and per nucleus:

$$\text{microscopic reaction rate} = r(E) \quad . \tag{1.19}$$

This diagram also illustrates the basic experimental procedure to find cross sections:

$$\sigma = \frac{\text{reaction/s and nucleus}}{\text{neutron/cm}^2\text{s}} = \frac{\text{reaction cross section (in cm}^2)}{\text{neutron and nucleus}} \quad . \tag{1.20}$$

The lower diagrams in Fig. 1-5 illustrate the "macroscopic cross section," which is introduced in the next section.

A neutron generally emerges from a scattering process with an energy E' different from its original value E. Thus, the detailed description of the scattering process requires a scattering cross section that depends on *both* energies. This cross section is called the "differential scattering cross section," and is written as $\sigma_s(E \rightarrow E')$ or $\sigma_s(E,E')$. A differential cross section has the dimensions $cm^2/U(E)$, since $\sigma_s(E \rightarrow E')\, dE'$ must have the dimension of a cross section. The total scattering cross section can be obtained from the differential cross section by integration over all possible final energies (holds also for σ_{in}):

$$\sigma_s(E) = \int_0^\infty \sigma_s(E \rightarrow E')\, dE' \quad . \tag{1.21}$$

The neutron always loses energy if its incident energy is much larger than the thermal energy of the scattering nucleus (kT).[e] Thus, the energy of the scattered neutron is, in this case, always less than or equal to its energy before the collision; one then has only "downscattering" for which Eq. (1.21) can be replaced by:

$$\sigma_s(E) = \int_0^E \sigma_s(E \rightarrow E')\, dE' \quad . \tag{1.22}$$

If, however, the neutron energy is comparable to kT (i.e., if the neutron is a "thermal" neutron), the thermal motion of the scattering nucleus must be considered, and the neutron may actually gain energy in a collision, or be "upscattered." As seen below, whether or not upscattering occurs makes a large difference in the mathematical complexity of the treatment of the energy dependence.

The differential scattering cross section is often called a "transfer cross section" and its ratio to the total scattering cross section is called "transfer kernel" (again, the subscript s may be replaced by S or in):

transfer kernel:

$$K_s(E \rightarrow E') = \frac{\sigma_s(E \rightarrow E')}{\sigma_s(E)} \quad ; \tag{1.23a}$$

transfer cross section:

$$\sigma_s(E \rightarrow E') = \sigma_s(E) \cdot K_s(E \rightarrow E') \quad . \tag{1.23b}$$

---

[e]Here, k is the Boltzmann constant, equal to 1.3804×10^{-16} erg/K and T is the absolute temperature of the system in degrees Kelvin. At room temperature, ($\simeq$293 K), one has $kT \simeq 0.025$ eV.

Neutrons emerge with fission neutron energies (say, E') from the fission process, caused by a neutron of energy E. This can, in principle, also be described by a transfer cross section:

$$\text{fission transfer cross section} = \nu(E)\sigma_f(E \rightarrow E') \quad , \tag{1.24}$$

where $\nu(E)$ denotes the average number of neutrons emitted in a fission reaction induced by energy E, and $\sigma_f(E \rightarrow E')$ is the cross section for emission of neutrons with energy E'. Fortunately for the theoretical treatment, the spectrum of fission neutrons depends only weakly on the inducing neutron energy, a dependency that can be neglected in most cases. Thus, one can approximate:

$$\sigma_f(E \rightarrow E') \simeq \chi(E')\sigma_f(E) \quad , \tag{1.25a}$$

where $\chi(E')\,dE'$ is the fraction of fission neutrons born with energy E' in dE'; $\chi(E')$ is called the "fission neutron spectrum." The differential cross section for the production of fission neutrons appears then factored into E- and E'-dependent quantities. A typical fission spectrum (as shown in Fig. 1-2) is usually normalized to unity.

If the dependence of the fission neutron spectrum on the inducing energy is taken into account, one has, instead of Eq. (1.25a), the more accurate formula:

$$\sigma_f(E \rightarrow E') = \chi(E,E')\sigma_f(E) \quad . \tag{1.25b}$$

With increasing E, the E' dependence of χ is shifted somewhat toward higher energies.

1-6 Reaction Rate Densities and Macroscopic Cross Sections

The reaction rate density is one of the key concepts in reactor physics as it combines the three basic concepts: atom densities, N, cross sections, σ, and neutron flux, ϕ, into a single and concrete concept. Reaction rate densities represent the building blocks for all neutron balance equations. The general definition of reaction rate densities, $R(\boldsymbol{r})$, includes integration over contributions at all energies:

$$R(\boldsymbol{r}) = N(\boldsymbol{r})\int_0^\infty \sigma(E)\phi(\boldsymbol{r},E)\,dE \quad ,$$

$$\left[\frac{\text{neutron reactions}}{\text{cm}^3\text{s}}\right] = \left[\frac{\text{nuclei}}{\text{cm}^3}\right]\cdot\left[\frac{\text{reaction-cm}^2}{\text{nucleus}}\right]\left[\frac{\text{neutrons}}{\text{cm}^2\text{s}}\right] \quad . \tag{1.26}$$

Reaction rates can be based directly on the flux, $\phi(\boldsymbol{r},E)$; they need not be calculated from angular reaction rates since

$$R(\boldsymbol{r}) = N(\boldsymbol{r})\int_E \sigma(E)\int_\Omega \phi(\boldsymbol{r},E,\boldsymbol{\Omega})\, d\Omega\, dE = N(\boldsymbol{r})\int_E \sigma(E)\phi(\boldsymbol{r},E)\, dE \quad . \tag{1.27}$$

In the following, the limits of integration are left off when integration is to be carried out over the entire range of the respective variable.

In reaction rate equations such as Eq. (1.26), the number density of the reacting nuclei, $N(\boldsymbol{r})$, is always multiplied by the microscopic cross section. The product number-density times the microscopic cross section is called the "macroscopic cross section" and is denoted by Σ; for example:

$$\Sigma_c(\boldsymbol{r},E) = N(\boldsymbol{r})\sigma_c(E) \quad . \tag{1.28}$$

The dimensions of the macroscopic cross section are:

$$\Sigma = \left[\frac{\text{nuclei}}{\text{cm}^3}\right]\left[\frac{\text{cm}^2}{\text{nucleus}}\right] = \left[\text{cm}^{-1}\right] \quad .$$

For atom densities and microscopic cross sections having an order of magnitude of $10^{23}/\text{cm}^3$ and 10^{-23} to 10^{-24} cm^2, respectively, typical macroscopic cross sections are of the order of 1 to 10^{-1} cm^{-1}.

The inverse of the macroscopic cross section can be interpreted as the mean-free-path (mfp) for neutrons to cause the interaction. Therefore, the probability that a neutron will cause an interaction in traveling a distance dx is

$$\Sigma\, dx = \frac{dx}{\lambda} \quad . \tag{1.29}$$

Thus,

$$\lambda = \Sigma^{-1} \tag{1.30}$$

is the neutron mfp for that reaction.

The lower left diagram in Fig. 1-5 illustrates the formation of the macroscopic cross section: The slice of the volume 1 cm^2 by dx contains $N\, dx$ nuclei. The chance for a reaction is given by the sum of all microscopic cross sections in an infinitesimal slice, dx:

$$\sigma N\, dx = \Sigma\, dx \quad . \tag{1.31}$$

The lower right diagram in Fig. 1-5 illustrates all microscopic cross sections in a thick slice of material. The number of cross sections is then so numerous that "overlapping" of cross-sectional areas becomes significant. Cross sections are located partially "behind" each other. Then the nuclei in front remove neutrons from the beam, and nuclei in the back are exposed to a lesser extent. This requires the consideration of the attenuation of the beam. Let $\phi_\Omega(x)$ denote the flux in the beam at x. In each *infinitesimal* slice, for which "overlapping" is absolutely negligible, the change in dx is given by:

$$d\phi_\Omega(x) = -\Sigma\phi_\Omega(x)\, dx \text{ or } \frac{d\phi_\Omega(x)}{dx} = -\Sigma\phi_\Omega(x) \quad . \tag{1.32}$$

Equation (1.32) readily yields the intensity reduction for the attenuating beam by solving the differential equation, Eq. (1.32):

$$\phi_\Omega(x) = \phi_\Omega(0)\exp(-\Sigma x) \quad , \tag{1.33}$$

i.e., the initial beam flux is reduced by the "beam attenuation factor," $\exp(-\Sigma x)$, which accounts for the effect of cross sections overlapping as depicted in Fig. 1-5.

The simple concept of reaction rates as described above has to be extended to include all the isotopes contained in a compound or mixture. The index i is consistently used throughout this book to denote isotopes:

$$R(\boldsymbol{r}) = \int_E \sum_i N_i(\boldsymbol{r})\sigma_i(E)\phi(\boldsymbol{r},E)\, dE = \int_E \Sigma(\boldsymbol{r},E)\phi(\boldsymbol{r},E)\, dE \quad , \tag{1.34}$$

where

$$\Sigma(\boldsymbol{r},E) = \sum_i N_i(\boldsymbol{r})\sigma_i(E) = \sum_i \Sigma_i(\boldsymbol{r},E) \tag{1.35}$$

is the macroscopic cross section of the entire conglomerate of isotopes. Equations (1.34) and (1.35) demonstrate that the macroscopic cross section is more than just a notation for the product of two quantities: It describes the reaction probability of macroscopic quantities of matter. As Eq. (1.35) shows, the macroscopic cross section of a mixture is determined by summing the macroscopic cross sections of each of the individual isotopic contributions. Both quantities (i.e., Σ and Σ_i) are macroscopic cross sections.

In many cases, one uses the chemical element cross sections instead of the isotope cross sections; e.g., σ_{Fe} denotes the microscopic cross section of iron. This can and should be done when the isotopic composition of a chemical element remains largely unchanged during its life in the reactor. The same simplification cannot be applied for uranium, plutonium, and some of the FPs where the detailed isotopic distribution must be described accurately. To account for elements in the notation, one simply extends the meaning of the index i from isotopes to chemical elements. Then $\Sigma_i(\boldsymbol{r},E)$ in Eq. (1.35) is the macroscopic cross section of isotope or element i at the point $\boldsymbol{r}$; e.g., for chromium in stainless steel, the quantity $\Sigma_{Cr}(\boldsymbol{r},E)$ denotes the chromium contribution to the macroscopic cross section of steel; it is *not* the macroscopic cross section of pure chromium. One rarely needs the macroscopic cross sections of pure substances. Therefore, no specific notation is introduced to denote macroscopic cross sections of pure isotopes or elements.

A concluding cautioning remark on the above subject appears to be in order: Although the definition of macroscopic cross sections is logically simple, their practical calculation is frequently a source for unsuspected errors, especially for heterogeneous arrangements of mixtures of compounds. Therefore, it is advisable for the beginning nuclear engineer to assure himself that for each volume the correct number of atoms of each kind is multiplied with the respective microscopic cross sections.

1-7 The Basic Theoretical Approach

The discussion of the engineering requirements of reactor physics makes it quite evident that there is a large variety of physics phenomena affecting the engineering, design, and operation of a reactor. Practically all of these phenomena are directly or indirectly based on or related to *reaction rates* of neutron/nuclear interactions. All these reaction rates have a single common root, the neutron flux with its space and energy dependency: $\phi(\boldsymbol{r},E)$. The general theoretical approach, pursued in reactor physics, therefore aims first at this common root of all reaction rates, the neutron flux. From the neutron flux and the appropriate material density and nuclear physics information, one can then calculate all required reaction rates.

The neutron flux with its space and energy dependencies is quite a complicated function. Its variations are even more complicated when its angular dependence affects its spatial variation, as is the case for heterogeneous material arrangements such as the configuration consisting of fuel rod, cladding, and coolant, which represents the inherent structure of a power reactor core. These angular effects are even stronger for control rods, especially for those of the cruciform shape, which is employed in some power reactors (e.g., BWRs).

In addition to space, energy, and direction, the neutron flux can also depend on time. One of its time variations reflects directly the long-term changes in material compositions present in the reactor, which were discussed in Sec. 1-2. Much larger time variations of the flux have to be described in reactor dynamics problems as required for safety investigations. Thus, the full description of neutron behavior involves seven dimensions: three spatial coordinates, the energy, two angles, and time.

Since the neutron flux is, in its detailed dependencies, such a complicated multidimensional function, one cannot just calculate it. One cannot even aim at calculating it directly with all its dependencies. One must at first devise a basic approach of how to go about solving this complicated problem.

Since the calculation of complicated multidimensional particle distributions is not unique to reactor physics, one can learn from experiences in other fields, particularly from kinetic gas theory. However, the particular configurations of interest in reactor physics and the special dependencies (of space energy, angles, and time) require the development of reactor-specific solution techniques. The development of these solution techniques and the investigation of typical results comprise the major body of theoretical reactor physics.

Devising a basic approach for the detailed calculation of the neutron flux requires a semiquantitative knowledge of its dependencies in individual dimensions. Therefore, several chapters (3 through 6) are devoted to the investigation of the flux dependencies in isolation. This provides the necessary general understanding of the neutron behavior and it reveals and suggests possibilities for devising practical approaches of various degrees of sophistication. Actually, the common element of nearly all approaches for treating complicated multidimensional problems is the composition of the full solution out of partial solutions as they can be derived for isolated problems. One usually proceeds in the following way.

The behavior is investigated in isolated dimensions and then attempts are made to determine the limits of the applicability of the isolated treatment. If one could strictly treat all dimensions in isolation, the entire problem would be "separable" in all variables. This is not possible for all practical purposes. But, it is practical to try to utilize as much information from the isolated treatment as possible. The utilization of this information generally involves some sort of effective parameters, which are calculated from an isolated solution and are subsequently used in reduced treatment or for the treatment of a different dimension. Examples are:

1. the "group" constants, which represent lumped information about the detailed energy variation of nuclear cross sections and a zero-dimensional neutron spectrum, to be used for the calculation of the space-dependent multigroup flux
2. "cell-averaged" cross sections, derived from information on the angular flux; the averages are then applied to the direct flux calculation, without considering the angular flux any further.

This basic structure of the overall approach is reflected in the structure of the presentation of the material in this book. The explanation of the fundamental concepts in this chapter is followed by a chapter on neutron balance equations, in which the later isolated investigations are prepared. Further, features common to most of the later solutions are also discussed extensively in Chapter 2. Chapter 3 concentrates on the pure space dependence of the flux, whereas the energy dependence is investigated in Chapters 4 and 5. Methods for detailed treatment of the angular dependency are briefly reviewed in Chapter 6. Chapter 7 is devoted to the combined treatment of space and energy dependency. Time dependencies of the flux are not covered in this text. They are the subject of reactor kinetics and dynamics, which are presented in Ref. 11 as a direct and consistent expansion of this text on reactor statics.

1-8 Summarizing Remarks

In this introductory chapter, the basic calculational approach and background fundamentals used in nuclear reactor core design have been outlined. The major role a reactor designer, engineer, or physicist plays in establishing a safe and efficient reactor was illustrated by the variety of information required in the design and operation of a power reactor. The relationship between the design activity described in this book, namely, the physics of nuclear reactor theory, and the other necessary engineering disciplines required to develop a reactor system was presented for purposes of perspective.

The major emphasis in this chapter, however, was on reviewing the fundamental concepts necessary to understand this subject known as nuclear reactor theory. The underlying phenomena—the fission process and the fundamental parameters associated with it—were defined and discussed. The importance of the neutron flux—the basic concept used to describe the average neutron population—was stressed throughout this chapter. This key quantity, along with its several versions and dependencies, forms the basis of much of reactor theory and a clear understanding of this concept is essential. As mentioned, the behavior of the neutron flux in reactor systems is explored in depth in Chapters 3 through 7 where its space, energy, and angular dependencies are investigated.

The importance of the neutron flux in reactor physics and the reason it receives so much attention is based on its use in calculating neutron/nucleus reaction rates. Two other parameters are important in this calculation: the atom density and the cross section. The probability that the various physical processes occur during neutron/nucleus interactions is determined by the neutron cross section. The various cross-section definitions are provided along with the mathematical and physical interpretation of the macroscopic cross-section concept. Extensive use is made throughout reactor theory of this concept, both in simple models as well as in the group constants in multidimensional reactor models. Therefore, a good understanding of this concept is essential. Additional information on cross sections is provided in later chapters where specific neutron reaction phenomena are discussed (e.g., the neutron scattering events described in Chapter 4).

Other neutron parameters related to the neutron flux, such as neutron current, neutron sources, and neutron density, were also defined and discussed briefly in this chapter. Their mathematical role in describing the net migration of neutrons in a reactor is made clearer in later chapters. Additional concepts and parameter definitions that are important in reactor theory are introduced and discussed in later chapters. For reactor theory application details, several excellent texts are available. References 7 through 10, 12, 13, and 14 describe some of the early reactor development approaches whereas the rather recent calculational techniques are presented in Refs. 15, 16, and 17.

Homework Problems

1. The energy distribution of fission neutrons shown in Fig. 1-2 is calculated from the analytical approximation

$$\chi(E) = \frac{2\alpha}{\sqrt{\pi}} (\alpha E)^{1/2} \exp(-\alpha E) \quad .$$

   a. Find E_{max}, the most frequent fission neutron energy in terms of α.
   b. Determine E_{max} from Fig. 1-2 (^{235}U) and find α.
   c. Find $\overline{E}$, the average fission neutron energy in terms of α, and give its approximate value for ^{235}U.
   d. Give $\chi(E)$ with $\overline{E}$ instead of α as the parameter.
2. A value of ~6.3/s has been found for the mean number of fission/kg U induced by cosmic ray neutrons. What is the approximate value of the initiating flux? (Find a representative fission cross section for ~5 MeV neutrons from Fig. 1-1).
3. In terms of beam fluxes and beam currents, let neutrons move along the x axis in both directions. Let $\phi^+(x)$ and $\phi^-(x)$ be the two components of the angular flux, $\phi(x,\mathbf{\Omega})$, with $\mathbf{\Omega}$ having only the two values: the plus or minus direction of the x axis. Find $\phi(x)$ and $J(x)$.

4. Suppose

$$2\pi\phi(z,\boldsymbol{\Omega}) = \phi(z,\mu) = C_0(z) + C_1(z)\mu$$

with

$$\mu = \cos\vartheta \quad ,$$

where ϑ measures the angular deviation from the z axis.
   a. Find $\phi(z)$ and $J(z)$.
   b. Express $\phi(z,\mu)$ in terms of $\phi(z)$, $J(z)$, and μ.
5. Suppose an angular flux in a slab geometry is given by

$$\phi(z,\mu) = \phi_0(\cos Bz + A\mu \sin Bz) \quad .$$

   a. Find $\phi(z)$ and $J(z)$.
   b. Rewrite $\phi(z,\mu)$ by eliminating A.
   c. Find J^+ and J^-, the partial currents in the upper and lower half of the solid angle ($\mu \geqslant 0$ and $\mu \simeq 0$).
   d. Check J^+ and J^- by comparing them with J.
6. Select and illustrate proper coordinate systems for the list of geometries given below. Write the angular flux, $\phi(\boldsymbol{r},E,\boldsymbol{\Omega})$, in these coordinates. Observe symmetries if applicable. Give the explicit formulas for the angular integral, space integral, the gradient, and the divergence. The geometries are:
   a. plane two-dimensional system
   b. rectangular three-dimensional system
   c. cylindrical system (infinitely high cylinder; α_r = azimuthal angle of $\boldsymbol{r}$)
   d. spherical system.
7. Find the macroscopic scattering cross section of $U0_2$, assuming a density of 10 g/cm^3 (= 10 000 kg/m^3) and $\sigma_s(\mathrm{U}) \simeq 8.9$ b and $\sigma_s(0) \simeq 3.75$ b (10^{-28} m^2).
8. Give a formula for the macroscopic cross sections for $U0_2$ fuel with $\gamma = N(^{235}\mathrm{U})/N(^{238}\mathrm{U})$ as the enrichment factor.
9. An effective thermal neutron absorber is cadmium. Determine the cadmium shielding thickness necessary to reduce a beam of thermal neutrons to one-tenth of its incident intensity. Assume $\rho_{\mathrm{Cd}} = 8.50$ g/cm^3 (8500 kg/m^3) and σ_a (Cd) = 2450 b.
10. An underlying assumption in neutron diffusion theory (and transport theory as well) is that the number of neutron/neutron collisions is negligible compared to collisions of neutrons with atomic nuclei. Convince yourself that this approximation is valid even for high-flux experimental reactor conditions (use $\phi_{th} = 10^{16}$/cm^{2}s, $\overline{E} = 0.025$ eV, and $\overline{v} = 2200$ m/s) and a reasonable value for the neutron/neutron cross section (use $\sigma_{nn} = 10$ b). Show that the number of

neutron/neutron collisions per second is given by $1/2\ vn \times n \times \sigma_{nn}$. What is the significance of the 1/2 in this reaction rate? Compare this reaction rate with the one in solid UO_2 under the same conditions.

11. Consider a point source emitting 10^{12} neutron/s in a vacuum. Calculate the flux and the current at a distance of 1 m from the source.

Review Questions

1a. Briefly explain the fission process.
b. Give the approximate magnitude of the total energy release of a fission reaction, and discuss the spatial location and the time of the release.
2a. What is η for fission reactions?
b. Give typical η values for ^{235}U, ^{233}U, and ^{239}Pu for E_{th} and for 2 MeV.
c. What follows from these η values with respect to possible reactor types?
3. What is the basic concept used to describe the average neutron population? Give all of its dependencies.
4. Give the definitions of flux, total flux, thermal flux, and group g flux by appropriate integrals.
5. Give the definitions of neutron current, total current, and group g current by appropriate integrals.
6. Explain the physical meaning of the current.
7. Express an isotropic angular flux in terms of the flux.
8. Give the angular vector $\boldsymbol{\Omega}$ for a spherical coordinate system.
9. Give σ_t in terms of the four most important components.
10. Give the differential scattering cross section, the scattering kernel, and form the total scattering cross section.
11. Give the fission neutron source (for a single fissionable isotope).
12a. Give the general expression for a macroscopic reaction rate density.
b. Why can it be formed with the flux rather than the angular flux?
13. Give the macroscopic cross section for a compound or a mixture of isotopes.
14a. Derive the attenuation factor for a beam of neutrons.
b. What is the mean-free-path in terms of a cross section?

REFERENCES

1. R. D. Evans, *The Atomic Nucleus,* McGraw-Hill Book Co., New York (1955).
2. I. Kaplan, *Nuclear Physics,* 2nd ed., Addison-Wesley Publishing Co., Inc., Reading, Massachusetts (1963).
3. O. Oldenburg and N. C. Rasmussen, *Modern Physics for Engineers,* McGraw-Hill Book Co., New York (1966).
4. S. Glasstone and A. Sesonske, *Nuclear Reactor Engineering,* 3rd ed., D. Van Nostrand Co., Princeton, New Jersey (1981).
5. M. M. El-Wakil, *Nuclear Power Engineering,* McGraw-Hill Book Co., New York (1962).
6. A. Sesonske, "Nuclear Power Plant Design Analysis," TID-26241, U.S. AEC Technical Information Center, Oak Ridge, Tennessee (1973).
7. A. M. Weinberg and E. P. Wigner, *The Physical Theory of Neutron Chain Reactors,* University of Chicago Press, Chicago (1958).
8. G. I Bell and S. Glasstone, *Nuclear Reactor Theory,* Van Nostrand Reinhold Co., New York (1970).
9. J. R. Lamarsh, *Introduction to Nuclear Reactor Theory,* Addison-Wesley Publishing Co., Reading, Massachusetts (1966).
10. S. Glasstone and M. C. Edlund, *The Elements of Nuclear Reactor Theory,* D. Van Nostrand Co., Princeton, New Jersey (1952).
11. K. O. Ott and R. J. Neuhold, *Introductory Nuclear Reactor Dynamics,* American Nuclear Society, La Grange Park, Illinois (1985).
12. R. L. Murray, *Nuclear Reactor Physics,* Prentice-Hall, Inc., Englewood Cliffs, New Jersey (1957).
13. H. S. Isbin, *Introductory Nuclear Reactor Theory,* Reinhold Publishing Corp., New York (1963).
14. R. V. Meghreblian and D. K. Holmes, *Reactor Analysis,* McGraw-Hill Book Co., New York (1960).
15. P. F. Zweifel, *Reactor Physics,* McGraw-Hill Book Co., New York (1973).
16. A. F. Henry, *Nuclear Reactor Analysis,* MIT Press, Cambridge, Massachusetts (1975).
17. J. J. Duderstadt and L. L. Hamilton, *Nuclear Reactor Analysis,* John Wiley & Sons, New York (1976).

Two

NEUTRON BALANCE EQUATIONS AND THE FUNDAMENTAL NEUTRONICS PROBLEMS

2-1 Introduction

The fundamental concepts of neutron flux, nuclear reaction rates, and sources described in the first chapter represent the "building blocks" for the development of a theoretical description of neutron behavior in an assembly of material such as a reactor. In general terms, the task is quite simple. Conservation laws, similar to the ones employed in heat transfer, fluid mechanics, or kinetic gas theory, can be applied to obtain neutron balance relationships for a volume of interest. The exact balance equations, i.e., the one for $\phi(\boldsymbol{r},E,\boldsymbol{\Omega})$ or $\phi(\boldsymbol{r},E,\boldsymbol{\Omega},t)$, provide the foundation for a topic generally referred to as neutron "transport" theory. To obtain a tractable solution, approximations are required in the general transport balance relation. Of particular practical importance are models that fall into a broad category usually referred to as "diffusion" theory. Both types of neutron balance are developed in this chapter along with the appropriate boundary conditions. Consistent with the major objective of this text, diffusion theory is emphasized with transport theory relegated to a supporting role. Methods of solution of these formulations are deferred until later chapters. Additional information and details about the neutron diffusion process can be found in Refs. 1 and 2 with transport theory presented in more advanced texts.[3,4]

Also developed and discussed in this chapter is probably the single most important reactor concept: *criticality*, which is the condition necessary to obtain a stable neutron population. This concept evolves naturally from the neutron balance formulation. A thorough understanding of this concept is required since the criticality constant of a reactor configuration is a frequently sought after reactor parameter. To appreciate the generality of the interrelationship between reactor criticality and the neutron balance equations, a generalized notation, the "operator" notation, is used extensively. Reactor criticality is also discussed in terms of neutron reaction rates in the context of the simple four- or six-factor formulas. However, with the common extensive use of computers, these four- or six-factor expressions for criticality are not explicitly used in calculations anymore. Nevertheless, in addition to their historical significance, they are still very valuable heuristic concepts.

In this chapter, attention focuses on the formulation of the time-independent or "static" neutronics problems. There is no need, in statics problems, to consider the delayed-neutron source explicitly. Only the total neutron source, i.e., the sum of the prompt and delayed contributions, impacts the static solutions. Therefore, one always considers the total fission neutron source in statics problems. Transient or time-dependent neutron behavior is treated elsewhere.[5]

2-2 The Static Balance Equations of Neutron Transport and Diffusion

The so-called "transport theory" correctly describes the average transport of neutrons; average in the sense that statistical fluctuations of the neutron population are not considered. The fundamental quantity in transport theory is the "angular flux." Analogous to other theoretical fields, the basis of transport theory is the balance equation for the particles involved, described here by the angle-dependent neutron flux, $\phi(\boldsymbol{r},E,\boldsymbol{\Omega})$. A neutron balance equation can also be written using the neutron density [see Eq. (1.4)], which leads to analogous expressions. But it is the neutron flux that is now in general use and is therefore emphasized throughout this text.

There are two approaches that are used to obtain a balance equation for the angular flux: a differential and an integral approach. Both approaches are briefly described in the following. Although the solution of the transport equation, i.e., the analytical or numerical treatment of the angular flux, is the subject of more advanced courses, it is very important for the nuclear engineer to be familiar with the transport theory balance equations since they represent the basis of all approximate methods. A very brief discussion of transport theory is presented in Chapter 6.

Further, the basic balance equation of the energy-dependent diffusion theory is derived in this section. Reduced formulations, such as the one-group or spatial diffusion equation can be derived readily from the energy-dependent formulation (see Secs. 2-2D and 2-2F).

2-2A The Differential Form of the Transport Equation (Boltzmann Equation)

The differential balance equation for the transport of particles (gas molecules) was first derived by Boltzmann. The equation for neutron transport is analogous to the "linear" Boltzmann equation, which describes the transport of a low-density gas in a medium with much higher density. Then the collisions between gas molecules themselves can be neglected. The collision density between gas molecules is proportional to the square of the density of the gas. This leads to a nonlinear transport equation, which represents a more complicated problem. The neutron densities even in high flux reactors are so low that neutron/neutron collisions can be neglected and the linear Boltzmann equation can be applied.

The approach to derive the Boltzmann equation is the same as the one always used to derive differential balance equations: One pretends to know $\phi(z,E,\mathbf{\Omega})$ and forms:

$$\phi(z + dz,E,\mathbf{\Omega}) = \phi(z,E,\mathbf{\Omega}) + \text{sum of changes in } dz \quad . \tag{2.1a}$$

Thus the difference $d\phi$ is obtained as a sum of the changes:

$$d\phi(z,E,\mathbf{\Omega}) = \phi(z + dz,E,\mathbf{\Omega}) - \phi(z,E,\mathbf{\Omega}) = \text{sum of changes in } dz \quad . \tag{2.1b}$$

For simplicity, only one space dimension is considered in the derivation. The neutron balance in an infinitesimal slice dz, illustrated in Fig. 2-1, can be readily written down since the total number of reactions occurring in dz is strictly proportional to the path length dl;

$$dl = \frac{dz}{\mu} \quad , \tag{2.2}$$

with

$$\mu = \cos\vartheta \quad . \tag{2.3}$$

No multiple collisions need to be considered along an infinitesimally small path. Thus, the changes in dz are given by the reaction rates or source densities times the path length.

The losses and gains (sources) that contribute to the change in dz are then obtained as:

$$\Sigma_t(z,E)\phi(z,E,\mathbf{\Omega})\,\frac{dz}{\mu}, \quad \text{total neutron loss in } dz \quad , \tag{2.4a}$$

$$\frac{1}{4\pi}S_f(z,E)\,\frac{dz}{\mu}, \quad \text{fission neutron source in } dz \quad , \tag{2.4b}$$

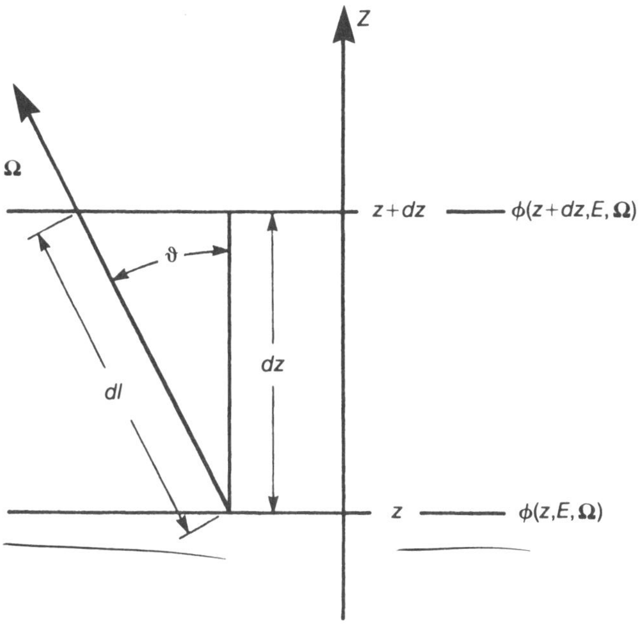

Fig. 2-1. Diagram for the derivation of the differential neutron balance equation for the angular flux (Boltzmann equation).

$$S(z,E,\mathbf{\Omega})\,\frac{dz}{\mu}, \quad \text{independent neutron source in } dz \quad , \qquad (2.4c)$$

and

$$\int_{\Omega'}\int_{E'} \Sigma_S(z,E'\rightarrow E,\mathbf{\Omega}'\rightarrow\mathbf{\Omega})\phi(z,E',\mathbf{\Omega}')\,dE'\,d\Omega'\cdot\frac{dz}{\mu} \quad ,$$

$$\text{scattering neutron source in } dz \quad . \qquad (2.4d)$$

By means of Eqs. (2.1) through (2.4), one can express ϕ at $z + dz$ in terms of ϕ at z. This procedure is continued for other spatial locations. Thus, by the consideration of a sequence of the balance equations in infinitesimal space increments, one can relate ϕ to its value at a boundary, where "boundary conditions" are needed as discussed in Sec. 2-2E.

In the actual theoretical treatment of neutron fluxes, one does not use the infinitesimal balance, Eqs. (2.1) through (2.4), directly; rather, one casts the infinitesimal balance, at first, into the form of a differential equation or integro-differential equation. These balance equations are the basis for further mathematical treatment, such as: analytical or partially analytical solutions, separation of variables, derivation of general theorems for solutions, numerical solutions, etc. If one subsequently applies the elegant and powerful mathematical tools for solving differential equations with finite difference methods, the expression may not appear to be too different from the original infinitesimal balance equation. However, the application of the mathematical tools provides the needed iterative techniques, information on proper size of intervals, convergence of iterative solutions, etc.

Adding all terms of Eq. (2.4) with the appropriate signs, i.e., gains are positive and losses are negative, and dividing by dz/μ yields an equation for $d\phi(z,E,\mathbf{\Omega})/dl$, the change of the angular flux per dl. The result is the

Boltzmann equation in plane geometry:

$$\mu\frac{\partial}{\partial z}\phi(z,E,\mathbf{\Omega}) = -\Sigma_t(z,E)\phi(z,E,\mathbf{\Omega})$$

$$+\int_{\Omega'}\int_{E'}\Sigma_S(z,E'\rightarrow E,\mathbf{\Omega}'\rightarrow\mathbf{\Omega})\phi(z,E'\mathbf{\Omega}')\,dE'\,d\Omega'$$

$$+\frac{1}{4\pi}S_f(z,E)+S(z,E,\mathbf{\Omega}) \quad . \tag{2.5}$$

Equation (2.5) is an integro-differential equation, since the unknown function appears differentiated as well as under an integral.

The quantity $d\phi(z,E,\mathbf{\Omega})$ in Eq. (2.1b) and, therefore, also the derivation in the Boltzmann equation, Eq. (2.5), have been introduced formally as the net result of all changes through nuclear reactions and sources occurring along the path dl. In a stationary state, where ϕ is independent of time, the net changes $d\phi$ cannot remain in dz and accrue; the entire surplus "leaks out" of dz, and the flux remains constant in time. This change of the angular flux through the leakage out of dz is described by the derivative term on the left side of Eq. (2.5).

Neutron leakage can also be considered a physical phenomenon that affects the neutron balance. One can therefore obtain the same neutron balance equation, Eq. (2.5), from requiring that the sum of *all* changes, i.e., neutron sources, nuclear reactions, and leakage, be zero at all points in space and at all energies.

It should be noted that the angular fission source, which must be used in Eq. (2.5), is independent of $\mathbf{\Omega}$. This is due to the isotropic nature of the average neutron emission in the fission process; thus, $S_f(z,E,\mathbf{\Omega})$ $d\Omega$, the average number of neutrons emitted into the interval $d\Omega$, is independent of $\mathbf{\Omega}$ and only proportional to the fraction of the total solid angle:

$$S_f(z,E,\mathbf{\Omega})\, d\Omega = S_f(z,E)\frac{d\Omega}{4\pi} = \frac{1}{4\pi}S_f(z,E)\, d\Omega \quad , \tag{2.6}$$

where $S_f(z,E)$ is the angular-integrated fission source:

$$S_f(z,E) = \int_\Omega S_f(z,E,\mathbf{\Omega})\, d\Omega \quad . \tag{2.7}$$

The fission source $S_f(z,E)$ is obtained from the flux, the fission cross section, and the number $\nu(E')$ and energy distribution of fission neutrons $\chi(E)$ (see Sec. 1-5):

$$S_f(z,E) = \chi(E)\int_{E'} \nu(E')\Sigma_f(z,E')\phi(z,E')\, dE' \quad . \tag{2.8}$$

Equation (2.8) holds for *one* fissionable isotope. If several isotopes are fissioned, their contributions must be combined. Since $\chi(E)$ is not appreciably different for different isotopes, the combination of several contributions still has the same form as Eq. (2.8) with $\nu\Sigma_f(z,E)$ accounting for the various contributions:

$$\nu\Sigma_f(z,E) = \sum_i \nu_i(E)\Sigma_{f,i}(z,E) \quad . \tag{2.9}$$

Note that $\nu\Sigma_f(z,E)$ on the left side of Eq. (2.9) is considered to be a single quantity.

The energy distribution of the emitted fission neutrons is normalized to one:

$$\int_E \chi(E)\, dE = 1 \quad , \tag{2.10}$$

since $\nu(E')$ describes the total (average) number of neutrons emitted in a fission reaction caused by a neutron with the energy E'.

In two or three space dimensions, the simple derivative in Eq. (2.5) is replaced by

$$\mu\frac{\partial}{\partial z}\phi(z,E,\mathbf{\Omega}) \rightarrow \mathbf{\Omega}\cdot\nabla\phi(\mathbf{r},E,\mathbf{\Omega}) \quad , \tag{2.11}$$

which gives the leakage rate in three dimensions. The substitution, Eq. (2.11), follows from the fact that the derivative along a path dl in the direction of $\mathbf{\Omega}$ is the component of the gradient in this direction, i.e., the dot product of $\mathbf{\Omega}$ and the gradient.

The substitution of z by $\boldsymbol{r}$ and of the leakage in Eq. (2.5) results in the general Boltzmann equation:

$$\boldsymbol{\Omega}\cdot\nabla\phi(\boldsymbol{r},E,\boldsymbol{\Omega}) = -\Sigma_t(\boldsymbol{r},E)\phi(\boldsymbol{r},E,\boldsymbol{\Omega})$$

$$+\int_{\Omega'}\int_{E'}\Sigma_S(\boldsymbol{r},E'\rightarrow E,\boldsymbol{\Omega}'\rightarrow\boldsymbol{\Omega})\phi(\boldsymbol{r},E',\boldsymbol{\Omega}')\,dE'\,d\Omega'$$

$$+\frac{1}{4\pi}S_f(\boldsymbol{r},E) + S(\boldsymbol{r},E,\boldsymbol{\Omega}) \quad . \tag{2.12}$$

Expressions for the neutron leakage term $\boldsymbol{\Omega}\cdot\nabla\phi(\boldsymbol{r},E,\boldsymbol{\Omega})$ in terms of various coordinates (i.e., plane, spherical, or cylindrical) can be found in Appendix A.

2-2B The Integral Transport Equation

An equation for the correct description of neutron transport can also be derived using a very different conceptual approach than the one leading to the differential form of the transport equation. This approach is often applied in deriving an *integral* equation for an unknown function rather than a *differential* equation. In the derivation, one pretends to know $\phi(\boldsymbol{r}',E',\boldsymbol{\Omega}')$ at *all* space points $\boldsymbol{r}'$ as a basis for the construction of $\phi(\boldsymbol{r},E,\boldsymbol{\Omega})$ as the integral of all contributions:

$$\phi(\boldsymbol{r},E,\boldsymbol{\Omega}) = \int\left[\begin{array}{l}\text{over all scattering and source events}\\ \text{that } \textit{directly} \text{ contribute to } \phi(\boldsymbol{r},E,\boldsymbol{\Omega})\end{array}\right] dl \quad . \tag{2.13}$$

"Directly" means contributions through free-flight transmission between the source at $\boldsymbol{r}'$ and the neutron's appearance at $\boldsymbol{r}$. Since the neutron passes through $\boldsymbol{r}$ with the direction $\boldsymbol{\Omega}$, it must have acquired this direction in its previous collision at $\boldsymbol{r}'$ (or it must have been generated at $\boldsymbol{r}'$ along $\boldsymbol{\Omega}$). Figure 2-2 illustrates the conceptual idea of Eq. (2.13). The line of integration is indicated as a dashed line.

Thus, instead of the simple neutron propagation through an infinitesimal slice—which leads to the Boltzmann equation—one considers, in the derivation of the integral transport equation, the similarly simple propagation of a neutron through free flight in the entire space.

The approach indicated in Eq. (2.13) leads to the integral transport equation for the angular flux. The explicit formulation of Eq. (2.13) can be readily worked out (see Chapter 6). An integral transport equation for the flux, rather than for the angular flux, can be derived if one assumes that the scattering source and the other sources are isotropic. This is a good approximation for thermal neutrons scattered on heavy nuclei. The approach to derive the integral transport equation for the

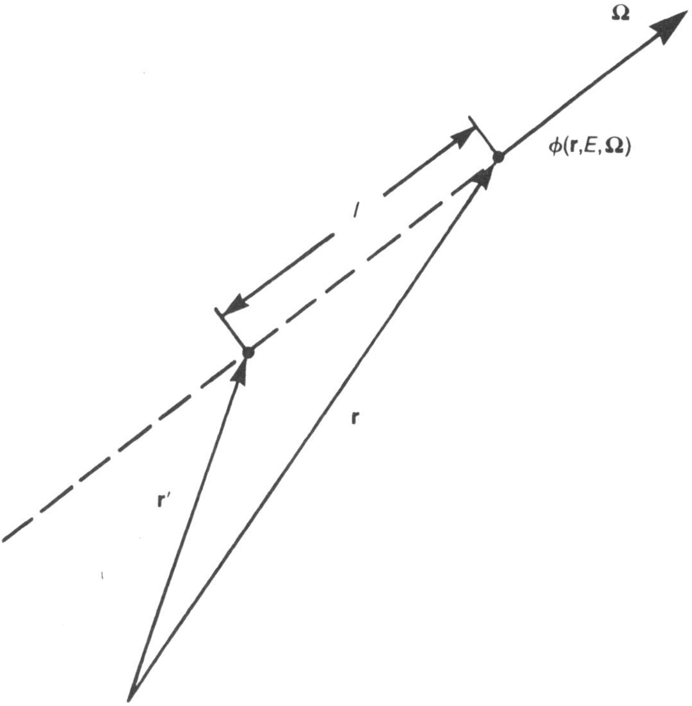

Fig. 2-2. Illustration of the approach to construct the angular flux in the integral transport equation.

neutron flux can be formulated as:

$$\phi(\boldsymbol{r},E) = \int_{V'} \left[\begin{array}{l} \text{over all scattering and source events} \\ \text{that directly contribute to } \phi(\boldsymbol{r},E) \end{array} \right] dV' \quad , \qquad (2.14)$$

with the integration being extended over the entire space.

Let $q(\boldsymbol{r}',E)$ denote the total source density at $\boldsymbol{r}'$, consisting of scattered, fission, and independent source neutrons:

$$q(\boldsymbol{r}',E) = \int_{E'} \Sigma_S(\boldsymbol{r}',E' \rightarrow E)\phi(\boldsymbol{r}',E')\, dE' + S_f(\boldsymbol{r}',E) + S(\boldsymbol{r}',E) \quad . \qquad (2.15)$$

The fraction of the isotropic source emitted into the solid angle increment $d\Omega$ is given by

$$q(\boldsymbol{r}',E)\frac{d\Omega}{4\pi} \quad . \qquad (2.16)$$

As shown in Fig. 2-3, the neutrons emitted out of the volume element dV' at $\boldsymbol{r}'$ into $d\Omega$ cover, after the distance l with

$$l = |\boldsymbol{r}' - \boldsymbol{r}| \quad , \tag{2.17}$$

the area

$$df = d\Omega \cdot l^2 \quad . \tag{2.18}$$

The probability that a neutron emitted at $\boldsymbol{r}'$ in the direction $\boldsymbol{r}' - \boldsymbol{r}$ is neither absorbed nor scattered during its flight from $\boldsymbol{r}'$ to $\boldsymbol{r}$ is, according to the basic definition of macroscopic cross sections (see Sec. 1-6), given by the "attenuation" factor:

$$\exp[-\Sigma_t(E)|\boldsymbol{r}' - \boldsymbol{r}|] \quad . \tag{2.19}$$

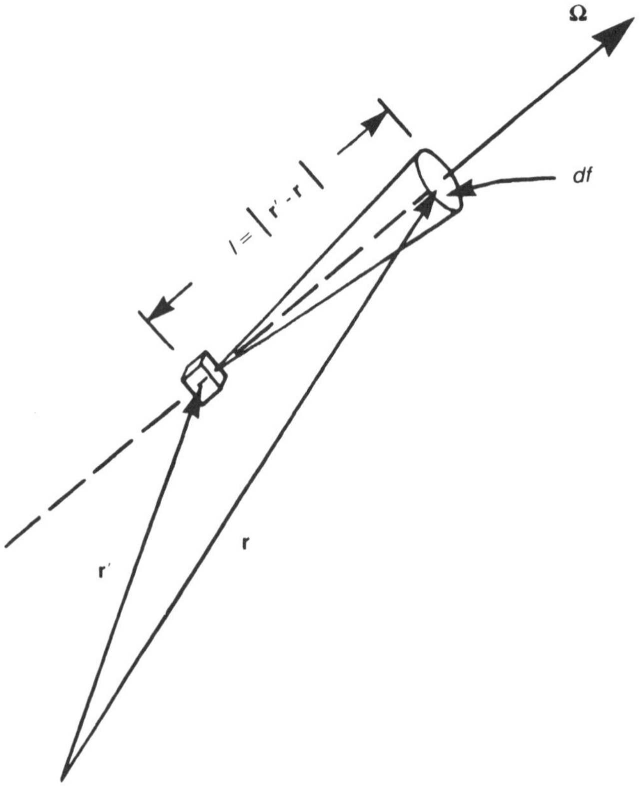

Fig. 2-3. Illustration of the integration in the derivation of the integral transport equation for the flux from isotropic sources.

In the formulation of the attenuation factor, Eq. (2.19), it is assumed the medium is homogeneous. The attenuation factor for inhomogeneous media is given by

$$\exp\left[-\int_0^l \Sigma_t(\boldsymbol{r}'',E)\,dl'\right] , \tag{2.20}$$

with l' denoting the distance from $\boldsymbol{r}'$ to $\boldsymbol{r}''$ along the direction $\boldsymbol{r}' - \boldsymbol{r}$.

The three elements—source, free flight, and attenuation—must be combined: $q(\boldsymbol{r}',E)\,dV'$, the isotropic source out of the volume element dV', is to be multiplied with $d\Omega/4\pi$ to obtain the fraction emitted into $d\Omega$. The result is divided by $df = d\Omega \cdot l^2$ in order to obtain at $\boldsymbol{r}$ a flux per square centimeter; then the attenuation factor has to be applied. The subsequent integration over space combines the contributions from all directions and thus yields the flux:

$$\phi(\boldsymbol{r},E) = \int_{V'} q(\boldsymbol{r}',E)\exp[-\Sigma_t(E)|\boldsymbol{r}' - \boldsymbol{r}|]\frac{dV'}{4\pi|\boldsymbol{r}' - \boldsymbol{r}|^2} . \tag{2.21}$$

Since $q(\boldsymbol{r}',E)$ depends on the flux $\phi(\boldsymbol{r}',E)$ according to Eq. (2.15), Eq. (2.21) is not an explicit prescription to calculate the flux. Rather, it is an integral equation for the flux $\phi(\boldsymbol{r},E)$, the integral transport equation.

The solutions of Eq. (2.21) are beyond the scope of this text and are not considered.

The often applied one-group transport equation for thermal neutrons is readily obtained by appropriate simplifications of Eq. (2.21). If Σ_s is independent of space and if the considered one-group neutrons are not directly produced in the medium V, the source, Eq. (2.15), assumes the form

$$q(\boldsymbol{r}) = \Sigma_S \phi(\boldsymbol{r}) . \tag{2.22}$$

The E' integration is eliminated in the transition to a one-group model (see Sec. 2-2D). One then obtains for the

one-group integral transport equation:

$$\phi(\boldsymbol{r}) = \Sigma_S \int_V \phi(\boldsymbol{r}')\exp(-\Sigma_t|\boldsymbol{r}' - \boldsymbol{r}|)\frac{dV'}{4\pi|\boldsymbol{r}' - \boldsymbol{r}|^2} . \tag{2.23}$$

Neutrons may enter the medium V through the boundary of V. They are then considered in the integration at the boundary.

2-2C The Diffusion Theory Equation

It is often possible to find a good approximation of the neutron flux without going through the costly and in most cases practically impossible procedure of finding the angular flux first. The simplest of these approximations is the "diffusion theory."

In Chapter 1 it was shown that reaction rate densities can legitimately be expressed in terms of the flux $\phi(\boldsymbol{r},E)$, i.e., one need not involve the angular flux $\phi(\boldsymbol{r},E,\boldsymbol{\Omega})$ or consider angular reaction rates. This suggests that one might be able to formulate a balance equation for the neutron flux alone, and that one does not have to go through the rigorous and complicated Boltzmann equation. The diffusion equation is derived independently although it can also be obtained by integrating the Boltzmann equation with respect to $\boldsymbol{\Omega}$ and properly simplifying the result.

In the derivation of a balance equation for the flux only, the same approach is employed as in deriving the Boltzmann equation, i.e., one considers the neutron gains and losses in an infinitesimal spatial volume. The derivation is carried out in three dimensions; thus a three-dimensional differential volume, $dV = \Delta x \Delta y \Delta z$, is considered, which is located at space point $\boldsymbol{r}$ as shown in Fig. 2-4.

The production rate of neutrons of energy E in dE and in volume dV at $\boldsymbol{r}$ has three contributions:

$$S(\boldsymbol{r},E)\Delta x \Delta y \Delta z = \text{independent neutron source in } dV \quad ,$$

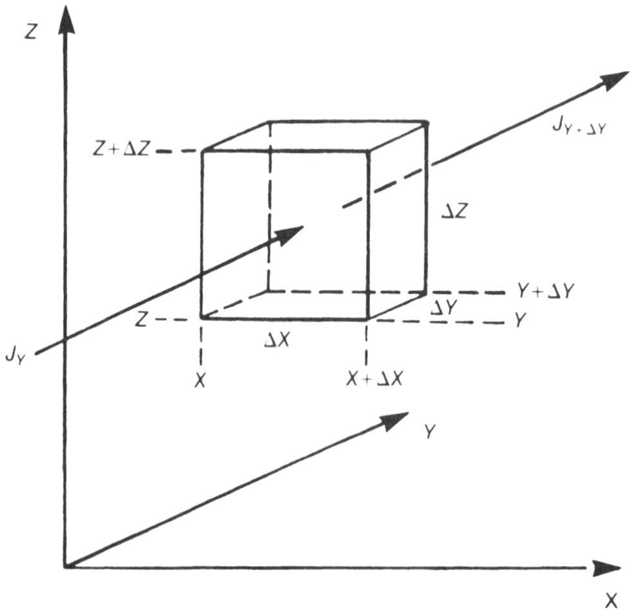

Fig. 2-4. Illustration of the derivation of the neutron leakage term in the diffusion theory.

$$\chi(E)\int_{E'} \nu\Sigma_f(\boldsymbol{r},E')\phi(\boldsymbol{r},E')\,dE'\Delta x\Delta y\Delta z = \text{fission source in } dV \quad ,$$

and

$$\int_{E'} \Sigma_S(\boldsymbol{r},E'\to E)\phi(\boldsymbol{r},E')\,dE'\;\Delta x\Delta y\Delta z$$

$$= \text{scattering source from neutrons of energy } E' \text{ in } dV \quad . \tag{2.24a}$$

The neutron losses consist of:

$$\Sigma_a(\boldsymbol{r},E)\phi(\boldsymbol{r},E)\Delta x\Delta y\Delta z = \text{ absorption losses in } dV \quad ,$$

$$\Sigma_S(\boldsymbol{r},E)\phi(\boldsymbol{r},E)\Delta x\Delta y\Delta z = \text{scattering losses in } dV \quad ,$$

and

$$\left.\begin{aligned}&[J_{x+\Delta x}(\boldsymbol{r},E) - J_x(\boldsymbol{r},E)]\Delta y\Delta z\\ +&[J_{y+\Delta y}(\boldsymbol{r},E) - J_y(\boldsymbol{r},E)]\Delta x\Delta z\\ +&[J_{z+\Delta z}(\boldsymbol{r},E) - J_z(\boldsymbol{r},E)]\Delta x\Delta y\end{aligned}\right\} = \begin{aligned}&\text{neutron leakage from } dV\\ &\text{in } x,\ y, \text{ and } z \text{ directions,}\\ &\text{respectively} \quad .\end{aligned} \tag{2.24b}$$

Equating the above stationary neutron gains to losses, as given in Eqs. (2.24), dividing by $\Delta x\Delta y\Delta z$, and taking the limit of $\Delta x\Delta y\Delta z$ to infinitesimally small values yields the balance equation in terms of $\phi(\boldsymbol{r},E)$:

$$S(\boldsymbol{r},E) + \chi(E)\int_{E'} \nu\Sigma_f(\boldsymbol{r},E')\phi(\boldsymbol{r},E')\,dE' + \int_{E'} \Sigma_S(\boldsymbol{r},E'\to E)\phi(\boldsymbol{r},E')\,dE'$$

$$= \Sigma_a(\boldsymbol{r},E)\phi(\boldsymbol{r},E) + \Sigma_S(\boldsymbol{r},E)\phi(\boldsymbol{r},E) + \nabla\cdot\boldsymbol{J}(\boldsymbol{r},E) \quad . \tag{2.25}$$

All terms in Eq. (2.25) can be obtained directly from integrating the Boltzmann equation, Eq. (2.5), with respect to Ω as shown in Chapter 6.

The leakage term is given by the vector gradient of the current and is not (yet) expressed in terms of the flux. Since the current is given by

$$\boldsymbol{J}(\boldsymbol{r},E) = \int_{\Omega} \boldsymbol{\Omega}\phi(\boldsymbol{r},E,\boldsymbol{\Omega})\,d\boldsymbol{\Omega} \quad , \tag{2.26}$$

one needs an approximate expression for the angular flux to express $\boldsymbol{J}$ in terms of the flux (see Chapter 6).

Since it is not possible to derive a rigorous neutron balance equation for the flux $\phi(\boldsymbol{r},E)$ alone, various procedures have been developed for an approximate calculation of the current $\boldsymbol{J}(\boldsymbol{r},E)$. The major ones are discussed in this text. The most important of these procedures is the diffusion theory, an approach that has been employed in various areas of physics (particularly in the theory of gases) prior to its application in nuclear reactor theory.

The diffusion approximation consists of assuming that the current can be found as the gradient of the flux,

Fick's Law of Diffusion:

$$\boldsymbol{J}(\boldsymbol{r},E) = -D(\boldsymbol{r},E)\nabla\phi(\boldsymbol{r},E) \quad . \tag{2.27}$$

Replacing the neutron current in Eq. (2.25) by the Fick's Law expression of Eq. (2.27) yields the

diffusion equation:

$$-\nabla\cdot D(\boldsymbol{r},E)\nabla\phi(\boldsymbol{r},E) + \Sigma_t(\boldsymbol{r},E)\phi(\boldsymbol{r},E) - \int_{E'} \Sigma_S(\boldsymbol{r},E'\rightarrow E)\phi(\boldsymbol{r},E')\,dE'$$

$$= \chi(E)\int_{E'} \nu\Sigma_f(\boldsymbol{r},E')\phi(\boldsymbol{r},E')\,dE' + S(\boldsymbol{r},E) \quad . \tag{2.28}$$

It seems plausible to assume that the migration of neutrons should qualify for the application of Fick's Law due to the large number of neutron collisions. Fick's Law essentially expresses the fact that the current is due to

> more neutrons moving from the areas of high neutron density to the areas of low density than the other way around.

However, Fick's Law is an approximation in that it assumes that the current develops out of the free back-and-forth diffusion of neutrons in such a way that the net result is determined by the relative density of neutrons at different locations. In regions of strong absorption, or at boundaries between materials with very different types of nuclear cross sections, or at a neutron source, the assumption of free diffusion does not apply. The angular distribution of the flux is directly influenced by interfaces and small areas of large cross sections. In essence, the approximate character of the diffusion theory in these cases is the price one pays for giving up the direct dependence of the current on the angular flux, Eq. (2.26).

The qualitative conclusion from this discussion can be proved quantitatively; namely, that diffusion theory yields accurate results only several ($\simeq 3$) mean-free-paths (mfps) away from interfaces or localized sources.

Practical cases in which the free diffusion of neutrons is inhibited and, therefore, the diffusion theory approximation of the current is inaccurate or invalid are indicated in the following.

Control blades or rods that have a high absorption cross section obviously interrupt the free back-and-forth diffusion and thus invalidate the application of diffusion theory.

An even more drastic example is a medium of very low density with dimensions that are small or comparable to the total mfp. Then, neutrons scatter primarily with the surrounding wall material, and there is no free back-and-forth diffusion within the medium.

A third case in which the neutron current is poorly represented by the gradient of the flux is the medium around a concentrated neutron source (such as a "point" source or a "plane" source). The source emits neutrons and the resulting part of the current is obviously unrelated to the gradient of the flux.

A fourth case is the boundary between a medium and a vacuum (see Sec. 2-2E, especially Fig. 2-6).

A further discussion of the limitations of the diffusion approximation is presented in Chapter 6.

2-2D Few-Group Diffusion Equations

A continuous energy dependence of the neutron flux has been included in the formulation of the balance equation of the neutron diffusion as presented in Sec. 2-2C. For practical reactor problems, one generally does not have to calculate the full continuous energy dependence together with the space dependence. It is normally sufficient to lump the energy-dependent flux into several components. The neutrons in each of these components are called a neutron "group" and the corresponding flux component is called the "group flux." The approach of subdividing the neutron population into groups, i.e., the flux into group fluxes, is called the "multigroup" approach. The general multigroup equations are presented in Chapter 7. In this section, the simple few-group diffusion equations are derived from the general energy-dependent equation, Eq. (2.28). If Eq. (2.28) is integrated over *all* energies, a diffusion equation is obtained that depends only on space. This equation is usually called the "one-group" diffusion equation although terms like "one-speed" are also sometimes used.

The term "one-group" does not uniquely identify the type of approximation applied. Three different one-group diffusion equations are derived in the following: equations for the total, the thermal, and the fast neutron flux, respectively.

Derivation of the One-Group Diffusion Equation for the Total Flux, $\phi(\boldsymbol{r})$. The derivation of the diffusion equation for the total flux requires the multiplication of Eq. (2.28) by dE and integration with respect to E from zero to infinity. This leads to the following components:

$$\int_E S(\boldsymbol{r},E)\,dE = S(\boldsymbol{r}) = \text{total source density} \tag{2.29}$$

and

$$\int_E \chi(E)\,dE \int_{E'} \nu\Sigma_f(\boldsymbol{r},E')\phi(\boldsymbol{r},E')\,dE' = \nu\Sigma_f(\boldsymbol{r})\phi(\boldsymbol{r}) \quad , \tag{2.30}$$

since

$$\int_E \chi(E)\,dE = 1 \quad ,$$

according to Eq. (2.10); $\phi(\boldsymbol{r})$ is the total flux defined in Eq. (1.9), and the quantity $\nu\Sigma_f(\boldsymbol{r})$ is the flux weighted average value of $\nu\Sigma_f(\boldsymbol{r},E)$:

$$\nu\Sigma_f(\boldsymbol{r}) = \overline{\nu\Sigma_f(\boldsymbol{r},E)}^E = \frac{\int_E \nu\Sigma_f(\boldsymbol{r},E)\phi(\boldsymbol{r},E)\,dE}{\int_E \phi(\boldsymbol{r},E)\,dE} \quad . \tag{2.31}$$

Note that $\nu\Sigma_f(\boldsymbol{r})$ is defined by Eq. (2.31) as a single quantity and not a product of two quantities. As the averaging formula, Eq. (2.31) yields just one number.

If one would like to find *two* average quantities $\overline{\nu}$ and $\overline{\Sigma_f}$ instead of the single one of Eq. (2.31), obviously two equations are needed. The second equation can be easily provided by the definition of an average Σ_f:

$$\Sigma_f(\boldsymbol{r}) = \overline{\Sigma_f(\boldsymbol{r},E)}^E = \frac{1}{\phi(\boldsymbol{r})}\int_E \Sigma_f(\boldsymbol{r},E)\phi(\boldsymbol{r},E)\,dE \quad . \tag{2.32}$$

Dividing Eq. (2.31) by the average Σ_f yields the average ν:

$$\overline{\nu} = \frac{\overline{\nu\Sigma_f(\boldsymbol{r},E)}^E}{\overline{\Sigma_f(\boldsymbol{r},E)}^E} \quad , \tag{2.33}$$

or explicitly:

$$\overline{\nu} = \frac{\int_E \nu\Sigma_f(\boldsymbol{r},E)\phi(\boldsymbol{r},E)\,dE}{\int_E \Sigma_f(\boldsymbol{r},E)\phi(\boldsymbol{r},E)\,dE} \quad . \tag{2.34}$$

The value of $\overline{\nu}$ resulting from Eqs. (2.33) and (2.34) may show a slight space dependence, but often one does not form $\overline{\nu}$ at all, but instead operates with the average $\overline{\nu\Sigma_f}$ as a single quantity given by Eq. (2.31).

A formula of the same type as Eq. (2.32) holds for neutron absorption, a, and scattering, S; e.g., for absorption:

$$\Sigma_a(\boldsymbol{r}) = \frac{1}{\phi(\boldsymbol{r})}\int_E \Sigma_a(\boldsymbol{r},E)\phi(\boldsymbol{r},E)\, dE \quad . \tag{2.35}$$

The integration of the two scattering terms in Eq. (2.28) yields

$$\begin{aligned}\int_E \Sigma_S(\boldsymbol{r},E)\phi(\boldsymbol{r},E)\, dE &- \int_E\int_{E'} \Sigma_S(\boldsymbol{r},E'\to E)\phi(\boldsymbol{r},E')\, dE'\, dE \\ &= \Sigma_S(\boldsymbol{r})\phi(\boldsymbol{r}) - \int_{E'}\int_E \Sigma_S(\boldsymbol{r},E'\to E)\, dE\ \phi(\boldsymbol{r},E')\, dE' \\ &= \Sigma_S(\boldsymbol{r})\phi(\boldsymbol{r}) - \Sigma_S(\boldsymbol{r})\phi(\boldsymbol{r}) = 0 \quad .\end{aligned} \tag{2.36}$$

Use has been made of the Sec. 1-5 definition of the total scattering cross section based on the differential cross-section concept, i.e.,

$$\Sigma_s(\boldsymbol{r},E') = \int_E \Sigma_s(\boldsymbol{r},E'\to E)\, dE \quad .$$

Thus, integration over all energies eliminates the scattering terms in the neutron balance equation. This is physically obvious since each neutron that disappears at energy E, i.e.,

$$\Sigma_S(\boldsymbol{r},E)\phi(\boldsymbol{r},E) \quad ,$$

reappears with certainty at some other energy E',

$$\int_E \Sigma_S(\boldsymbol{r},E\to E')\phi(\boldsymbol{r},E)\, dE \quad ,$$

so that the net result on the *total* neutron balance is zero. In other words, the number of "departures" equals the number of "arrivals."

The integral of the energy-dependent leakage term is expressed in terms of an average diffusion constant, $D(\boldsymbol{r})$:

$$\int_E D(\boldsymbol{r},E)\nabla\phi(\boldsymbol{r},E)\, dE = D(\boldsymbol{r})\nabla\phi(\boldsymbol{r}) \quad . \tag{2.37}$$

The average diffusion constant cannot be obtained from Eq. (2.37) in the same way as, for example, Σ_f was in Eq. (2.32) since one cannot divide by a vector (here $\nabla\phi$) and obtain a scalar number. Actually, Eq. (2.37) represents three equations, one for each component of $\nabla\phi$:

$$\int_E D(\boldsymbol{r},E)\frac{\partial}{\partial x}\phi(\boldsymbol{r},E)\, dE = D_x(\boldsymbol{r})\frac{\partial}{\partial x}\phi(\boldsymbol{r}) \quad , \tag{2.38a}$$

$$\int_E D(\boldsymbol{r},E)\frac{\partial}{\partial y}\phi(\boldsymbol{r},E)\,dE = D_y(\boldsymbol{r})\frac{\partial}{\partial y}\phi(\boldsymbol{r}) \quad , \tag{2.38b}$$

and

$$\int_E D(\boldsymbol{r},E)\frac{\partial}{\partial z}\phi(\boldsymbol{r},E)\,dE = D_z(\boldsymbol{r})\frac{\partial}{\partial z}\phi(\boldsymbol{r}) \quad . \tag{2.38c}$$

Equations (2.38) may be abbreviated in vector notation

$$\int_E D(\boldsymbol{r},E)\begin{pmatrix}\nabla_x\\ \nabla_y\\ \nabla_z\end{pmatrix}\phi(\boldsymbol{r},E)\,dE = \begin{pmatrix}D_x & 0 & 0\\ 0 & D_y & 0\\ 0 & 0 & D_z\end{pmatrix}\begin{pmatrix}\nabla_x\\ \nabla_y\\ \nabla_z\end{pmatrix}\phi(\boldsymbol{r}) \quad . \tag{2.39}$$

Thus, $D(\boldsymbol{r})$ actually is a diagonal matrix. The three components of **D** are, for the most part, not very different. One therefore normally uses a scalar diffusion constant, except in special cases for which a significant directional dependence of **D** is indicated.

Combining all terms then yields the

one-group diffusion equation for the total flux:

$$-\nabla\cdot D(\boldsymbol{r})\nabla\phi(\boldsymbol{r}) + \Sigma_a(\boldsymbol{r})\phi(\boldsymbol{r}) = \nu\Sigma_f(\boldsymbol{r})\phi(\boldsymbol{r}) + S(\boldsymbol{r}) \quad . \tag{2.40}$$

Thus, the neutron balance of the total flux is governed by the interplay of diffusion, absorption, and sources.

Derivation of the One-Group Equation for the Thermal Neutron Flux. For the treatment of thermal neutrons or thermal reactor problems, one often employs a one-group diffusion equation for the thermal neutrons only, i.e., for neutrons having energies comparable to the energy of the thermal motion of the scattering medium. Then the energy integrations in the definitions of the one-group quantities [Eqs. (2.29) to (2.37)] are extended only over the thermal group, i.e., from 0 to E_{th} (E_{th} is usually chosen between 0.4 and 2 eV). The neutrons of higher energies are treated in some other way; compare, for example, the discussion in Chapter 7 of two-group equations.

If the energy integration is extended over only part of the range, the two scattering terms in Eq. (2.28) will not cancel. The first term gives (note that there is no inelastic scattering for thermal neutrons):

$$\int_0^{E_{th}} \Sigma_S(\boldsymbol{r},E)\phi(\boldsymbol{r},E)\,dE = \Sigma_s^{th}(\boldsymbol{r})\phi_{th}(\boldsymbol{r}) \quad , \tag{2.41}$$

with ϕ_{th} being the thermal flux of Eq. (1.9) and Σ_s^{th} being the average scattering cross sections for thermal neutrons. The second term can be split into two parts (the space dependence is temporarily omitted in the notation):

$$\int_0^{E_{th}}\int_0^{\infty}\Sigma_S(E'\to E)\phi(E')\,dE'\,dE$$
$$=\int_0^{E_{th}}\int_0^{E_{th}}\Sigma_S(E'\to E)\phi(E')\,dE'\,dE$$
$$+\int_0^{E_{th}}\int_{E_{th}}^{\infty}\Sigma_S(E'\to E)\phi(E')\,dE'\,dE \quad . \tag{2.42}$$

If E_{th} is chosen high enough so that virtually no upscattering occurs beyond E_{th}, one can extend the limit of the E integration in the first integral on the right side of Eq. (2.42) to infinity without making an appreciable error. The result of this integral is the total scattering cross section in the thermal range, i.e., the same term as obtained in Eq. (2.41).

The second term on the right side of Eq. (2.42) describes the scattering from the fast into the thermal range. Combining Eqs. (2.41) and (2.42) gives

$$\Sigma_s^{th}(\boldsymbol{r})\phi(\boldsymbol{r})-\int_0^{E_{th}}\int_0^{\infty}\Sigma_S(\boldsymbol{r},E'\to E)\phi(\boldsymbol{r},E')\,dE'\,dE$$
$$=\Sigma_s^{th}(\boldsymbol{r})\phi(\boldsymbol{r})-\Sigma_s^{th}(\boldsymbol{r})\phi(\boldsymbol{r})-S_S^{th}(\boldsymbol{r})=-S_S^{th}(\boldsymbol{r}) \quad . \tag{2.43}$$

Thus, the scattering within the thermal group cancels (as does all "within group" scattering if one does not have upscattering beyond the group boundary). Thus, the number of "arrivals" of scattered neutrons equals the number of "departures" in the thermal group. The residue of the scattering terms in Eq. (2.43) represents the scattering source of thermal neutrons; $S_S^{th}(\boldsymbol{r})$.

The other terms in the diffusion equation are condensed in the same way as for the total flux; the resulting average group constants hold for the thermal neutrons [see, for example, Eq. (2.41)]. The only change is in the fission source, which can be neglected for thermal neutrons since

$$\int_0^{E_{th}}\chi(E)\,dE\simeq 0 \quad .$$

Combining all terms gives the

one-group diffusion equation for thermal neutrons:

$$-\nabla\cdot D_{th}(\boldsymbol{r})\nabla\phi_{th}(\boldsymbol{r})+\Sigma_a^{th}(\boldsymbol{r})\phi_{th}(\boldsymbol{r})=S_S^{th}(\boldsymbol{r})+S^{th}(\boldsymbol{r}) \quad , \tag{2.44}$$

which formally agrees with Eq. (2.40), except that the fission source is replaced by a scattering source from above-thermal energies.

The one-group equation for fast neutrons can be derived in the same way as the previous one-group equations, Eqs. (2.40) and (2.44). A subscript 1 is used to denote the fast group, as it is the "first" group in a two- or multigroup scheme; a subscript 2 indicates the thermal group in a two-group scheme. The result is

$$-\nabla\cdot D_1(\boldsymbol{r})\nabla\phi_1(\boldsymbol{r}) + \Sigma_{a1}(\boldsymbol{r})\phi_1(\boldsymbol{r}) + \Sigma_{S12}(\boldsymbol{r})\phi_1(\boldsymbol{r}) = \nu\Sigma_{f1}(\boldsymbol{r})\phi_1(\boldsymbol{r}) + \nu\Sigma_{f2}(\boldsymbol{r})\phi_2(\boldsymbol{r}) + S_1(\boldsymbol{r}) \quad . \tag{2.45}$$

The left side of Eq. (2.45) contains as an additional loss term the scattering of neutrons from the fast to the thermal group (group 1 to group 2 in a two-group scheme):

$$\Sigma_{S12}(\boldsymbol{r})\phi_1(\boldsymbol{r}) = S_S^{th}(\boldsymbol{r}) \quad , \tag{2.46}$$

which equals the scattering source in Eq. (2.44). The right side contains the fission source from fast and thermal neutron fission:

$$\nu\Sigma_{f1}(\boldsymbol{r})\phi_1(\boldsymbol{r}) = \text{source from "fast" fission} \quad ,$$
$$\nu\Sigma_{f2}(\boldsymbol{r})\phi_2(\boldsymbol{r}) = \text{source from thermal neutron fission} \quad . \tag{2.47}$$

Since Eqs. (2.44) and (2.45) are obtained from energy integrals over two intervals, which together make up the total interval used in Eq. (2.40), one should expect that the sum of Eqs. (2.44) and (2.45) equals Eq. (2.40). This can be readily verified for all reaction rates and sources. However, the sum of the diffusion terms equals the diffusion term in Eq. (2.40) only approximately due to the fact that approximate scalar diffusion constants are used in all cases.

Equations (2.44) and (2.45) taken together are the *two-group diffusion equations*, a coupled set of differential equations for the two unknown group fluxes $\phi_1(\boldsymbol{r})$ and $\phi_2(\boldsymbol{r})$, i.e., for the fast and thermal flux, respectively:

$$-\nabla\cdot D_1\nabla\phi_1(\boldsymbol{r}) + \Sigma_{a1}\phi_1(\boldsymbol{r}) + \Sigma_{S12}\phi_1(\boldsymbol{r}) = \nu\Sigma_{f1}\phi_1(\boldsymbol{r}) + \nu\Sigma_{f2}\phi_2(\boldsymbol{r}) + S_1(\boldsymbol{r}) \quad ,$$
$$-\nabla\cdot D_2\nabla\phi_2(\boldsymbol{r}) + \Sigma_{a2}\phi_2(\boldsymbol{r}) = \Sigma_{S12}\phi_1(\boldsymbol{r}) + S_2(\boldsymbol{r}) \quad . \tag{2.48}$$

All fission neutrons are assumed to be generated in the fast group.

A more detailed description of the neutron energy dependency is normally required for several reasons. Recall that neutrons are born by fission at relatively high energies (~2 MeV) and slow down to very low energies (~0.025 eV), where the majority of new fissions occur. The strong dependency of the nuclear cross sections on incident neutron energy over this large energy span necessitates a more realistic energy description for any reactor calculational model. The approach that is almost universally employed to account for this energy dependency is the *multigroup* method or model. This model is basically an extension of the two-group model into a generalized many group format.

The multigroup theory is described in detail in Chapter 7 after the information on the continuous energy dependence of the flux has been developed in Chapters 4 and 5. This information is needed for the "generation" of the group constants. Here only a preliminary formulation is presented.

In this multigroup model, the continuous neutron energy range in the diffusion model of Eq. (2.28) is partitioned into G discrete energy intervals. This subdivision results in the neutron flux $\phi(\boldsymbol{r},E)$ being represented by a series of G group fluxes, $\phi_g(\boldsymbol{r})$. These multigroup fluxes $\phi_g(\boldsymbol{r})$ are defined as the integral of $\phi(\boldsymbol{r},E)$ over the energy of each group (E_g to E_{g-1}) and represent the flux of all neutrons in the g'th group whose energy E is between $E_g < E < E_{g-1}$ (note that the group index increases as the energy decreases):

$$\phi_g(\boldsymbol{r}) = \int_{E_g}^{E_{g-1}} \phi(\boldsymbol{r},E)\, dE = \int_g \phi(\boldsymbol{r},E)\, dE \quad . \tag{2.49}$$

Applying this energy discretization to Eq. (2.28) results in the multigroup diffusion equation written for a space point $\boldsymbol{r}$ and energy group g, a generalization of the two-group equations, Eqs. (2.48):

$$\begin{aligned} -\nabla\cdot D_g(\boldsymbol{r})\nabla\phi_g(\boldsymbol{r}) + \Sigma_{tg}(\boldsymbol{r})\phi_g(\boldsymbol{r}) - \sum_{g'}\Sigma_{sg'g}\phi_{g'}(\boldsymbol{r}) \\ = \chi_g\sum_{g'}\nu\Sigma_{fg'}\phi_{g'}(\boldsymbol{r}) \quad . \end{aligned} \tag{2.50}$$

The multigroup formulation represents a set of G equations describing the diffusion and slowing down of neutrons as they progress from group to group. Each term is a reaction rate; even the leakage term of Eq. (2.50) can be interpreted as such. These equations are coupled through the group transfer cross sections and through the fission source term.

A detailed description of this multigroup model is provided in Chapter 7 along with a brief sketch of the procedure for its numerical solution. The group cross sections are to be generated for each group using a procedure in which reaction rates are conserved (see Sec. 7-3 for details).

While in theory it is possible to employ a great number of energy groups, in practice a few groups are usually sufficient. For most thermal reactor calculations, 2 to 4 groups are traditionally employed; a larger number is necessary for fast reactors (10 to 100 groups).

2-2E Boundary Conditions for Transport and Diffusion Equations

Differential equations must be completed by boundary conditions in order to have unique solutions. Differential equations represent a complete formulation of a problem only when accompanied by boundary conditions.

Furthermore, since the solution of differential equations generally requires the coefficients to be continuous, a larger domain involving discontinuous coefficients, e.g., in the form of different compositions in different regions, must be broken down into subdomains (regions), with continuous coefficients in each subdomain. Solutions are obtained in each subdomain, and the subdomain solutions are connected by interface conditions to obtain the total solution. Eventually, the solution, connected at all interfaces, is subjected to the outer boundary conditions, which then yields the desired solution.

The *Boltzmann equation* is an integro-differential equation of first order; thus *one* boundary condition and *one* condition for each interface are required.

Interface condition: $\phi(\boldsymbol{r},E,\Omega)$ is continuous at region interfaces, since the "interface" has neither a finite neutron absorption nor emission capability:

$$\phi(\boldsymbol{r}_b^+,E,\boldsymbol{\Omega})=\phi(\boldsymbol{r}_b^-,E,\boldsymbol{\Omega}) \quad , \tag{2.51}$$

where $\boldsymbol{r}_b^+$ and $\boldsymbol{r}_b^-$ indicate the two sides of the interface.

The *boundary condition* at the outer boundary ($\boldsymbol{r}_v$), with a vacuum outside of a convex medium, is

$$\phi(\boldsymbol{r}_v,E,\boldsymbol{\Omega})\equiv 0 \text{ for } \boldsymbol{\Omega} \text{ pointing inward} \quad , \tag{2.52}$$

i.e., no neutrons move from the vacuum into the system since a vacuum contains neither a source nor a scattering material. Equation (2.52) also means that no neutrons emitted into the vacuum from the reactor will ever come back; it is therefore sometimes called the "black body" boundary condition.

The *integral transport equation* does not contain a derivative. Therefore, no boundary conditions are required.

It is instructive to realize from physical considerations why the integral transport equation does not need a boundary condition, for example, at the interface with a vacuum. One can calculate

$$\phi(\boldsymbol{r}_v,E,\boldsymbol{\Omega}_{\text{inward}}) = \int \begin{pmatrix}\text{all contributions in the direction} \\ \text{of } \boldsymbol{\Omega} \text{ out of the vacuum}\end{pmatrix} dl \quad . \tag{2.53}$$

There *are* neutrons in the vacuum, namely those that have been emitted from the reactor, but there are neither scattering centers nor sources in the vacuum. Thus, there is *no* contribution to the integral and the condition that must be imposed on the Boltzmann equation is obtained by integration in the case of the integral transport equation:

$$\phi(\boldsymbol{r}_v,E,\boldsymbol{\Omega}_{\text{inward}})\equiv 0 \;[\equiv \text{to Eq. (2.52)}] \quad .$$

In the same way, one can derive the interface condition for the angular flux.

The *diffusion theory* balance equation is a differential equation of second order; thus, it requires two boundary conditions as well as two conditions on each interface if present in the system. Normally, outside the boundaries or the external surfaces of reactors, vacuum is assumed. Reactor interfaces on the other hand are surfaces between regions with different compositions.

The flux and the current are originally defined as integrals of the angular flux. Thus, the corresponding interface and boundary conditions can be derived from these definitions and applied to the interface and boundary condition for the angular flux, Eqs. (2.51) and (2.52).

For the interface conditions, one obtains from the equality of the angular fluxes at both sides of the interface ($\boldsymbol{r}_b^+$ and $\boldsymbol{r}_b^-$ denote the space coordinates at the two sides of an interface):

$$\phi(\boldsymbol{r}_b^+,E) = \int_\Omega \phi(\boldsymbol{r}_b^+,E,\boldsymbol{\Omega})\,d\Omega$$

$$= \int_\Omega \phi(\boldsymbol{r}_b^-,E,\boldsymbol{\Omega})\,d\Omega = \phi(\boldsymbol{r}_b^-,E) \quad , \tag{2.54}$$

i.e., the flux is continuous at interfaces. Also:

$$\boldsymbol{J}(\boldsymbol{r}_b^+,E) = \int_\Omega \boldsymbol{\Omega}\phi(\boldsymbol{r}_b^+,E,\boldsymbol{\Omega})\,d\Omega$$

$$= \int_\Omega \boldsymbol{\Omega}\phi(\boldsymbol{r}_b^-,E,\boldsymbol{\Omega})\,d\Omega = \boldsymbol{J}(\boldsymbol{r}_b^-,E) \quad , \tag{2.55}$$

i.e., the current is continuous at interfaces. Note that from Fick's Law [Eq. (2.27)] and from Eq. (2.55) it follows that the first derivative of the flux cannot be continuous at a boundary between different materials if D has different values on the two sides of the interface since $\boldsymbol{J}$ is continuous. An example of the flux and current boundary condition in diffusion theory is illustrated in Fig. 2-5.

The boundary condition at a vacuum represents a particular problem in the framework of diffusion theory. Since the detailed angular

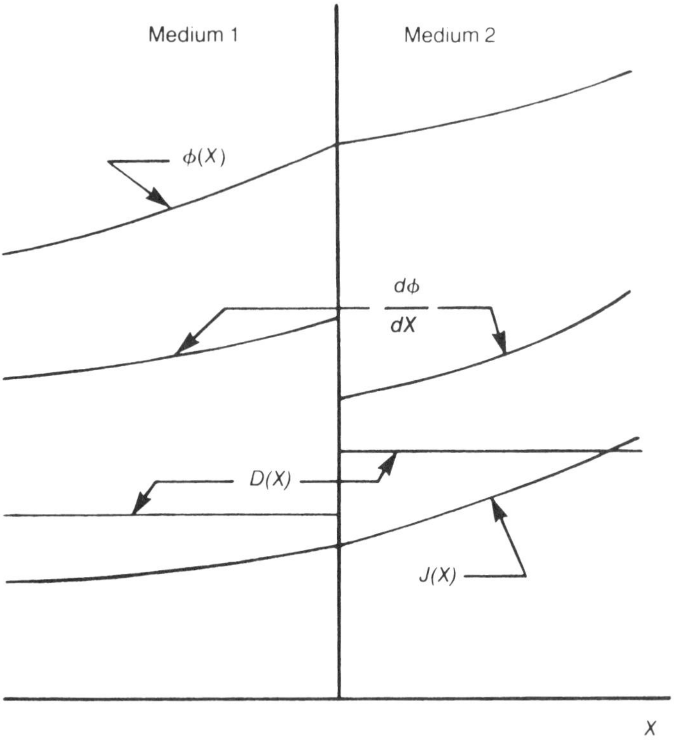

Fig. 2-5. Illustration of the space dependencies of $\phi(X)$, $J(X)$, $D(X)$, and $d\phi(X)/dX$ near an interface between two media in diffusion theory.

dependence is not incorporated in the model, the correct physical boundary condition,

$$\phi(\boldsymbol{r}_v, E, \boldsymbol{\Omega}_{\text{inward}}) \equiv 0 \quad ,$$

cannot be realized. Therefore, the correct boundary condition, Eq. (2.52), must be replaced by an appropriate approximation. The derivation of the diffusion theory boundary condition is presented in the following only for the one-group diffusion equation.[a] Furthermore, since the curvature of the outer boundary is normally small, the angular distribution is considered only in plane geometry.

[a]For the boundary conditions of the multigroup diffusion equation see Chapter 7.

It is shown in Chapter 6 that the angular distribution in one-group diffusion theory in plane geometry is given by

$$[\phi(x,\mu)]_{\text{diff}} = \frac{1}{2}\phi(x) + \frac{3}{2}J(x)\mu \quad , \tag{2.56}$$

with $\mu = \cos\vartheta$ and ϑ being the angular deviation from the normal to the slab. Figure 2-6 shows μ times the angular distribution, Eq. (2.56), as a dashed parabola as it approximates a correct angular distribution given as solid line [note: $\phi(\boldsymbol{r}_v,\boldsymbol{\Omega})$ is zero for incoming directions]. The diffusion theory angular distribution in Fig. 2-6 is based on letting the net incoming current equal zero. It is obvious that one can approximate the physically correct boundary condition only crudely by letting only the *net* incoming part of the current equal zero.

Before deriving the analytical form of the zero net-incoming-current, it is verified that the angular flux of Eq. (2.56) yields the flux as well as the current after the appropriate integration with respect to solid angle:

$$\int_{-1}^{1} \phi(x,\mu)\, d\mu = \frac{1}{2}\phi(x)\cdot 2 + 0 = \phi(x) \tag{2.57}$$

and

$$\int_{-1}^{1} \mu\phi(x,\mu)\, d\mu = 0 + \frac{3}{2}J(x)\cdot\frac{2}{3} = J(x) \quad . \tag{2.58}$$

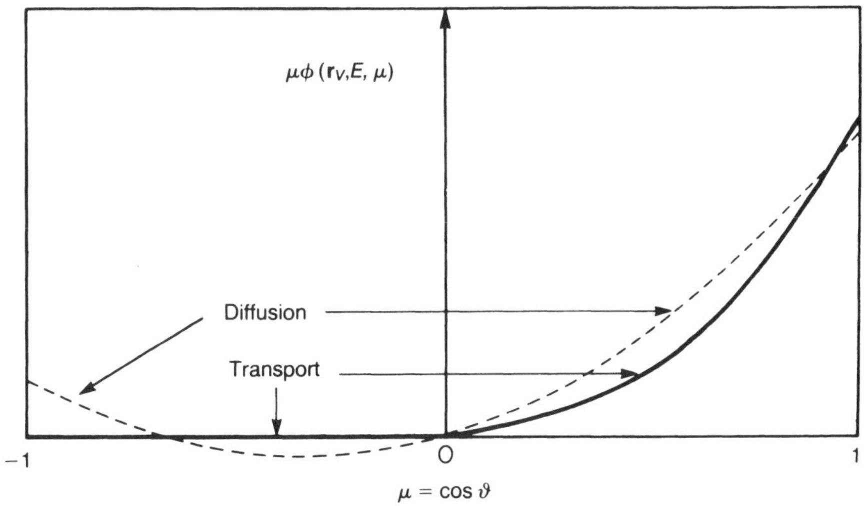

Fig. 2-6. Angular fluxes times μ at a vacuum boundary in transport theory (solid line) and in diffusion theory (dashed line).

To obtain the incoming part of the current, the range of μ integration is from -1 to 0 only.[b] Then both terms of the angular flux contribute:

$$J_{\text{in}} = \int_{-1}^{0} \mu\phi(x_v,\mu)\, d\mu$$

$$= \frac{1}{2}\phi(x_v)\int_{-1}^{0} \mu\, d\mu + \frac{3}{2}J(x_v)\int_{-1}^{0} \mu^2\, d\mu \tag{2.59}$$

and

$$J_{\text{in}}(x_v) = -\frac{1}{4}\phi(x_v) + \frac{1}{2}J(x_v) = 0 \quad . \tag{2.60}$$

Inserting the Fick's Law relationship [Eq. (2.27)]:

$$J(x) = -D\,\frac{d\phi(x)}{dx}$$

yields as the boundary condition:

$$\phi(x_v) + 2D\,\frac{d\phi(x_v)}{dx} = 0 \quad . \tag{2.61}$$

It is shown in Chapter 6 that the diffusion constant can be expressed by the inverse of a cross section, the "transport" cross section, Σ_{tr}. The corresponding mfp is denoted by λ_{tr}; it is related to D as:

$$D = \frac{\lambda_{tr}}{3} \tag{2.62}$$

and thus

$$\phi(x_v) + \frac{2}{3}\lambda_{tr}\left(\frac{d\phi}{dx}\right)_{x_v} = 0 \quad . \tag{2.63}$$

The transport path, λ_{tr}, will be defined below [see, however, Eq. (2.68)].

Since $\left(\frac{d\phi}{dx}\right)_{x_v}$ is generally negative for a medium with a vacuum at the right side, one can write the boundary condition in the following form, which holds on both sides of the interface:

$$\phi(x_v) - \frac{2}{3}\lambda_{\text{tr}}\left|\frac{d\phi}{dx}\right|_{x_v} = 0 \quad . \tag{2.64}$$

---

[b]If the vacuum is at the right of the medium.

Equation (2.64), the condition of vanishing *incoming* neutron current, represents the relationship between the flux and the flux gradient at the boundary. This boundary condition can be interpreted as the condition of *vanishing flux* at an extrapolated boundary, x_{ex}, with

$$x_{ex} = x_v + \frac{2}{3}\lambda_{tr} \quad . \tag{2.65a}$$

If ϕ is extrapolated linearly into the vacuum, Eq. (2.64) can be written as:

$$\phi(x_{ex}) = \phi(x_v + \frac{2}{3}\lambda_{tr}) = 0 \quad . \tag{2.65b}$$

The equality of Eqs. (2.64) and (2.65) for a linear flux extrapolation may be readily shown. A Taylor expansion of Eq. (2.65) is truncated automatically after the second term because of the linearity of the extrapolation:

$$\phi(x_v + \frac{2}{3}\lambda_{tr}) = \phi(x_v) - \frac{2}{3}\lambda_{tr}\left|\frac{d\phi}{dx}\right|_{x_v} = 0 \quad . \tag{2.66}$$

Figure 2-7 illustrates this boundary condition for a case with vacuum on the left side. The solid line gives the actual flux, which is constant in the vacuum around a slab of material.

In the more accurate transport theoretical derivation of the diffusion theory boundary condition, the factor 2/3 in Eq. (2.64) is replaced by 0.71:

$$\phi(x_v) - 0.71\lambda_{tr}\left|\frac{d\phi}{dx}\right|_{x_v} = 0 \quad . \tag{2.67a}$$

Thus,

$$\phi(x_v + 0.71\lambda_{tr}) = \phi(x_{ex}) = 0 \tag{2.67b}$$

and

$$x_{ex} = x_v + 0.71\lambda_{tr} \quad . \tag{2.67c}$$

It will be shown in Chapter 6 that the transport cross section is given by:

$$\lambda_{tr}^{-1} = \Sigma_{tr} = \Sigma_t - \bar{\mu}_s\Sigma_s \quad , \tag{2.68}$$

with $\bar{\mu}_s$ being the average cosine in elastic scattering. Based on the simple elastic scattering considerations in a scattering medium of mass number A, $\bar{\mu}_s = 2/3A$ [see Eq. (4.22)].

Two additional comments to the vacuum boundary condition are appropriate:

1. The linear extrapolation of the flux into the vacuum (shown by the dotted-dashed line in Fig. 2-7) is not meant to actually represent the flux distribution in this area. It is merely a different interpretation of the actual boundary condition, Eq. (2.67a), which only involves quantities at the boundary.

Physically, there is no material between the actual and the extrapolated boundary. Thus, the flux cannot be "calculated" from diffusion theory in this domain, since the application of diffusion theory requires the presence of material (otherwise, all macroscopic cross sections are zero and D is infinite). The following artifice may be employed to circumvent this difficulty if the application of the boundary condition, Eq. (2.67b), is desired, instead of Eq. (2.67a). If one substitutes Eq. (2.67b) for Eq. (2.67a) as a boundary condition, the actual boundary does not appear in the treatment any more and one automatically fills the extrapolated volume with material. This alters the physical problem and leads therefore to an inaccurate result. In many cases, however, this error is small or it is unimportant because the flux is small near the boundary.

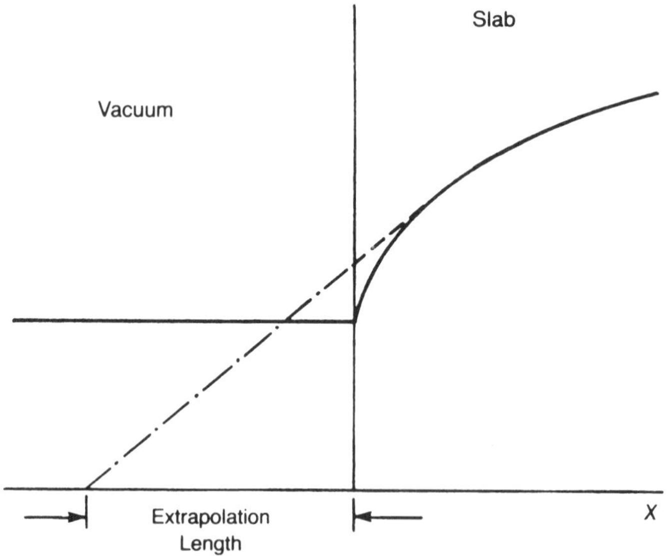

Fig. 2-7. Typical flux and flux extrapolation near a slab/vacuum interface. The solid line illustrates a flux calculated from transport theory. The dashed line corresponds to a diffusion theory flux. The dotted-dashed line represents the extrapolation of the diffusion theory flux.

2. The practical importance of the vacuum boundary condition is greatly reduced by the fact that reactor cores are surrounded by several large regions of material: reflectors or blankets, thermal shields and other vessel internals, and the reactor vessel. Furthermore, large regions of concrete are imbedded in the reactor vessel as a biological shield to provide the necessary protection for personnel and the population. Thus, the boundary condition can, in principle, be applied at a surface where the flux is *very* small and any errors in the boundary condition will have no impact on the reactor.

2-2F The One-Group Diffusion Equation for a One-Region Assembly

The simplest neutron balance equation is the one that describes a one-region assembly, surrounded by air or a vacuum, in one-group diffusion theory. One-region assemblies are, for example, bare one-region cores or cubes consisting of natural uranium fuel and a graphite moderator. Although the actual applicability of this simplest neutronics model is limited to one-region assemblies, it is very useful in heuristic or semiquantitative considerations.

In a one-region system with a uniform composition distribution, all macroscopic cross sections are independent of space; i.e.:

$$D(\boldsymbol{r}) = D \quad ,$$

$$\Sigma_a(\boldsymbol{r}) = \Sigma_a \quad ,$$

and

$$\nu\Sigma_f(\boldsymbol{r}) = \nu\Sigma_f \quad . \tag{2.69}$$

Suppose the source S is either zero or its spatial dependence is proportional to the flux:

$$S(\boldsymbol{r}) = S f_\phi(\boldsymbol{r}) \quad , \tag{2.70}$$

with $f_\phi(\boldsymbol{r})$ being the spatial shape of the flux:

$$\phi(\boldsymbol{r}) = \hat{\phi} f_\phi(\boldsymbol{r}) \quad . \tag{2.71}$$

The flux shape factor is normalized to one:

$$\int_V f_\phi(\boldsymbol{r})\, d\boldsymbol{r} = 1 \quad . \tag{2.72}$$

Then, the amplitude factor in Eq. (2.71) describes the integrated flux in the assembly:

$$\int_V \phi(\boldsymbol{r})\, d\boldsymbol{r} = \hat{\phi} \quad , \tag{2.73}$$

with $\hat{\phi}$ being the "integrated flux," which is used below.

With constant coefficients, Eq. (2.40) assumes the form

$$-D\nabla^2\phi(\boldsymbol{r}) + \Sigma_a\phi(\boldsymbol{r}) = \nu\Sigma_f\phi(\boldsymbol{r}) + Sf_\phi(\boldsymbol{r}) \quad , \tag{2.74}$$

or without the independent source

$$-D\nabla^2\phi(\boldsymbol{r}) + \Sigma_a\phi(\boldsymbol{r}) = \nu\Sigma_f\phi(\boldsymbol{r}) \quad . \tag{2.75}$$

Dividing by D and combining the absorption and fission terms yields:

$$\nabla^2\phi(\boldsymbol{r}) + \frac{\nu\Sigma_f - \Sigma_a}{D}\phi(\boldsymbol{r}) = 0 \quad . \tag{2.76a}$$

The factor in front of ϕ contains the information on the "material" that makes up the system. It is identified below in Sec. 2-6B as "material buckling," B_m^2:

$$B_m^2 = \frac{\nu\Sigma_f - \Sigma_a}{D} \quad . \tag{2.76b}$$

For simplicity, let $\phi(\boldsymbol{r})$ be zero at the extrapolated boundary [Eq. (2.67b)]:

$$\phi(\boldsymbol{r}_{ex}) = 0 \quad . \tag{2.76c}$$

The neutron balance equation, Eq. (2.76a), is of the same form as the wave equation (Helmholtz equation):

$$\nabla^2\phi(\boldsymbol{r}) + B^2\phi(\boldsymbol{r}) = 0 \quad . \tag{2.77}$$

As shown in the examples below, Eq. (2.77), with the boundary condition of Eq. (2.76c), has a positive solution only for a special value of B^2, the so-called "geometrical buckling," B_{geo}^2:

$$B^2 = B_{geo}^2 \quad . \tag{2.78}$$

The value of the geometrical buckling is uniquely given by the boundary conditions and cannot be obtained from the physics quantities in Eqs. (2.75) and (2.76).

The relations of Eqs. (2.76) and (2.77) are discussed extensively in Secs. 2-4 and 2-6. Here, Eq. (2.77) is merely used to eliminate $\nabla^2\phi$ from Eq. (2.75). This yields the homogeneous one-group, one-region diffusion equation:

$$(DB^2 + \Sigma_a)\phi(\boldsymbol{r}) = \nu\Sigma_f\phi(\boldsymbol{r}) \quad . \tag{2.79}$$

Integration with respect to space leads to the simplest neutron balance equation, a zero-dimensional, one-group balance equation:

$$(DB^2 + \Sigma_a)\hat{\phi} = \nu\Sigma_f\hat{\phi} \quad , \tag{2.80}$$

or with an independent source as given by Eq. (2.70):

$$(DB^2 + \Sigma_a)\hat{\phi} = \nu\Sigma_f\hat{\phi} + S \quad , \tag{2.81}$$

with B^2 being the geometrical buckling of a given one-region geometry. The meaning and the value of B^2 is further discussed in Secs. 2-4 and 2-6.

This rather simple one-group, one-dimensional reactor model described by Eq. (2.81) hardly portrays the real multigroup, multiregion reactor world. It contains nevertheless many of the necessary features, and is therefore used in the next several sections as an example of the fundamental physical and mathematical nature of the type of neutronics problems encountered in reactor design.

2-3 Operators in Reactor Applications

Although conceptually quite simple, the energy-dependent Boltzmann equation and even the diffusion equations are so lengthy that one can easily lose track of their meaning in the flood of all the symbols if they are used explicitly in lengthy derivations. Furthermore, many derivations and relations that involve neutron balance equations are independent of the specific approximation. There is, therefore, an incentive to express these equations in an appropriate shorthand that is independent of the specific approximation. Such a shorthand is available. It is the concept of "operators," which can be directly applied to neutron balance equations. The definition of operators is briefly reviewed in the following section and these concepts are subsequently applied in reactor theory problems.

2-3A Definition of Operators

Operators express certain mathematical operations or prescriptions to be carried out with a function or a vector. Applying an operator to a function yields another function. There are various kinds of operators. For example, $\frac{d}{dx}$, $\frac{d}{dt}$, and $\frac{d^2}{dx^2}$ are differential operators. Their application produces the derivative, e.g.:

$$\frac{d}{dx}f(x) = f'(x) \quad . \tag{2.82a}$$

The simplest matrix operator has two rows and two columns:

$$\begin{pmatrix} a_{11} & a_{12} \\ a_{21} & a_{22} \end{pmatrix} .$$

Its application to a vector yields another vector:

$$\begin{pmatrix} a_{11} & a_{12} \\ a_{21} & a_{22} \end{pmatrix} \begin{pmatrix} x_1 \\ x_2 \end{pmatrix} = \begin{pmatrix} y_1 \\ y_2 \end{pmatrix} . \tag{2.82b}$$

The direct continuous analog to a matrix is a "kernel," $K(E' \rightarrow E)$, the central part of an integral operator,

$$\int K(E' \rightarrow E) \ldots \, dE' .$$

Its application to an appropriate function yields another function:

$$\int_{E'} K(E' \rightarrow E)\phi(E') \, dE' = S_s f_s(E) . \tag{2.83a}$$

Here, $K(E' \rightarrow E)$ may be the scattering kernel [see Eqs. (1.23)]; $f(E)$ is then the energy distribution of the scattered neutrons. Also fission can be expressed by a kernel:

$$K(E' \rightarrow E) = \nu\Sigma_f(E')\chi(E) ;$$

its application to the flux yields the fission source:

$$\chi(E)\int_{E'} \nu\Sigma_f(E')\phi(\boldsymbol{r},E') \, dE' = S_f(\boldsymbol{r},E) . \tag{2.83b}$$

The simplest operator is the multiplication operator, e.g., $\Sigma_a(\boldsymbol{r},E)$.

The application of operators to functions or vectors that yield other functions or vectors of the same basic type can be called a "transformation." One can also consider operators as quantities that perform a mapping of one space upon another. It is not necessary here to make use of these mathematical properties of operators.

Some other basic properties of operators that are needed in the following are briefly reviewed here.

An operator can only be applied to an appropriate class of functions. This class of functions is characterized by three properties:

1. The function must have the appropriate structure and dependence. For example, $\frac{d}{dt}$ is to be applied to a function of time

such as $p(t)$ or $\phi(\boldsymbol{r},t)$, while $\begin{pmatrix} a_{11} & a_{12} \\ a_{21} & a_{22} \end{pmatrix}$ is to be applied to a vector with two components.

2. The result of the application of the operator must exist. For example, $p(t)$ must have a first derivative. Or, in order for $\frac{d^2}{dx^2}\phi(x)$ to be formed, the second derivative of $\phi(x)$ must exist; etc.
3. Their scalar product must exist. (This is no problem for vectors, but it is of relevance for integrals, since they may not have a finite value.)

For vectors, the scalar product is often called the "dot product," e.g.,

$$(\boldsymbol{y},\boldsymbol{x}) = (y_1 \; y_2)\begin{pmatrix} x_1 \\ x_2 \end{pmatrix} = y_1x_1 + y_2x_2 \quad , \tag{2.84}$$

with $(y_1 \; y_2)$ being a row and $\begin{pmatrix} x_1 \\ x_2 \end{pmatrix}$ a column vector. One of the vectors may already be transformed by a matrix; the corresponding scalar product is given by:

$$(y,\mathbf{A}x) = (y_1 \; y_2)\begin{pmatrix} a_{11} & a_{12} \\ a_{21} & a_{22} \end{pmatrix}\begin{pmatrix} x_1 \\ x_2 \end{pmatrix} \quad . \tag{2.85}$$

If the functions are continuous, then the scalar product is the product of the functions, integrated over the respective domain, e.g.,

$$(\psi,\phi) = \int_{-a}^{a} \psi(x)\phi(x)\, dx \quad , \tag{2.86}$$

or with a differential operator included, e.g.,

$$(\psi,\nabla_x^2\phi) = \int_{-a}^{a} \psi(x)\frac{d^2}{dx^2}\phi(x)\, dx \quad , \tag{2.87a}$$

or with an integral operator included, e.g.,

$$(\psi,\mathbf{K}\phi) = \int_E \psi(E)\int_{E'} K(E\rightarrow E')\phi(E')\, dE'\, dE \quad . \tag{2.87b}$$

The entirety or class of functions to which an operator can be applied forms a "functional space." In the case of a matrix, this is simply a finite dimensional "vector space." For differential and integral operators, this space normally has an infinite number of dimensions; it is then a "Hilbert space."

In the next section, these mathematical concepts are applied to neutron balance equations. The operators will find extensive application in an area called "perturbation theory," which deals with small changes in otherwise "unperturbed" systems such as nuclear reactors. Perturbation theory is presented in the companion text on reactor dynamics.[5]

2-3B Neutron Balance Equations in Operator Form

In the operator formulation of the energy-dependent diffusion equation or the Boltzmann equation [see Eqs. (2.28) and (2.12)], one generally identifies three terms that describe different physical effects:

$$S(\boldsymbol{r},E) = \text{independent source}$$

and

$$\mathbf{F}\Phi = \chi(E)\int_0^\infty \nu\Sigma_f(\boldsymbol{r},E')\phi(\boldsymbol{r},E')\,dE' = \text{fission source} \quad , \tag{2.88}$$

where $\mathbf{F}$ is the fission operator. The remainder of Eq. (2.28) is combined into one term, which is abbreviated as $\mathbf{M}\Phi$ with $\mathbf{M}$ as the "migration and loss operator"; i.e., in diffusion approximation:

$$\mathbf{M}\Phi = -\nabla\cdot D(\boldsymbol{r},E)\nabla\phi(\boldsymbol{r},E) + \Sigma_t(\boldsymbol{r},E)\phi(\boldsymbol{r},E) - \int_{E'}\Sigma_S(\boldsymbol{r},E'\rightarrow E)\phi(\boldsymbol{r},E')\,dE' \quad . \tag{2.89}$$

Equations (2.88) and (2.89) define the operators $\mathbf{F}$ and $\mathbf{M}$, respectively. The general flux that appears in the operator equation is denoted by Φ without an argument, since the operator formulation should be applicable to any approximation.

This gives the energy-dependent diffusion equation in the form:

$$\mathbf{M}\Phi = \mathbf{F}\Phi + S \quad . \tag{2.90}$$

The Boltzmann equation gives a different definition of $\mathbf{M}$, which can be readily obtained from Eq. (2.12). Neutron migration and loss can also be expressed in one of the simpler models described in Secs. 2-2D and 2-2F. In one-group diffusion theory, the operators and S for all neutrons are given by:

$$\mathbf{M} = -\nabla\cdot D(\boldsymbol{r})\nabla + \Sigma_a(\boldsymbol{r}) \quad ,$$

$$\mathbf{F} = \nu\Sigma_f(\boldsymbol{r}) \quad , \tag{2.91}$$

and

$$S = S(\boldsymbol{r}) \quad .$$

In the simplest neutronics model, Eq. (2.81), all operators are reduced to numbers (multiplication operators):

$$\mathbf{M} = DB^2 + \Sigma_a \quad ,$$

$$\mathbf{F} = \nu\Sigma_f \quad , \qquad (2.92)$$

and

$$S = S \quad .$$

The formal identity of all neutron balance equations shows the advantage of the use of operator notation. By means of operators, the neutron balance equation may be expressed in the general form, Eq. (2.90), which is independent of the specific approximation employed. For example, Φ then denotes:

the angular flux $\phi(\boldsymbol{r},E,\boldsymbol{\Omega})$
the flux $\phi(\boldsymbol{r},E)$
the total flux $\phi(\boldsymbol{r})$
or
the integrated flux $\hat{\phi}$,

depending on whether **M** and **F** are the operators of

the Boltzmann equation
the energy-dependent diffusion equation
the one-group diffusion equation
or
the one-group space-integrated diffusion equation.

2-4 Fundamental Neutronics Problems

2-4A Introduction

The first concepts in the development of a new area like reactor physics have their specific merit in providing an initial understanding of the phenomena. Being the first of their kind, these early concepts and approaches are normally not very elegant and often not precisely defined. With further development of an area, more accurate, refined, consistent, and elegant concepts and approaches evolve. Eventually, the formulation and conceptual understanding assume a mature form. Ideally, this viewpoint of historical development of an area should do justice to the early concepts and provide better insight into the new conceptual understanding.

Modern understanding of nuclear reactor physics conceptually embraces the theory of both thermal and fast reactors with development oriented toward obtaining solution methods suitable to computer application. Approximate concepts, such as the four- and six-factor formulas, then receive positions of historical interest without a *quantitative* practical application in reactor design. Nevertheless, these simplified "factorized" concepts of reactor criticality serve a very valuable purpose in the *qualitative* understanding and comprehension of reactor physics phenomena. Therefore, a discussion of this factorized method of estimating reactor criticality is presented in Sec. 2-5.

The modern conceptual understanding of reactor behavior is more mathematical but far from being complicated. As an analogy, one can look at the so-called "new math," which provides an early conceptual understanding on the basis of set theory. In the "old math," set theory was first encountered in graduate education and then only by students majoring in mathematics.

An important facet of the modern conceptual understanding of reactor physics is based on the realization that neutron balance equations do not, per se, have physically realizable solutions. There are essentially four possibilities one can encounter:

1. physically *realizable* solutions[c]
2. physically *allowed* solutions,[c] which may or may not be physically realizable
3. *mathematical* solutions, which may be physically not allowed[d]
4. the *trivial* solution ($\phi \equiv 0$), which is normally not sought after; in some cases, however, it can be the physically realized solution.

All four types of solutions are discussed in this section. It should be noted that if one presumes all of the equations encountered will have only physically realizable solutions, one automatically relinquishes the possibility of fully understanding the theoretical treatment of the problem.

Many of the neutronics problems that are solved for reactor design considerations only have solutions that are physically *allowed* but *not realizable.* Problems that have physically allowed solutions can be modified in various ways, if so desired, such that physically realizable solutions are obtained (e.g., by proper adjustment of the fuel enrichment). Math-

---

[c]A physically *realizable* solution is one that can be realized in nature; a physically *allowable* flux must be nonnegative and nonsingular.

[d]A solution that is partly negative or singular may mathematically solve the problem, but it is "physically not allowed."

ematical solutions that are physically not allowed have to be suppressed; this is done automatically by most computer codes for flux calculations. For some applications, one is interested in mathematical solutions that are not allowed physically (e.g., for special perturbation considerations; compare Sec. 2-4E). One normally wants to avoid the trivial solution (except when it is physically realized); special measures are required to achieve this goal.

The significance of the distinction between physically realizable and allowed solutions is:

> Physically realizable solutions are obtained for problems that describe a system that is, in principle and in a certain approximation, physically realizable, while physically allowed solutions (that are not realizable) are obtained from problems that contain an artificial (nonphysical) modification and thus cannot be realized. These modifications often have the form of a nonphysical parameter that plays the role of an eigenvalue. Frequently, the analyst decides on the kind of artifice suitable for his purpose. In most cases, the artifice is necessary since a characterization of an off-critical system (e.g., its degree of offcriticality) is sought *without* the changes needed to establish a physical solution (e.g., by adding a source or by initiating a transient). The desired characterization often requires the determination of a "static solution" to the problem. A static solution that is not physically realizable means that it cannot be established as a "steady state" of a physical system.

It should be noted that this distinction is a principle one and that accuracy questions are disregarded.

2-4B Source-Sink Problems

In the source-sink problem, it is assumed that all sources, including the fission source, are given sources, *independent of the flux:*

$$S_t(\boldsymbol{r},E) = S_f(\boldsymbol{r},E) + S(\boldsymbol{r},E) \quad , \tag{2.93}$$

with S_t being the total source. The balance equation then has the form:

$$\mathbf{M}\Phi = S_t \quad . \tag{2.94}$$

The source may be localized, e.g., as a "point source" or a "plane source," or it may be distributed as a fission source of a *given* strength. For the solution of source-sink problems, it holds in general that *source-sink problems have a unique physically realizable solution.* This mathematical statement is physically obvious. The simple neutronics model is used to illustrate the statement:

$$(DB^2 + \Sigma_a)\phi = S_t \tag{2.95a}$$

and

$$\phi = \frac{S_t}{DB^2 + \Sigma_a} \quad . \tag{2.95b}$$

All terms in Eq. (2.95a) are positive, and since S_t is a physical neutron source, it follows that ϕ is realizable.

2-4C Subcritical Reactors with an Independent Source

The neutron balance equation for a subcritical reactor, including an independent source, is the following:

$$(\mathbf{M} - \mathbf{F})\Phi = S \quad . \tag{2.96}$$

The fact that the reactor is subcritical means that there are fewer fission neutrons produced ($\mathbf{F}\Phi$) than lost ($\mathbf{M}\Phi$).

The same physical intuition as applied in the source-sink problem yields, heuristically, the theorem:

> An independent source in a subcritical reactor yields a unique physically realizable neutron flux; thus, Eq. (2.96) has a physically realizable solution.

For example,

$$(DB^2 + \Sigma_a - \nu\Sigma_f)\phi = S \quad . \tag{2.97}$$

The positive terms on the left side of Eq. (2.97) are numerically larger than the negative ones since the reactor is subcritical. Thus, Eq. (2.97) yields a positive flux, since S is positive.

Solving Eq. (2.97) for ϕ yields the magnitude of the flux as it is established in a subcritical reactor through "multiplication" of the independent source neutrons by means of nuclear fission:

$$\phi = \frac{S}{DB^2 + \Sigma_a - \nu\Sigma_f} \quad . \tag{2.98a}$$

Equation (2.98a) is the simplest form of a so-called "source multiplication formula" that has the general form:

$$\phi = S \cdot M_S \quad , \tag{2.98b}$$

with M_S as the "source multiplication factor" (see, e.g., Secs. 3-4D and 3-4E).

As indicated by Eq. (2.98a), the resultant flux is proportional to the source. This shows that *in the limit of $S \rightarrow 0$, the only physically realizable solution is the trivial solution* (compare Sec. 2-4D). Note that the solution of Eqs. (2.96) or (2.97) is always unique.

The denominator of Eq. (2.98a) determines the magnitude or factor of the source multiplication. If, for example, the neutron loss rate, $(DB^2 + \Sigma_a)\phi$, is 10% larger than the fission rate, $\nu\Sigma_f\phi$, one obtains from the modified Eq. (2.98a),

$$\nu\Sigma_f\phi = \frac{S}{\dfrac{DB^2 + \Sigma_a}{\nu\Sigma_f} - 1} = \frac{S}{\dfrac{1}{k} - 1} = \frac{S}{-\rho} \quad , \tag{2.98c}$$

that the fission source is ten times larger than the independent source. The simplest form of the multiplication constant k (or k_{eff}) and of the reactivity ρ, which will be introduced in Sec. 2-4E, appear in the denominator of Eq. (2.98c).

Equations (2.98) also show that by increasing $\nu\Sigma_f$ or by decreasing $(DB^2 + \Sigma_a)$, one can let the denominator approach very small values. The flux in the respective steady states would then be very large.

2-4D The Critical Reactor

The "critical" reactor is characterized by its capability to sustain a stationary (i.e., steady-state) chain reaction *without* an independent source. Thus, the neutron balance equation is given by:

$$(\mathbf{M} - \mathbf{F})\Phi = 0 \quad . \tag{2.99}$$

In addition to the desired solution, i.e., the flux in a critical reactor, this equation obviously also has a trivial solution:

$$\Phi \equiv 0 \quad . \tag{2.100}$$

For off-critical balance equations, the trivial solution is the *only* solution of $(\mathbf{M} - \mathbf{F})\Phi = 0$:

subcritical reactor:

$$(\mathbf{M} - \mathbf{F})_{\text{sub}}\Phi = 0 \quad , \tag{2.101a}$$

in which $\Phi \equiv 0$ in the absence of an independent source, as demonstrated in the previous section;

supercritical reactor:

$$(\mathbf{M} - \mathbf{F})_{\text{sup}}\Phi = 0 \quad , \tag{2.101b}$$

in which $\Phi \equiv 0$ is the only possible *stationary* solution in a supercritical reactor; i.e., since in a supercritical reactor there are more neutrons produced than lost, one can obtain a stationary solution in such a system only if not a single neutron is available for multiplication; thus ϕ must be zero in the entire system.

Only in the critical case does Eq. (2.99) have a nontrivial solution for which ϕ is nonnegative and physically realizable. This requires mathematically that $(\mathbf{M} - \mathbf{F})$ be "singular": A singular operator does not have an inverse. Again, the one-group separated problem is used as an example. Equation (2.99) then has the form:

$$(DB^2 + \Sigma_a - \nu\Sigma_f)\phi = 0 \quad , \tag{2.102}$$

which has a nontrivial solution ($\phi \not\equiv 0$) only if the operator is singular, i.e., if the inverse does not exist. It would be infinite (and thus does not exist) in this example:

$$(\mathbf{M} - \mathbf{F})^{-1} = (DB^2 + \Sigma_a - \nu\Sigma_f)^{-1} \quad . \tag{2.103}$$

If a reactor can sustain a finite flux level in the absence of an independent source, it is called a "critical reactor." The criticality condition is: *The operator* $\mathbf{M} - \mathbf{F}$ *must be singular for a critical reactor*, that is, the parentheses of Eq. (2.102) must be zero. This may be rewritten as:

$$\frac{\nu\Sigma_f}{DB^2 + \Sigma_a} = 1 \quad , \tag{2.104}$$

which is the criticality condition for a one-group, one-region reactor.

Equation (2.104) is often converted into two factors,[e] one describing the multiplication properties of the medium and the other the leakage characteristic of the assembly:

$$\frac{\nu\Sigma_f}{\Sigma_a} \cdot \frac{\Sigma_a}{DB^2 + \Sigma_a} = \frac{\nu\Sigma_f}{\Sigma_a} \cdot \frac{1}{1 + L^2B^2} = k_\infty P = 1 \quad . \tag{2.105}$$

Here, k_∞ is the multiplication constant of that medium, the "medium multiplication constant." The notation k_∞ comes from its interpretation as the multiplication constant of an infinite medium (see Sec. 2-5). Further,

$$L^2 = \frac{D}{\Sigma_a} = \text{square of the "diffusion length"} \quad , \tag{2.106}$$

---

[e]For further factorization of multiplication constants, see Sec. 2-5.

while P is the nonleakage probability, i.e., the fraction of the total neutron loss due to absorption in the reactor:

$$P = \frac{\Sigma_a \phi}{(DB^2 + \Sigma_a)\phi} = \frac{1}{1 + L^2 B^2} \quad . \tag{2.107}$$

To make the operator ($\mathbf{M} - \mathbf{F}$) singular, i.e., to make the reactor critical, requires certain measures—first in the design of the reactor, and then in the control of it. The three standard measures are:

1. search for critical enrichment [variation of $\nu\Sigma_f$ in Eq. (2.103)]
2. search for critical geometry [variation of B^2 in Eq. (2.103)]
3. search for critical control absorption [variation of Σ_a in Eq. (2.103)].

A more detailed discussion of the general types of solutions and procedures, including searches for criticality, is presented in Sec. 2-6B.

2-4E The Static Off-Critical Reactor Problem and the Static Multiplication Constant

A legitimate and very important question is how far away a given system is from criticality; in other words, What is the degree of off-criticality (OC) of a given system? The balance equation is:

$$(\mathbf{M} - \mathbf{F})_{OC}\Phi = 0 \quad . \tag{2.108}$$

From Eqs. (2.101), it follows that the above equation has only one solution, namely $\Phi \equiv 0$. Thus, by solving Eq. (2.108) correctly, one will not learn anything about the degree of off-criticality.

One therefore has to alter the balance equation, Eq. (2.108), to obtain a nontrivial solution. But the system should not be altered, i.e., the operators **M** and **F** should not be changed.

One possible alteration would be to introduce a source:

$$(\mathbf{M} - \mathbf{F})_{OC}\Phi = S \quad . \tag{2.109}$$

The magnitude of the resulting flux, i.e., the degree of source multiplication, is then a measure of the degree of off-criticality [compare the source multiplication formula, Eq. (2.98)]. However, this procedure is reasonably applicable only in a *subcritical* reactor. The static solution for a supercritical system with an independent source results in a negative neutron flux, as can be seen from Eq. (2.97), since the term in parentheses on the left side of Eq. (2.97) is negative if the reactor is supercritical. Thus, an approach to generally define the degree of off-criticality by Eq. (2.108) would involve physically nonallowed solutions. Therefore,

Eq. (2.109) cannot be applied for determining in general the degree of off-criticality.

In actuality, the applied procedure to obtain a nontrivial solution is to introduce a modifying factor, λ, in front of the fission operator $\mathbf{F}$; i.e., instead of Eq. (2.108), one solves the modified problem:

$$(\mathbf{M} - \lambda\mathbf{F})\Phi_\lambda = 0 \quad . \tag{2.110}$$

The modifying factor λ is varied in order to make the operator $(\mathbf{M} - \lambda\mathbf{F})$ singular; mathematically speaking, one finds the "proper" value, the "eigenvalue" λ, which makes the operator singular. Then a nontrivial solution is obtained for this eigenvalue problem, which resulted from an artificial alteration of the original problem. The resulting flux, Φ_λ, the so-called "λ mode" is positive and thus physically allowed. However, since Eq. (2.108) is altered by a nonphysical operation (multiplication of $\mathbf{F}$ by λ), *the solution is physically not realizable.*

The off-critical eigenvalue problem, Eq. (2.110), has a physically allowed but not a physically realizable solution. Although Φ_λ is physically not realizable, it may practically be a close approximation to the realized fluxes.

The eigenvalue λ can be interpreted in various ways. The most common interpretation is found by integrating Eq. (2.110) with respect to space and energy and solving for $\frac{1}{\lambda}$:

$$\frac{1}{\lambda} = \frac{\int_V \int_E \mathbf{F}\Phi_\lambda \, dE \, dV}{\int_V \int_E \mathbf{M}\Phi_\lambda \, dE \, dV} = k \ (= k_{\text{eff}}) \quad , \tag{2.111a}$$

where

$$k = \left(\frac{\text{total neutron production rate}}{\text{total neutron loss rate}} \right)_{\Phi_\lambda} = k_{\text{eff}} \tag{2.111b}$$

is the "multiplication constant" for an off-critical reactor in the eigenstate Φ_λ.

In most of the older literature, the denotation k_{eff} is used for the static multiplication constant. The index "eff" was applied historically to distinguish it from the medium multiplication constant, k_∞; k_{eff} was considered descriptive of the multiplication of neutrons that were "effective" in a finite system. The refined definition, Eqs. (2.111), shows that, strictly speaking, k_{eff} is *not* "effective" in the sense of "acting" since it describes the multiplication of neutrons in the artificial eigenstate Φ_λ.

On the contrary, a nuclear engineer must make every effort to prevent a $k>1$ from becoming accidentally effective or acting in a reactor. A further reason to abandon the index "eff" is that it is too cumbersome if further indices are needed to denote k in different states [compare, e.g., Eq. (2.115)]. Therefore, the classical and famous term k_{eff} is appearing less in modern reactor theory literature.

By inserting Eq. (2.111) into Eq. (2.110), one obtains the balance equation of the off-criticality problem that defines k:

$$(\mathbf{M} - \frac{1}{k}\mathbf{F})\Phi = 0 \quad . \tag{2.112}$$

In part of the literature, the criticality factor defined herein as k is introduced as an eigenvalue, then also denoted by the symbol λ. Here, λ' is used to denote this kind of eigenvalue:

$$(\mathbf{M} - \frac{1}{\lambda'}\mathbf{F})\Phi_\lambda = 0 \quad , \tag{2.113}$$

with $\lambda' = k$, defined by Eq. (2.113). Certainly, Eqs. (2.110) and (2.113) have the same solutions, Φ_λ, and they differ only in the notation of the eigenvalue.

The degree of off-criticality is also measured by the static reactivity, ρ ($= \rho^{stat}$):

$$\rho = 1 - \lambda = 1 - \frac{1}{k} = \frac{k-1}{k} = \frac{\Delta k}{k} \quad . \tag{2.114}$$

In addition to the reactivity, one needs the concept of "reactivity increments":

$$\delta\rho = \rho_1 - \rho_0 = \frac{k_1 - k_0}{k_1 \cdot k_0} = \frac{\Delta k}{k_1 k_0} \quad . \tag{2.115}$$

Equation (2.115) is often, but erroneously, considered the "general" definition of the reactivity from which the "special" value, Eq. (2.114), can be obtained if k_0 assumes the special value $k_0 = 1$. Another source of confusion is the habit of many authors to denote ρ as well as $\delta\rho$ with $\Delta k/k$. But normally no numerical error results from this erroneous usage of $\Delta k/k$ since $\delta\rho$ is correctly calculated from appropriate computer programs as defined by Eq. (2.115).

There is another interpretation of λ that is often used. Since λ multiplies the fission source, one interprets $\lambda \neq 1$ as an artificial change in ν:

$$\nu' = \lambda\nu \quad . \tag{2.116}$$

As an example, a subcritical facility with a $k = 0.2$, i.e., $\lambda = 5$, may have an actual ν of $\nu = 2.5$. The ν' is equal to 12.5, i.e., the system would be "critical" if 12.5 neutrons were emitted in each fission event. This example also suggests that in a highly subcritical system, one may strongly deform the actual physical situation by applying an eigenvalue. In near-critical systems, the deformation is minor. If $k \simeq 1.11$, ν is only reduced by ~10%.

Eigenvalue problems normally have more than one solution. But they have only one nonnegative solution, the "fundamental mode." The eigenvalue of the fundamental mode is the smallest or largest, depending on the way it is introduced, i.e.,

$$(\mathbf{M} - \lambda\mathbf{F})\Phi = 0, \text{ the smallest } \lambda \text{ is the fundamental eigenvalue (the largest } k) \quad , \qquad (2.117a)$$

$$(\mathbf{M} - \frac{1}{\lambda'}\mathbf{F})\Phi = 0, \text{ the largest } \lambda' \text{ is the fundamental eigenvalue} \quad , \qquad (2.117b)$$

and

$$\nabla^2\Phi + B^2\Phi = 0, \text{ the smallest } B^2 \text{ is the fundamental eigenvalue} \quad . \qquad (2.117c)$$

These equations are not eigenvalue problems per se. Equation (2.117c), for example, is just a differential equation with a continuous variety of solutions. What makes it an eigenvalue problem, having special solutions (the eigenfunctions), is the requirement that the solution satisfy two (homogeneous) boundary conditions; e.g., in one dimension, that equation is solved by $\cos Bx$ (see Sec. 3-3). But if the solution is required to be zero at both ends of a symmetrical slab, one particular value of B^2 (the eigenvalue) is needed to achieve that (see Fig. 2-8).

Eigenfunctions that belong to the eigenvalues beyond the fundamental eigenvalues are called "higher eigenfunctions" or "higher modes." Generally, higher modes are mathematical solutions but not physically allowed solutions. However, higher modes can appear in physically allowed solutions together with the fundamental mode as shown in the following example.

Figure 2-8 illustrates a flux distribution in an asymmetric reactor. The cosine function is the fundamental mode. The next higher mode, the sine function, is partially negative. It can be used in conjunction with the fundamental mode to describe approximately the distorted flux. In the earlier times of reactor physics, higher modes were applied in practice to describe perturbations, as in the example of Fig. 2-8. Now, numerical solutions with computers have replaced the explicit combinations

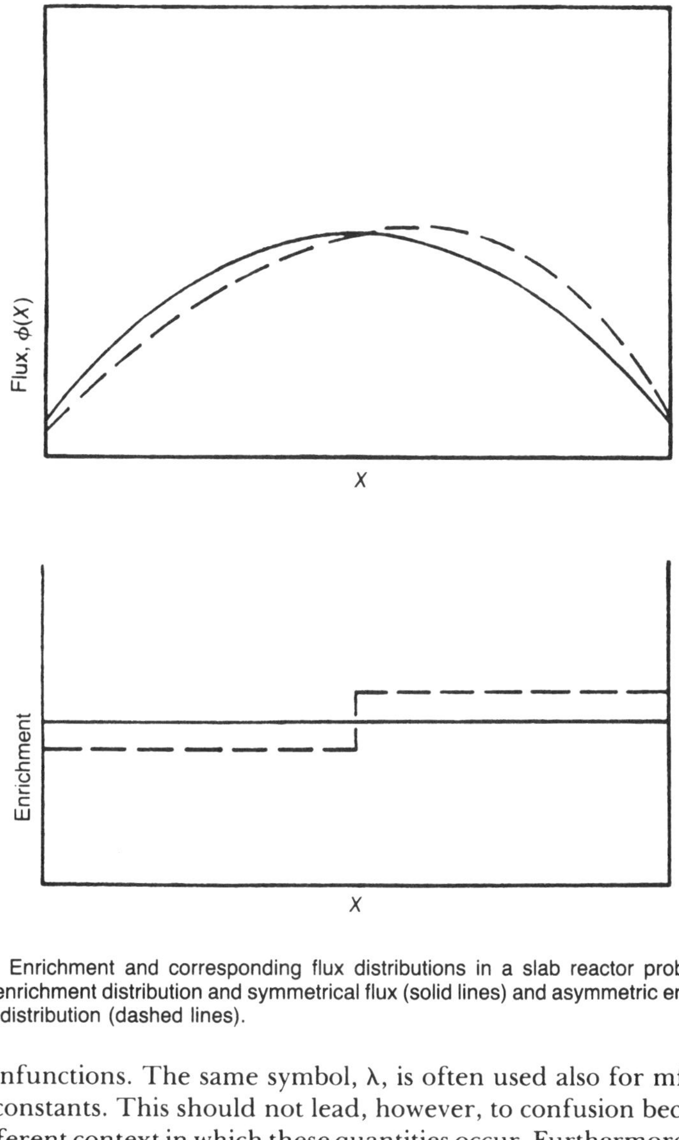

Fig. 2-8. Enrichment and corresponding flux distributions in a slab reactor problem with uniform enrichment distribution and symmetrical flux (solid lines) and asymmetric enrichment and flux distribution (dashed lines).

of eigenfunctions. The same symbol, λ, is often used also for mfps and decay constants. This should not lead, however, to confusion because of the different context in which these quantities occur. Furthermore, decay constants and mfp normally have a descriptive index.

2-4F The Time-Dependent Flux and the Dynamic Reactivity

The last category of problems comprises time-dependent problems, especially reactor kinetics. If the sum of all neutron sources and losses in a volume element are not in balance as they are in steady-state problems, the neutron flux depends on time: $\Phi = \phi(\boldsymbol{r},E,t)$. The lack of neutron balance leads to a change in the neutron density with time: i.e., a new term appears on the right side of the time-dependent neutron balance equation:

$$(\mathbf{F}_p - \mathbf{M})\Phi + S_{\text{del}} = \frac{1}{v}\frac{\partial \Phi}{\partial t} \quad . \tag{2.118}$$

In time-dependent problems, the fission source needs to be separated in the "prompt" fission neutron source (described by the operator $\mathbf{F}_p$) and the delayed neutron source, S_{del}. The right side of Eq. (2.118) gives the change in the number of neutrons [$= \phi/v$, with $v = v(E)$ being the velocity of neutrons of energy E].

The off-criticality in the dynamics problem can be described by a "dynamic reactivity," ρ_{dyn}, which is slightly different from the corresponding static reactivity, $\rho = 1 - \lambda$.

Solutions of time-dependent problems are physically realizable solutions. The term ρ_{dyn} is actually driving a transient. Time-dependent problems are treated in reactor dynamics, for example, see Ref. 5.

The burnup and buildup of fuel isotopes and fission products represent another category of time-dependent problems. The variations are so slow in time that the flux can be calculated from a sequence of stationary problems; i.e., the right side of Eq. (2.118) can be neglected and the delayed neutrons can be combined with the prompt neutrons.

2-5 Estimates of Criticality for Thermal Reactors

In a fast reactor, there is only one reasonably accurate way to find or to estimate the static multiplication constant, namely, to solve the λ eigenvalue problem given by Eq. (2.110):

$$(\mathbf{M} - \lambda\mathbf{F})\Phi_\lambda = 0 \quad .$$

The solution of this equation requires a detailed treatment of the energy dependence, which is practically approximated by the multigroup model. This approach is discussed in Chapter 7.

In a thermal reactor, one can estimate k by separating the eigenvalue expression of Eq. (2.111a) into different physical phenomena. At first, k is factored into a material and a geometry related quantity:

$$k = \frac{1}{\lambda} = \frac{\displaystyle\iint \mathbf{F}\Phi_\lambda \, dE \, dV}{\displaystyle\iint \mathbf{M}\Phi_\lambda \, dE \, dV} = \left(\frac{R_p}{R_a + R_L}\right)_{\Phi_\lambda} = \left(\underbrace{\frac{R_p}{R_a}}_{k_\infty} \cdot \underbrace{\frac{R_a}{R_a + R_L}}_{P}\right)_{\Phi_\lambda} \quad , \tag{2.119}$$

i.e.,

$$k = k_\infty \cdot P \quad , \tag{2.120}$$

where R_p, R_a, and R_L are, respectively, the production, absorption, and leakage rates, and k_∞ is the medium multiplication constant (for a homogeneous medium or for a unit cell). In Eq. (2.120), the factor P is referred to as the "nonleakage probability."

If the medium multiplication constant is defined as in Eq. (2.120), i.e., by factoring the multiplication factor of the finite medium in $k_\infty \cdot P$, then the reaction rates in k_∞ must be calculated with the flux Φ_λ of the finite medium:

$$k_\infty = \left(\frac{R_p}{R_a}\right)_{\Phi_\lambda} \quad . \tag{2.121}$$

A multiplication constant for an infinite medium may also be defined as an eigenvalue in a true infinite-medium eigenvalue problem (see Sec. 2-6B):

$$(\mathbf{M}_\infty - \lambda_\infty \mathbf{F}_\infty)\Phi_\infty = 0 \quad . \tag{2.122}$$

Note that $\mathbf{F}$ is the same in finite and infinite media. Then $1/\lambda_\infty$ can be expressed as a ratio of reaction rates calculated with Φ_∞:

$$\frac{1}{\lambda_\infty} = \left(\frac{R_p}{R_a}\right)_{\Phi_\infty} = k^\infty = \text{infinite-medium } k \quad , \tag{2.123}$$

where the notation k^∞ is introduced to distinguish it from k_∞ as defined by Eq. (2.121). If, however, k_∞ is used in the context of Eq. (2.120), i.e., for calculating a finite-medium k, then definition Eq. (2.121) must be applied to be consistent with the factorization in Eq. (2.119). To distinguish the results of Eqs. (2.121) and (2.123), the latter one is called the "infinite-medium k" and the classical term "k_∞" is applied to denote the result of Eq. (2.121), since Eq. (2.120) represents the most frequent usage of a medium multiplication constant.

Historically, no distinction between the k_∞ concepts as defined by Eqs. (2.121) and (2.123) have been made; the computational procedure for finding k_∞ involved several approximations, which probably introduced inaccuracies larger than the difference between the two k_∞ concepts. Thus, a distinction between these two concepts was not meaningful.

Generally, k_∞ is factored further into quantities representing special physical phenomena. These are obtained by considering the fact that neutrons are born at high energy and are then slowed down to thermal energies where, through fissions, they produce new neutrons that appear at high energies, etc. This process is referred to as a "chain reaction." Figure 2-9 illustrates the chain reaction for a thermal reactor.

In this approach, then, k_∞ is factored into *four* factors: η, f, p, and ε:

$$\eta = \text{the } \eta \text{ factor} \quad ,$$

$$f = \text{the "thermal utilization" factor} \quad ,$$

$$p = \text{the "resonance escape probability" factor} \quad ,$$

and

$$\varepsilon = \text{the "fast fission" factor} \quad . \tag{2.124}$$

These four factors can be expressed as ratios of reaction or transfer rates. The "classical" definition of these factors (see p. 82 of Ref. 1) is then expressed by the following ratios:

$$k_\infty^{cl} = \underbrace{\frac{R_p(\text{th})}{R_a(\text{th,f})}}_{\eta} \cdot \underbrace{\frac{R_a(\text{th,f})}{R_a(\text{th})}}_{f} \cdot \underbrace{\frac{R_{tr}(\text{res}\rightarrow\text{th})}{R_{tr}(\text{fast}\rightarrow\text{res})}}_{p} \cdot \underbrace{\frac{R_{tr}(\text{fast}\rightarrow\text{res})}{R_p(\text{th})}}_{\varepsilon} \quad , \tag{2.125}$$

where

$$R_p(\text{th}) = \text{production rate from thermal fission}$$

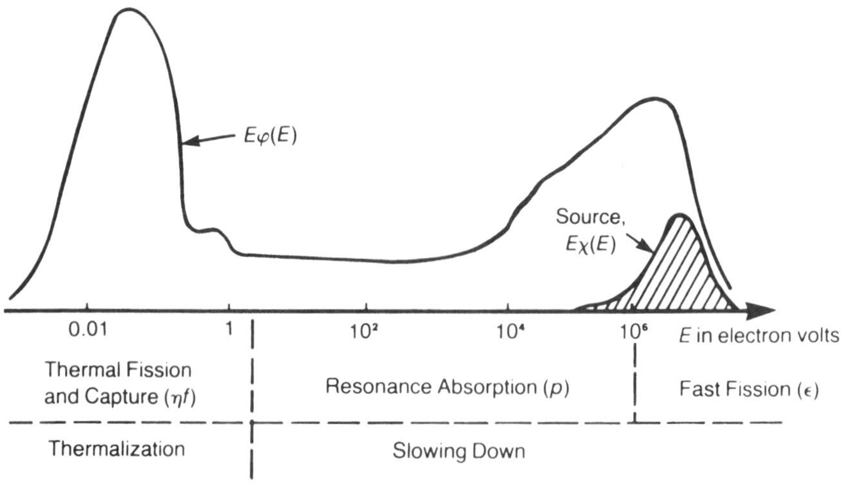

Fig. 2-9. Illustration of processes involved in a "chain reaction" for a thermal reactor along the energy chain.

R_a(th) = absorption rate of thermal neutrons
R_a(th,f) = absorption rate of thermal neutrons in fuel
R_{tr}(res→th) = transfer rate of neutrons from the resonance energy group to the thermal energy group
R_{tr} (fast→res) = transfer rate of neutrons from the fast energy group to the resonance energy group . (2.126)

The basis is the consideration of three rather broad energy groups: a thermal, a resonance, and a fast group (see Fig. 2-9). First, the fast fission factor ε (defined as the number of neutrons arriving below the ^{238}U fission threshold per neutron from thermal fission) is introduced. The resonance escape probability, p, is obtained as the ratio of the total number of neutrons leaving the resonance group to those that enter it from the fast group. The quantities η and f are given by the indicated ratios.

Historically, the four-factor formula given by Eq. (2.125) is the expression most generally used. A different definition[f] can be obtained by factoring k_∞ consistently into four factors, η, ε′, p', and f, in a way that is a direct expansion of the original definition of k_∞ in Eq. (2.119):

$$k_\infty = \frac{R_p}{R_a} = \underbrace{\frac{R_p(\mathrm{th})}{R_a(\mathrm{th,f})}}_{\eta} \cdot \underbrace{\frac{R_p}{R_p(\mathrm{th})}}_{\varepsilon'} \cdot \underbrace{\frac{R_a(\mathrm{th})}{R_a}}_{p'} \cdot \underbrace{\frac{R_a(\mathrm{th,f})}{R_a(\mathrm{th})}}_{f} \quad , \qquad (2.127)$$

where all quantities are defined reaction rate ratios (to be calculated with the finite-medium ϕ_λ). The fast fission factor, ε′, and resonance escape probability factor, p', are defined differently than the comparable quantities in Eq. (2.125), whereas η and f are the same.

Conceptually, Eq. (2.127) is simpler, in that k_∞ is the direct ratio of neutron production rate, R_p, to neutron absorption rate, R_a. The quantity ε′ is sometimes used in fast reactor applications[g] (e.g., for breeding gain estimates) because of its concise definition.

The equivalence of the two formulations of the four-factor formulas, Eqs. (2.125) and (2.127), can only be shown for an infinite medium.

A six-factor formula for estimating k for a finite thermal reactor can be obtained by separating the leakage into two probability components that describe the leakage out of the thermal group and the sum of the two other groups:

P_{th} = thermal leakage

[f]See p. 586 of Ref. 4.

[g]It should be noted that the total factored k_∞ expression or concept is not used in fast reactor design.

and

$$P_f = \text{fast leakage (resonance and fast fission groups)} \quad . \tag{2.128}$$

The static multiplication factor is then estimated by

$$k = k_\infty P_{th} P_f = \eta \varepsilon p f P_{th} P_f \quad , \tag{2.129}$$

or ε' and p' instead of ε and p.

If all reaction rates are calculated with Φ_λ, then k_∞ is consistent with its introduction in Eq. (2.119). However, instead of Φ_λ, one normally uses only a crude approximation, which is obtained by piecing together approximate components of the neutron energy spectrum. In this way, one obtains an estimate of k_∞. Specifically, typical procedures to find the four factors of k_∞ are as follows:

1. The fast fission factor ε was obtained through integral transport considerations (see Sec. 6-3E), without an explicit calculation of a spectrum.
2. The resonance escape probability factor p was normally calculated with a slowing down spectrum in an infinite medium (compare Chapter 5).
3. The thermal utilization factor f was calculated from effective one-group cross sections and spatial weighting factors (disadvantage factors) to account for heterogeneities.
4. The terms P_{th} and P_f were estimated on the basis of a separation of the space and energy dependence of the flux.

The conceptual applicability of the four- and six-factor formulas requires that the reactor core consist of a single medium so that a single k_∞ can describe the multiplication properties. Then, the neutron flux can be separated and the nonleakage probabilities can be estimated independent of the core composition.

Modern power reactors have several enrichments in the core; in addition, the material composition varies with burnup. Therefore, the application of factored formulas for criticality determination has long been replaced by the direct numerical solution of the λ eigenvalue problem. In this treatment, one normally applies the multigroup approach discussed in Chapter 7.

2-6 Separation of Space and Energy Dependencies

2-6A Balance Equations Resulting from Flux Separation

In a bare one-region reactor, the neutron flux, $\phi(\mathbf{r},E)$, is separable in its space and energy dependence:

$$\phi(\boldsymbol{r},E) = \phi(\boldsymbol{r}) \cdot \varphi(E) \quad . \tag{2.130}$$

Weinberg and Wigner[6] refer to this fact as a "fundamental theorem of reactor theory."

The relative normalization of the space- and energy-dependent factors in Eq. (2.130) is arbitrary. For most applications in this text, the relative normalization is chosen in such a way that the space-dependent factor is the total neutron flux:

$$\phi(\boldsymbol{r}) = \int_0^\infty \phi(\boldsymbol{r},E)\, dE \quad . \tag{2.131}$$

Then, $\varphi(E)$, the neutron spectrum, is normalized to one:

$$\int_0^\infty \varphi(E)\, dE = 1 \quad . \tag{2.132}$$

The dimensions of the flux have been split in this separation in the following way:

$$\phi(\boldsymbol{r},E)\left[\frac{\text{neutrons}}{\text{cm}^2\text{s U}(E)}\right] = \phi(\boldsymbol{r})\left[\frac{\text{neutrons}}{\text{cm}^2\text{s}}\right]\varphi(E)\left[\frac{1}{U(E)}\right] \quad . \tag{2.133}$$

Occasionally it is more convenient to use a normalization different from one, e.g., $\phi(\boldsymbol{r}_s,E)$, the spectrum at the space point $\boldsymbol{r}_s$, may be normalized to the total flux value at $\boldsymbol{r}_s$.

The term "spectrum" comes from the Latin word "spectare," to see. It was first used for the description of the visible light intensities as a function of the wave length. The term "spectrum" is now used to denote the *energy* dependence of all kinds of particle and wave intensities. The name "spectrum" was also introduced later into mathematics, where it denotes the entirety of eigenvalues of a given problem. This is suggested by the fact that the spectrum of emitted light is obtained from the quantum-mechanical treatment of an atom as the "spectrum of eigenvalues" of the operator that describes the energy characteristics of the atom. Also, the Fourier decomposition of a continuous spectrum into waves of pure frequencies suggested the term "eigenvalue spectrum" for the set of pure frequencies. The term "spectrum" is used occasionally for decompositions other than with respect to energy; e.g., mass spectrometry. However, in its reactor application, the term "spectrum" has been used almost universally to denote the energy distribution of the flux (neutron and photon).

Inserting the separation into the one-region, energy-dependent diffusion equation, modified by an eigenvalue λ, yields:

$$-D(E)\nabla^2\phi(\boldsymbol{r})\varphi(E) = -\Sigma_t(E)\phi(\boldsymbol{r})\varphi(E) + \int_0^\infty \Sigma_s(E'\rightarrow E)\varphi(E')\,dE' \cdot \phi(\boldsymbol{r}) + \lambda\chi(E)\int_0^\infty \nu\Sigma_f(E')\varphi(E')\,dE' \cdot \phi(\boldsymbol{r}) \quad . \tag{2.134}$$

Dividing Eq. (2.134) by $D(E)\varphi(E)\phi(\boldsymbol{r})$ separates out the space and energy dependencies:

$$-\frac{\nabla^2\phi(\boldsymbol{r})}{\phi(\boldsymbol{r})} = \frac{1}{D(E)\varphi(E)}\left[-\Sigma_t(E)\varphi(E) + \int_0^\infty \Sigma_s(E'\rightarrow E)\varphi(E')\,dE' + \lambda\chi(E)\int_0^\infty \nu\Sigma_f(E')\varphi(E')\,dE'\right] \quad . \tag{2.135}$$

The left side is a function of space, the right side a function of energy; the quantities can only be equal if both sides *do not depend* on the respective variables, i.e., both sides are equal to a constant:

$$-\frac{\nabla^2\phi(\boldsymbol{r})}{\phi(\boldsymbol{r})} = B^2 \tag{2.136}$$

or

$$\nabla^2\phi(\boldsymbol{r}) + B^2\phi(\boldsymbol{r}) = 0 \quad . \tag{2.137}$$

Also, the right side of Eq. (2.135) must be equal to B^2. This yields the balance equation for $\varphi(E)$:

$$[D(E)B^2 + \Sigma_t(E)]\varphi(E) - \int_0^\infty \Sigma_s(E'\rightarrow E)\varphi(E')\,dE' = \lambda\chi(E)\int_0^\infty \nu\Sigma_f(E')\varphi(E')\,dE' \quad . \tag{2.138}$$

Equations (2.137) and (2.138) reflect the following very important result:

> If the neutron flux is separable, two individual equations can be obtained from the original balance equation; one for the space and one for the energy dependence, respectively. The space-dependent factor is a solution of the wave equation (or Helmholtz equation).

The quantity B^2, which appears in both Eqs. (2.137) and (2.138), is called "buckling." This nomenclature indicates the fact that B^2 is related to the geometrical curvature of the flux through Eq. (2.136): B^2 is proportional to the relative second derivative of the flux, which describes the curvature.

As stated at the beginning of this section, the flux is separable in a bare one-region reactor. In a realistically reflected, multiregion core, the flux is only partially separable. One generally finds that $\phi(\boldsymbol{r},E)$ is not separable around interfaces, but $\phi(\boldsymbol{r},E)$ is separable far away from interfaces. This can only happen if the corresponding medium is sufficiently large. Because of this dependence on the distance from interfaces, the separated spectrum is often called the "asymptotic spectrum." Thus, in a multiregion system, an asymptotic spectrum can realistically exist, but only within a large region. If separability is assumed in the entire system, one makes a certain error. This error can be tolerated in some cases, e.g., in survey-type criticality calculations.

Envisioning the asymptotic spectrum as a solution of the integral transport equation, which is independent of boundary conditions, shows that it is independent of the size of the core, and it is the same in a reflected core as in an unreflected core of the same material, provided the regions are large enough. The same asymptotic spectrum in a critical core is also established in a subcritical neutron multiplying medium of the same composition as the core far away from the independent source that sustains the flux. Compare, for example, Secs. 2-6C and 3-2 for further discussion of the asymptotic nature of the separated neutron spectrum.

The different types of solutions that may be obtained from the pair of equations (2.137) and (2.138) are further examined in the following section.

2-6B General Types of Solutions for the Separated Balance Equations

There are about six types of solutions one can derive from the pair of equations (2.137) and (2.138), depending on what information is desired and depending on the class of the problem of interest.

The Multiplication Constant k for a Bare Homogeneous Medium (λ Eigenvalue Search). Given the geometry and the composition of the bare medium, the multiplication constant can be determined in two steps:

1. Complete Eq. (2.137) with appropriate boundary conditions for the description of the given geometry and solve the resulting eigenvalue problem. The solution[h] simultaneously yields the

[h]Compare Chapter 3.

spatial flux distribution and the geometrical B^2 eigenvalue, called the "geometrical buckling," B^2_{geo}:

$$B^2 = B^2_{geo} \quad . \tag{2.139}$$

2. Insert B^2_{geo} into Eq. (2.138) and obtain an eigenvalue problem in λ; the solution[i] simultaneously yields the desired eigenvalue λ ($= 1/k$) and the neutron spectrum for this eigenvalue problem, $\varphi_\lambda(E)$.

Note that the computational procedure could not start with solving the energy-dependent equation first since Eq. (2.138) contains two eigenvalues that cannot be determined simultaneously.

The Infinite-Medium k. The multiplication constant of a medium assumed to be infinitely large is found in the same way as for the finite system. The geometrical buckling resulting from Eq. (2.137) for an infinitely large system is zero, as it can also be seen from any one of the formulas for B^2_{geo}, found in Sec. 3-3. Thus, λ_∞ is obtained as the eigenvalue of Eq. (2.138) with $B^2 = 0$:

$$\Sigma_t(E)\varphi(E) - \int_{E'} \Sigma_s(E' \to E)\varphi(E')\, dE'$$

$$= \lambda_\infty \chi(E) \int_{E'} \nu\Sigma_f(E')\varphi(E')\, dE' \quad . \tag{2.140}$$

The resulting

$$\varphi(E) = \varphi_\infty(E) = \Phi_\infty \tag{2.141}$$

is the spectrum in the λ_∞ eigenvalue problem. Integration with respect to energy (see Sec. 2-2D) yields for the infinite-medium k:

$$\text{infinite-medium } k = \frac{\displaystyle\int_0^\infty \nu\Sigma_f(E)\varphi_\infty(E)\, dE}{\displaystyle\int_0^\infty \Sigma_a(E)\varphi_\infty(E)\, dE} = \left(\frac{R_p}{R_a}\right)_{\Phi_\infty} \quad . \tag{2.142}$$

As discussed above, this definition is somewhat different from the definition of Eq. (2.121):

$$k_\infty = \left(\frac{R_p}{R_a}\right)_{\Phi_\lambda} \text{, medium multiplication rate} \tag{2.143}$$

---

[i]Compare, for example, Sec. 4-3.

due to the fact that the spectrum in the finite system is used in Eq. (2.143).

If k_∞ is used in an application, it is logically always the definition in Eq. (2.143) that is applied (see the discussion of k_∞ in Sec. 2-5). The actual differences between the two values k_∞ and the infinite-medium k are small since the deviation of $\varphi(E)$ and $\varphi_\infty(E)$ largely cancels out of the ratio of Eqs. (2.138) and (2.139). Therefore, the notation k_∞ is used in the following for all approximations of the infinite-medium multiplication rate.

Material Buckling and Fundamental Spectrum. The spectra calculated in the λ eigenvalue problems are physically not realized (see Sec. 2-4E). A very important problem is the calculation of separated spectra that actually occur physically; these are the asymptotic spectra that are found far away from interfaces in a critical reactor or far away from sources in a subcritical medium.

The fission neutron source for physically realized spectra is unmodified by an eigenvalue, λ. One therefore has to omit λ in Eq. (2.138):

$$\chi(E)\int_{E'} \nu\Sigma_f(E')\varphi_m(E')\,dE' + \int_{E'} \Sigma_s(E'\to E)\varphi_m(E')\,dE' - \Sigma_t(E)\varphi_m(\mathrm{E})$$

$$= D(E)B_m^2\varphi_m(E) \quad , \tag{2.144}$$

which yields as the largest eigenvalue B_m^2, the "material buckling," and for the corresponding eigenfunction, the "fundamental spectrum," the so-called "B^2 mode," $\varphi_m(E)$. Equation (2.144) is the general definition of the material buckling. The most important application of the fundamental spectrum is as a weighting spectrum in the formation of group constants (see Chapter 7).

The situation is again illustrated in the simplest example: Integrating Eq. (2.144) with respect to energy yields the one-group model; $\nu\Sigma_f$, Σ_a, and D denote the corresponding one-group constants:

$$(\nu\Sigma_f - \Sigma_a)\varphi_m = B_m^2 D\varphi_m \quad , \tag{2.145}$$

with

$$\varphi = \int_0^\infty \varphi(E)\,dE \quad .$$

Equation (2.145) has a nontrivial solution if B_m^2 has the value

$$B_m^2 = \frac{\nu\Sigma_f - \Sigma_a}{D} = \frac{k_\infty - 1}{L^2} \quad . \tag{2.146}$$

The fact that λ has been set equal to one in Eq. (2.144) does not mean that the balance equation has been specialized for a critical reactor.

Actually, Eq. (2.144) holds for any neutron-multiplying medium, independent of its k_∞ value; k_∞ can be larger than, equal to, or smaller than one. The size of the medium does not enter the problem (except that the medium should be large to physically establish separability).

Suppose $B_m^2<0$. Then Eq. (2.144) yields the asymptotic spectrum as it is physically realized far away from a source that maintains a finite stationary flux. For a negative material buckling, k_∞ is smaller than one:

$$B_m^2<0 \text{ and thus } k_\infty<1 \quad . \tag{2.147}$$

The left side of Eq. (2.144) gives the balance of fission and absorption, and the right side, with $B_m^2<0$, represents a source. This term can be interpreted as a "leakage source," which provides neutrons through leakage to a volume element since the balance of fission and absorption is negative.

If $B_m^2>0$, the neutron balance of each point is positive and the right side of Eq. (2.144) must eliminate the surplus neutrons through leakage out of each volume element. For this positive neutron balance, one has $k_\infty>1$:

$$B_m^2>0 \text{ and thus } k_\infty>1 \quad . \tag{2.148}$$

A fundamental spectrum in a system with $k_\infty>1$ can be physically realized with or without an independent source. A source is required if the system is too small to be critical. If no source is needed to establish a fundamental spectrum, the reactor is necessarily critical. This leads directly into the next problem.

Equation (2.146) should not be interpreted as a prescription to find B_m^2 from a k_∞ value. Actually, the situation is just the reverse: Eq. (2.144) is to be solved first, giving B_m^2 and $\varphi_m(E)$. Only then can the one-group quantities for Eq. (2.145) (including L^2) be formed. Then one can determine k_∞ from B_m^2 and L^2 by

$$k_\infty = 1 + L^2 B_m^2 \quad . \tag{2.149}$$

Critical Geometry for a Bare Homogeneous Medium. The first step of the solution to this problem is the determination of the material buckling for the given composition. Since $\lambda = 1$ for the critical reactor, Eq. (2.138) is converted into Eq. (2.144), which yields B_m^2 and the fundamental spectrum. Both B^2 values in Eqs. (2.137) and (2.138) must be the same. Therefore, the spatial equation must be solved for "trial boundary conditions," which gives $B^2_{\text{geo(trial)}}$. By comparison of $B^2_{\text{geo(trial)}}$ with B_m^2, one obtains instructions on how to vary the boundary for the next trial solution. If $B^2_{geo(trial)} > B^2_m$, the system is subcritical and the dimensions must be increased for the next trial solution; if $B^2_{geo(trial)} < B^2_m$, the dimensions must be decreased. This procedure is called a "critical dimension search." The iteration converges toward B^2_m:

$$B^2_{geo(trial)} \rightarrow B^2_m \quad , \tag{2.150}$$

with the converged result

$$B^2_{geo} = B^2_m \quad . \tag{2.151}$$

If Eq. (2.137) can be solved analytically and thus B^2_{geo} be expressed analytically as a function of the dimensions of the system, one does not have to iterate and the critical dimensions can be found by setting the analytical expression for B^2_{geo} equal to B^2_m, for example, for a slab problem (compare Sec. 3-3). The critical height can be found by solving Eq. (2.152) for H':

$$B^2_{geo} = \left(\frac{\pi}{H'}\right)^2 = B^2_m \quad . \tag{2.152}$$

Critical Composition for a Bare Homogeneous System. In a system of given geometry, the spatial eigenvalue problem is to be solved first to find the corresponding geometrical buckling. Inserting B^2_{geo} into Eq. (2.138) and setting $\lambda = 1$ because of the desired criticality would lead to an equation without an explicit eigenvalue. A nontrivial solution, however, requires an eigenvalue, either λ (if $B^2 = B^2_{geo}$ is inserted) or B^2_m (if λ is set equal to unity). The composition is varied in a trial and error fashion; for each trial composition, an eigenvalue problem is solved. The variation of the composition for the next trial solution is found by some sort of an extrapolation or interpolation scheme based on the deviation of the eigenvalue from the target value. After the convergence of the problem, $B^2_m = B^2_{geo}$ and $\lambda = 1$ as desired. The search yields the critical composition or enrichment.

Search at Given Multiplication Constants. In the same way as criticality is achieved in the computational schemes discussed above, one can achieve (within limits) any preset eigenvalue $\lambda = 1/k$ as the converged result. Practically, this type of search is very important, since providing a power reactor with sufficient fuel for a, say, 1-yr operation requires the initial k to be larger than 1, e.g., 1.10.

Neutron Slowing Down Spectrum. The neutron spectrum during the slowing down process is important for the calculation of the absorption of neutrons in cross-section resonances (see Chapter 5) and for the formation of the respective group constants (see Chapter 7).

For the treatment of this slowing down problem, one does not consider the entire energy range and thus does not consider the complete chain reaction. If the thermal range (in a thermal reactor) is not treated, the strength of the fission source and the multiplication constant of the system cannot be known. The only fact known is that there is a fission source, proportional to $\chi(E)$; but the value of the factor that multiplies $\chi(E)$ is unknown. The spectrum in the slowing down region is therefore obtained from an inhomogeneous problem with an arbitrary source magnitude rather than from an eigenvalue problem. Thus, the balance equation in the slowing down region, with leakage described in the diffusion approximation, is given by:

$$[D(E)B^2 + \Sigma_t(E)]\varphi(E) - \int_E^\infty \Sigma_s(E' \to E)\varphi(E')\,dE' = \chi(E)\cdot s_0 \quad , \tag{2.153}$$

with an arbitrary source s_0 (e.g., $s_0 = 1$ neutron/cm)[j] and $B^2 = B^2_{\text{geo}}$ for the given size of the system. The solution of Eq. (2.153) is proportional to the arbitrary source; s_0 is to be determined separately (see Sec. 4-2).

In addition to this equation, one considers in slowing down problems the corresponding one for an infinite medium. Setting B^2 to zero gives the balance equation for the slowing down region in an infinite system:

$$\Sigma_t(E)\varphi(E) - \int_{E'} \Sigma_s(E' \to E)\varphi(E')\,dE' = s_0\chi(E) \quad . \tag{2.154}$$

The diffusion approximation had an impact only on the description of leakage. Thus, the slowing down equation in an infinite medium contains no approximation (provided the system is homogeneous). Equation (2.154), therefore, correctly describes the neutron slowing down spectrum in an infinite homogeneous medium.

Practically all neutrons in a fast reactor have energies in the slowing down range. Thus, the treatment of the slowing down problem is equivalent to the complete solution, which also yields the λ eigenvalue. In Sec. 4-2 it is shown how the λ eigenvalue can be extracted from the solution of Eq. (2.153) and the value of s_0.

The coupled relationship between B^2_{geo}, B^2_m, and λ can be seen in a comprehensive manner by considering the one-energy group model for

---

[j]This rather strange dimension for a "source" follows from the fact that the flux $\phi(\boldsymbol{r})$ has been canceled out of the balance equation for the spectrum.

a bare reactor. For this seemingly "monoenergetic" problem, Eq. (2.138) appears in the condensed form:

$$(DB^2 + \Sigma_a - \lambda\nu\Sigma_f)\varphi = 0 \quad . \tag{2.155}$$

To obtain a nontrivial solution for $\lambda = 1$, B^2 must be varied such that the operator in the parentheses becomes singular, and no inverse exists. This requires the parentheses to be zero, which yields the material buckling:

$$B_m^2 = \frac{k_\infty - 1}{L^2} \quad . \tag{2.156}$$

If B^2 in Eq. (2.155) is set equal to the B_{geo}^2 value of a given geometry, and if the composition is given too, one must retain the eigenvalue λ in order to allow for a nontrivial solution. One obtains the λ eigenvalue for a given geometry and a given composition, again by requiring the parentheses to be zero, as

$$k = \frac{1}{\lambda} = \frac{k_\infty}{1 + L^2 B_{\text{geo}}^2} \quad . \tag{2.157}$$

Analogously, the problem can be solved for a given geometry and a variable composition. The eigenvalue λ may be either equal to one or different from one if so desired.

2-6C Separation of Problems with Distributed Independent Sources

Problems with independent sources are not easily amenable to flux separation. However, since certain balance equations that are often applied in practical approximations are of the same form as the equations resulting from flux separation, the discussion of this area can shed light on some practically important approximations.

The previous two subsections dealt with flux separation and solutions for *homogeneous* problems, for which all terms in the balance equation contain the neutron flux. The other important class of problems comprises the cases in which a flux-independent source is present: *inhomogeneous* problems. The independent source may be a "distributed source" or it may be localized, e.g., as a "point" or "plane" source (see Sec. 2-4B).

The balance equations and boundary conditions for a distributed source are given by:

$$(\mathbf{M} - \mathbf{F})\Phi = S = S(\boldsymbol{r},E) \quad . \tag{2.158}$$

If the source is distributed only within the system (and not outside of it), homogeneous boundary conditions have to be used; e.g., in the diffusion approximation, one can use the approximate boundary condition

$$\phi(\boldsymbol{r}_v,E) = 0 \quad , \tag{2.159}$$

if the system is surrounded by air, where $\boldsymbol{r}_v$ describes the vacuum boundary.

Note that Eq. (2.158) does not contain an eigenvalue. In general, it has a unique nontrivial solution, except when the operator in the bracket is singular. In the latter case (which corresponds to a critical reactor with an independent source), no solution exists; this means physically that no finite neutron flux exists in a critical reactor with an independent source.

The inhomogeneous equation, Eq. (2.158), is not separable in space and energy for physically realistic independent sources. One can readily show that separability would require the distributed source proportional to the total flux, $\phi(\boldsymbol{r})$, which is normally not the case for flux-independent sources. Alternately, separation would be possible if the energy dependence were proportional to $D(E)\varphi(E)$:

$$S(\boldsymbol{r},E) = S(\boldsymbol{r})\frac{D(E)}{D}\varphi(E) \quad , \tag{2.160}$$

which is equally unrealistic (D is the one-group diffusion constant). The latter case gives as a balance equation for the space dependence of the total flux in the presence of a source:

$$\nabla^2\phi(\boldsymbol{r}) + B_m^2\phi(\boldsymbol{r}) = -\frac{1}{D}S(\boldsymbol{r}) \quad , \tag{2.161}$$

with

$$D = \int_E D(E)\varphi(E)\, dE \quad . \tag{2.162}$$

The same equation as Eq. (2.161) is obtained if one integrates the original inhomogeneous equation, Eq. (2.158), with respect to energy and introduces one-group constants. These group constants actually depend on space, as shown and discussed in Sec. 7-2A. If one neglects this space dependence and substitutes approximate space-independent group constants, one obtains the common one-group balance equation. The discussion of the flux separation reveals the underlying assumption that would yield Eq. (2.161) exactly from flux separation.

2-6D Separation of Problems with Localized Independent Sources

If the source is localized in space, e.g., if it is an "external" source, the medium into which the source neutrons are emitted does not contain an independent source. The balance equation is then homogeneous:

$$(\mathbf{M} - \mathbf{F})\Phi = 0 \quad . \tag{2.163}$$

The presence of the source is accounted for in the boundary conditions, which are then "inhomogeneous" [see Eqs. (2.164) to (2.167)].

In diffusion theory, the boundary condition at a localized source can best be expressed by a neutron conservation condition. If the source is located at the interface between the medium and vacuum, only source neutrons enter the system. If the source is located at the center of a symmetric system, the current near the source consists only of source neutrons; neutrons that pass through the source cancel their contribution to the current due to the symmetry of the problem. Thus, the neutron conservation in both cases requires that the surface integral of the current be equal to the volume integral of the source:

$$\oint \boldsymbol{J}(\boldsymbol{r}_s,E)\cdot d\boldsymbol{f} = \int_{V_s} S(\boldsymbol{r},E)\, dV \quad , \tag{2.164}$$

with $d\boldsymbol{f}$ describing an element of the surface area around the source, V_s the source volume, and $S(\boldsymbol{r},E)$ the source density within the source. Special examples are:

the sphere source (in a system of spherical geometry):

$$4\pi r_s^2 J(r_s,E) = S(E)\cdot V_s \quad ; \tag{2.165}$$

the point source:

$$\lim_{r_s\to 0}[4\pi r_s^2 J(r_s,E)] = S_p(E) \quad , \tag{2.166}$$

with $S_p(E)$ being the neutron emission rate of the point source; and

the plane source ($x_s^{\pm}$ being the two sides of the source plane):

$$\pm J(x_s^{\pm},E) = S(x_s,E)/2 \quad . \tag{2.167}$$

Also, the boundary condition of asymmetrically located sources can be derived from the principle of neutron conservation. The source neutrons then provide only part of the neutron current. Examples are suggested as homework problems at the end of this and the next chapter.

Note that no eigenvalues appear in Eq. (2.163) since this equation describes the flux as it develops out of an external source. A homogeneous equation together with an inhomogeneous boundary condition is mathematically also an inhomogeneous problem that has a unique solution, provided that the operator is *not* singular (i.e., the reactor is not critical).

The separation of the balance equation, Eq. (2.163), proceeds in the same way as without an external source: The spatial flux factor is the solution of the wave equation, Eq. (2.137), and the spectrum follows from Eq. (2.138), with $\lambda = 1$. If $\lambda = 1$ in the energy-dependent balance equation (although the system is subcritical), B^2 must be treated as an eigenvalue and the solution is again the fundamental spectrum and B^2 equals B_m^2.

Since the B^2 values must be the same, B_m^2 must be inserted into Eq. (2.137). The resulting equation (in one planar dimension),

$$\frac{d^2}{dx^2}\phi(x) + B_m^2\phi(x) = 0 \quad , \tag{2.168}$$

must be solved with an inhomogeneous boundary condition. However, the boundary conditions, Eqs. (2.165), (2.166), and (2.167), depend on energy, and Eq. (2.168) contains only the spatial flux component. The solution of such a problem requires the superposition of several terms, each term containing an energy- and a flux-dependent factor:

$$\phi(x,E) = \sum_k a_k\phi_k(x)\varphi_m^k(E) \quad . \tag{2.169}$$

The $\varphi_m^k(E)$ represent the set of eigenfunctions of Eq. (2.138) (with $\lambda = 1$), and the corresponding eigenvalues $B_{m,k}^2$ are the higher material bucklings. The $B_{m,k}^2$ are to be inserted into Eq. (2.168); the resulting solutions represent the set of functions $\phi_k(x)$. The coefficients are found by adjusting Eq. (2.169) to the actual boundary condition, Eqs. (2.165), (2.166), and (2.167). A two-group example is presented in Secs. 2-7 and 3-4E.

The higher flux components normally decrease much faster—with increasing distance from the source—than the factor of the fundamental spectrum. This leads to an "asymptotic flux," being separated in space and energy:

$$\phi^{as}(x,E) = a_o\phi_o(x)\varphi_m^o(E) \quad . \tag{2.170}$$

This flux is physically realized at some distance from the source plane. Therefore, $\varphi_m^o(E) = \varphi_m(E)$ is called the "asymptotic spectrum."

Simplified reactor theory, based on one-group cross sections calculated with the asymptotic spectrum, is often called "asymptotic reactor theory." Most of the one-group formulas discussed above can be considered as part of asymptotic reactor theory.

2-7 Flux Separability Applications for Two Groups

The separation of the neutron flux in its space and energy dependencies leads to the fundamental spectrum $\varphi_m(E)$, which is physically realized far away from boundaries, and to the material buckling, B_m^2, representing the intrinsic neutron balance as well as the corresponding spatial flux curvature. The separated flux is the "asymptotic solution," Eq. (2.170). The deviations from the asymptotic flux are analytically given by the higher modes as generally indicated in Eq. (2.169). These concepts and characteristics of the flux solution are illustrated for two energy groups in this section and in Sec. 3-6. The explicit evaluation of the space dependency is deferred to Sec. 3-6. The vector notation introduced above is applied.

2-7A Separation of the Two-Group Diffusion Equation

For two energy groups, the flux appears as a vector with two components:

$$\phi(x,E) = \begin{pmatrix} \phi_1(x) \\ \phi_2(x) \end{pmatrix} \quad . \tag{2.171}$$

The two-group diffusion equations, Eqs. (2.48), are then written as a single operator equation to conform with the vector character of the flux. For the application in this section, Eqs. (2.48) are simplified for a homogeneous one-region problem in slab geometry:

$$-D_1\frac{d^2}{dx^2}\phi_1(x) + (\Sigma_{a1} + \Sigma_{S12})\phi_1(x) = \nu\Sigma_{f1}\phi_1(x) + \nu\Sigma_{f2}\phi_2(x) \quad ,$$

$$-D_2\frac{d^2}{dx^2}\phi_2(x) + \Sigma_{a2}\phi_2(x) - \Sigma_{S12}\phi_1(x) = 0 \quad . \tag{2.172}$$

Abbreviating Eq. (2.172) in operator notation gives

$$\mathbf{M\Phi} = \mathbf{F\Phi} \quad , \tag{2.173}$$

with

$$\mathbf{M} = \begin{pmatrix} -D_1\frac{d^2}{dx^2} & 0 \\ 0 & -D_2\frac{d^2}{dx^2} \end{pmatrix} + \begin{pmatrix} \Sigma_{a1} + \Sigma_{S12} & 0 \\ -\Sigma_{S12} & \Sigma_{a2} \end{pmatrix} \tag{2.174}$$

and

$$\mathbf{F} = \begin{pmatrix} \nu\Sigma_{f1} & \nu\Sigma_{f2} \\ 0 & 0 \end{pmatrix} \quad . \tag{2.175}$$

For the intended separation of variables, **M** has been split in two components, the leakage operator,

$$\mathbf{L} = -\begin{pmatrix} D_1 & 0 \\ 0 & D_2 \end{pmatrix} \frac{d^2}{dx^2} = -\mathbf{D}\frac{d^2}{dx^2} \quad , \tag{2.176}$$

and the non-leakage component **N**,

$$\mathbf{N} = \begin{pmatrix} \Sigma_{a1} + \Sigma_{S12} & 0 \\ -\Sigma_{S12} & \Sigma_{a2} \end{pmatrix} \quad . \tag{2.177}$$

Separation of the two-group flux in Eq. (2.171) identifies a spectrum vector $\boldsymbol{\varphi}$ and a common space dependency $\phi(x)$:

$$\boldsymbol{\Phi} = \begin{pmatrix} \phi_1(x) \\ \phi_2(x) \end{pmatrix} = \begin{pmatrix} \varphi_1 \\ \varphi_2 \end{pmatrix} \phi(x) \quad . \tag{2.178}$$

Inserting Eq. (2.178) into (2.173), rearranging terms, and dividing by $\phi(x)$ yields

$$-D_1\, \varphi_1 \frac{1}{\phi(x)} \frac{d^2}{dx^2}\phi(x) = \nu\Sigma_{f1}\varphi_1 + \nu\Sigma_{f2}\, \varphi_2 - (\Sigma_{a1} + \Sigma_{S12})\varphi_1 \quad ,$$

$$-D_2\, \varphi_2 \frac{1}{\phi(x)} \frac{d^2}{dx^2}\phi(x) = \Sigma_{S12}\, \varphi_1 - \Sigma_{a2}\, \varphi_2 \quad .$$

Dividing the first equation by $D_1\varphi_1$ and the second by $D_2\varphi_2$ gives the one-dimensional two-group analog to Eq. (2.135):

$$-\frac{1}{\phi(x)} \frac{d^2}{dx^2}\phi(x) = \frac{1}{D_1\varphi_1}[\nu\Sigma_{f1}\varphi_1 + \nu\Sigma_{f2}\varphi_2 - (\Sigma_{a1} + \Sigma_{S12})\, \varphi_1]$$

$$-\frac{1}{\phi(x)} \frac{d^2}{dx^2}\phi(x) = \frac{1}{D_2\varphi_2}(\Sigma_{S12}\varphi_1 - \Sigma_{a2}\, \varphi_2) \quad . \tag{2.179}$$

Again, both sides have to be equal to a constant, which is denoted by B^2. This gives the separated two-group equations, consisting of the wave equation for $\phi(x)$:

$$\frac{d^2}{dx^2}\phi(x) + B^2\phi(x) = 0 \quad , \tag{2.180}$$

and the two-group matrix equation,

$$[B^2\mathbf{D} + \mathbf{N} - \mathbf{F}]\, \boldsymbol{\varphi} = 0 \quad , \tag{2.181}$$

$$\begin{pmatrix} B^2D_1 + \Sigma_{a1} + \Sigma_{S12} - \nu\Sigma_{f1} & -\nu\Sigma_{f2} \\ -\Sigma_{S12} & B^2D_2 + \Sigma_{a2} \end{pmatrix} \begin{pmatrix} \varphi_1 \\ \varphi_2 \end{pmatrix} = 0 \qquad (2.182)$$

for the energy dependence.

2-7B Fundamental and Higher Material Bucklings and Spectra

The homogeneous equation, Eq. (2.181), has a nontrivial solution only for special values of B^2, the B^2 eigenvalues, obtained by making the operator in the bracket singular. This requires the corresponding determinant to be set equal to zero:

$$\begin{vmatrix} B^2D_1 + \Sigma_{a1} + \Sigma_{S12} - \nu\Sigma_{f1} & -\nu\Sigma_{f2} \\ -\Sigma_{S12} & B^2D_2 + \Sigma_{a2} \end{vmatrix} = 0 \quad . \qquad (2.183)$$

Equation (2.183) is a quadratic equation for B^2:

$$D_1D_2(B^2)^2 + [D_1\Sigma_{a2} + D_2(\Sigma_{a1} + \Sigma_{S12} - \nu\Sigma_{f1})]B^2$$
$$+ [\Sigma_{a2}(\Sigma_{a1} + \Sigma_{S12} - \nu\Sigma_{f1}) - \nu\Sigma_{f2}\Sigma_{S12}] = 0 \quad . \qquad (2.184)$$

The two solutions are

$$B_{1,2}^2 = -\frac{p}{2} \pm \left(\frac{p^2}{4} + q\right)^{1/2} \quad , \qquad (2.185)$$

with

$$p = \frac{\Sigma_{a2}}{D_2} + \frac{\Sigma_{a1} + \Sigma_{S1} - \Sigma_{f1}}{D_1} \qquad (2.186)$$

and

$$q = \frac{1}{D_1D_2}[\nu\Sigma_{f2}\Sigma_{S12} - \Sigma_{a2}(\Sigma_{a1} + \Sigma_{S12} - \nu\Sigma_{f1})] \quad . \qquad (2.187)$$

The plus sign in Eq. (2.185) gives the largest eigenvalue, which is the material buckling, B_m^2:

$$B_m^2 = -\frac{p}{2} + \left(\frac{p^2}{4} + q\right)^{1/2} \quad ; \qquad (2.188)$$

the negative sign gives the higher material buckling eigenvalue. For two groups, both material bucklings are real. For more than two groups, one or more pairs of the higher B_m^2 eigenvalues may be complex.

The material buckling is positive if there is intrinsically an over-production of neutrons. Then $k_\infty > 1$ as stated in Eq. (2.148). Apparently, B_m^2 is positive if q is positive. The bracket in Eq. (2.187) expresses the physical condition for $q>0$. Neglecting the "resonance" absorption (Σ_{a1}) and the fast fission ($\nu\Sigma_{f1}$) gives the simple condition:

$$q>0 \text{ if } \frac{\nu\Sigma_{f2}}{\Sigma_{a2}} = \eta > 1 \quad . \tag{2.189}$$

More completely, $q>0$ and thus

$$B_m^2 > 0 \text{ if } \frac{\nu\Sigma_{f2}}{\Sigma_{a2}} = \eta > 1 + \frac{\Sigma_{a1} - \nu\Sigma_{f1}}{\Sigma_{S12}} \quad ; \tag{2.190}$$

i.e., η must be larger than one plus the relative net losses in the fast group.

Having the B_m^2 eigenvalues allows a completion of the solution of Eq. (2.182). Inserting the B_m^2 eigenvalues in one of the two equations of Eq. (2.182)—the other equation is automatically satisfied per Eq. (2.183)—yields, for the corresponding eigenfunctions, the ratios:

$$\left.\frac{\varphi_1}{\varphi_2}\right]_m = \frac{B_m^2 D_2 + \Sigma_{a2}}{\Sigma_{S12}} \left(\begin{matrix}\text{fundamental}\\ \text{spectrum}\end{matrix}\right) \tag{2.191}$$

and

$$\left.\frac{\varphi_1}{\varphi_2}\right]_{(2)} = \frac{B_2^2 D_2 + \Sigma_{a2}}{\Sigma_{S12}} (\text{higher mode}) \quad . \tag{2.192}$$

A theorem on these eignvalue problems states that there exists a non-negative fundamental mode and this is the only non-negative eigenfunction. Thus,

$$\left.\frac{\varphi_1}{\varphi_2}\right]_m > 0$$

and

$$\left.\frac{\varphi_1}{\varphi_2}\right]_{(2)} < 0 \quad . \tag{2.193}$$

The free factor in φ is disposed by normalization:

$$\int_0^\infty \varphi(E)\, dE = \varphi_1 + \varphi_2 = 1 \quad . \tag{2.194}$$

Two examples are presented here to illustrate these concepts. The macroscopic group constants are given in Table 2-I. Applying the inequality (2.190) shows which sign to expect for B_m^2 in these two cases.

Apparently, B_m^2 should be positive in case 1, representing a thermal

TABLE 2-I
Macroscopic Group Constants for Two-Group Example

	Case 1	Case 2
Σ_{a1}	0.003/cm	0.003/cm
Σ_{S12}	0.127/cm	0.127/cm
D_1	1.3 cm	1.3 cm
Σ_{a2}	0.100/cm	0.100/cm
$\nu\Sigma_{f2}$	0.110/cm	0.040/cm
D_2	0.25 cm	0.25 cm

reactor core-type material, but in case 2, B_m^2 is negative. The numerical evaluation gives:

Case 1:

$$B_m^2 = 0.0059/\text{cm}^2$$

$$B_2^2 = -0.5059/\text{cm}^2$$

and

$$B_m^2 = -0.0547/\text{cm}^2$$

Case 2:

$$B_2^2 = -0.4453/\text{cm}^2 \quad .$$

The corresponding normalized eigenspectra are obtained as

Case 1:

$$\begin{pmatrix} \varphi_1 \\ \varphi_2 \end{pmatrix}_m = \begin{pmatrix} 0.4441 \\ 0.5559 \end{pmatrix} ; \begin{pmatrix} \varphi_1 \\ \varphi_2 \end{pmatrix}_{(2)} = \begin{pmatrix} -0.2634 \\ 1.2634 \end{pmatrix} \quad .$$

Case 2:

$$\begin{pmatrix} \varphi_1 \\ \varphi_2 \end{pmatrix}_m = \begin{pmatrix} 0.2960 \\ 0.7040 \end{pmatrix} ; \begin{pmatrix} \varphi_1 \\ \varphi_2 \end{pmatrix}_{(2)} = \begin{pmatrix} -0.4533 \\ 1.4533 \end{pmatrix} \quad .$$

As expected φ^m is non-negative and $\varphi^{(2)}$ has one negative component.

2-7C The Space-Dependent Flux Components

In separating the original balance equation, a B^2 was introduced that then appeared with the same value in both the energy and the spatial equation. The resulting B^2 values and spectra for two-group problems were given in the previous section. The same B^2 values need to be inserted in the spatial equation, Eq. (2.180). Its solution for a symmetrical slab problem gives the following results:

$$\text{Case 1:}\quad \phi^{(m)}(x) = \cos B_m x \quad (B_m^2 > 0)$$

$$\phi^{(2)}(x) = \cosh|B_2|x, \ (B_2^2 < 0) \quad .$$

$$\text{Case 2:}\quad \phi^{(m)}(x) = \cosh|B_m|x \ \ (B_m^2 < 0)$$

$$\phi^{(2)}(x) = \cosh|B_2|x \ \ (B_2^2 < 0)$$

or

$$\phi^{(m)}(x) = e^{-|B_m|x}$$

$$\phi^{(2)}(x) = e^{-|B_2|x} \quad .$$

The B_m component yields the "asymptotic flux," i.e.,

$$\mathbf{\Phi}^{as}(x) = \begin{pmatrix} \varphi_1 \\ \varphi_2 \end{pmatrix}_m \cos B_m x \quad ,$$

in case 1. The second component provides the adjustment needed to satisfy the boundary condition. The complete solution is discussed in Sec. 3-6 for several examples.

It should also be noted that the geometrical buckling, B_{geo}^2, plays no role in these analytical solutions and the resulting flux distributions. It is B_m^2 which quantifies the intrinsic neutron balance and thus the corresponding curvature, applying the argumentation presented in Sec. 3-4D. Only when the slab approaches the critical dimension does the B_{geo}^2 value approach B_m^2. But it is still B_m^2 which determines the intrinsic neutron balance and thus the curvature as it has been for all slab dimensions as well as during the approach to critical.

2-8 Summarizing Remarks

In the previous sections, some fundamental features of reactor theory were presented. Derived first were the neutron balance relationships along with necessary boundary conditions describing the behavior of neutrons based on the gains and losses of the neutron populations within a control volume. The relationship between the detailed transport theory representation and diffusion theory were discussed and some specific approximations were examined. The similarities between the various neutron balances were made more visible by recasting the equations into common operator notation. From this reformulation, the concept of reactor criticality naturally evolved in a general form independent of the particular approximation employed in the balance equation. Also, the conceptual approach for the quantification of off-criticality was described in the same general form. The resulting multiplication constant was compared with the traditional factored models for k_{eff}. Next, the important concept of flux separation was presented as a valuable approach toward gaining insight in and understanding of basic neutron flux solution and calculation procedures. Finally, this concept of flux separation was also applied to inhomogeneous problems. The general types of solutions with independent sources become more transparent because of the flux separation.

In the next several chapters, the technique of separating the space and energy dependencies of the neutron flux is extensively exploited in order to obtain analytical or numerical solutions for parts of the neutron balance equation. From the separated solutions, lumped quantities are derived as a basis for the treatment of more complicated problems or for a more accurate treatment of the general space-energy-dependent problem.

Homework Problems

1. Derive the integral transport equation, Eq. (2.13), for anisotropic scattering and an isotropic fission source and an anisotropic independent source.
2. Simplify the one-group integral transport equation, Eq. (2.23), to apply to a slab between $-(H/2) \leq z \leq H/2$, with H being the height of the slab. Assume vacuum outside the slab, but include an isotropic source of neutrons entering the medium from both sides.
3. Derive and present the one-group integral transport equation for the angular flux with isotropic scattering for the fuel part of a three-medium slab problem. Use Eq. (2.13) as the basis. Note, the neutrons are thermalized in the moderator, leading to a source of thermal neutrons, $S(z)$.

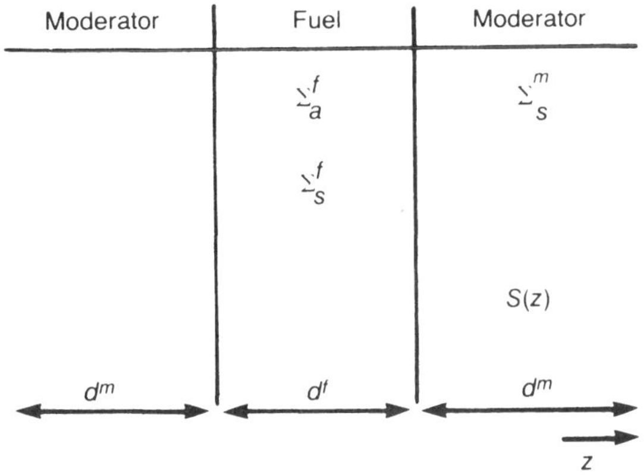

Assume d^m to be large enough so that the outer moderator boundary need not be considered.

4. Apply the general equation developed in problem 1 to the following case.

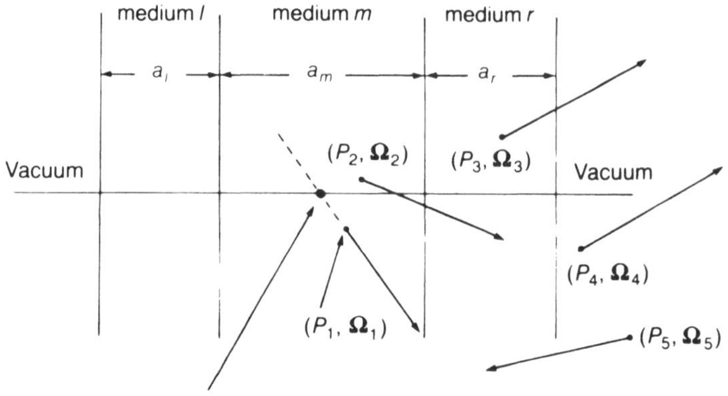

Present the integral transport equation for the sets of points and angles indicated above. Note the relations of points and angles to the source.

$$\Sigma_t \text{ (medium } l) = \Sigma_l \quad ,$$

$$\Sigma_t \text{ (medium } m) = \Sigma_m \quad ,$$

and

$$\Sigma_t \text{ (medium } r) = \Sigma_r \quad ,$$

with

$\Sigma_f \equiv 0$ in all three media.

5. *One-Group Problems.* Simplify the balance equations, Eqs. (2.137) and (2.138), for the λ eigenvalue problem so that they represent the corresponding pair of equations for plane geometry $[\phi(x)]$ and one energy group (φ). Assume as boundary conditions $\phi(\pm a) = 0$. Use these one-group values: $D = 0.4$ cm, $\Sigma_a = 0.0044$/cm, and $\nu\Sigma_f = 0.00584$/cm.

   a. Find the material buckling; describe the general procedure.

b. Find $k = \frac{1}{\lambda}$ for $a = 50$ cm; describe the general procedure.
c. Find the critical dimension; describe the general procedure.
d. Increase the critical dimensions by 5%; i.e., increase a to $a' = 1.05a$. Find the additional control rod absorption (in terms of an added $\delta\Sigma_a$) to make the system critical. Give also the new Σ_a, say, $\Sigma_a' = \Sigma_a + \delta\Sigma_a$. Describe the general procedure.
e. Take the critical dimension. Modify the original composition by increasing $\nu\Sigma_f$ such that the resulting k equals 1.05. Describe the general procedure.

6. *Energy-Dependent Problems.* Take the separated balance equations and boundary conditions for the λ eigenvalue problem.
   a. Describe the procedure for finding the material buckling.
   b. Describe the procedure for finding k.
   c. Describe the procedure for finding the critical dimensions.
   d. Describe the procedure for finding the critical composition.
7. *Source Problem.* Consider a subcritical slab $(-a \le x \le a)$ with a plane source at x_s. Find the boundary conditions at the source plane for $x_s = -a$; x (off center); 0; and $+a$.

Review Questions

1a. Describe in words the approach applied in deriving the one-dimensional Boltzmann equation.
b. Briefly present the derivation of the Boltzmann equation in one dimension without an independent source.
2a. Describe the approach applied in deriving the integral transport equation for the flux.
b. How many neutrons from an isotropic source $q(\mathbf{r},E)$ go in a solid angle element $d\Omega$?
c. Give the formulas for the two attenuation effects involved.
3a. Describe the approach applied in deriving the one-dimensional diffusion approximation.
b. What is the difference compared to the corresponding derivation of the Boltzmann equation?
c. Give the one-dimensional energy-dependent diffusion equation.
4a. Give Fick's Law.
b. Present the assumption on which Fick's Law is based (in general terms and in terms of the angular distribution).
c. List four cases in which Fick's Law is probably a poor approximation.

5. Derive the one-group diffusion equation for all neutrons from the energy-dependent diffusion equation.
6. Present the treatment of the scattering terms for the one-group equation for thermal neutrons.
7a. Why is the diffusion "constant" in the three-dimensional one-group diffusion equation actually a diagonal matrix?
b. Show it in formulas.
8. Why are no boundary conditions needed for the integral transport equation?
9a. Derive the interface conditions and the vacuum boundary condition for the Boltzmann equation from the integral transport equation.
b. Give the results.
10a. Derive the interface conditions of the diffusion theory from the corresponding conditions of the Boltzmann equation.
b. Sketch the behavior of the diffusion theory flux, $\phi(x)$, around an interface; explain the key aspect.
11a. Give the angular distribution approximation of the one-dimensional angular flux, $\phi(x,\mu)$, as used in diffusion theory.
b. Derive from it $\phi(x)$ and $J(x)$ by proper integrations.
12a. Describe and sketch the frequently used approach to find a diffusion theory boundary condition at a vacuum boundary.
b. Give the result in terms of ϕ and $d\phi/dx$.
c. How is this result often interpreted?
13. What is the difference between the two boundary conditions

$$\phi(x_v) - \frac{2}{3}\lambda_{tr} \left| \frac{d\phi}{dx} \right|_{x_v} = 0$$

and

$$\phi(x_v + \frac{2}{3}\lambda_{tr}) = 0$$

physically and in an application? When is the difference unimportant?
14. Give the more accurate diffusion theory boundary condition as derived from transport theory.
15. Give one reactor physics example of each of the following types of operators: multiplication, differentiation, matrix, and integral operators.

16. Give the four corresponding scalar products.
17. Write the stationary energy-dependent diffusion equation for an off-critical system in operator form and define the operators.
18. What are the two basic advantages in using operator notation?
19a. Which four general types of solutions can one encounter in reactor problems?
  b. Give an example of each type.
20a. What is a source-sink problem?
  b. What kind of solutions do source-sink problems have?
  c. Demonstrate the solution for the one-group integrated neutronics model.
21. What kind of solutions do subcritical problems have with an independent source? With $S = 0$? Demonstrate it with the one-group integrated neutronics model.
22. Give source multiplication formulas in the one-group integrated neutronics model for the flux and for the fission source.
23a. What are the two kinds of solutions for the critical reactor equation?
  b. What is the condition on the operator $(\mathbf{M} - \mathbf{F})$ in $(\mathbf{M} - \mathbf{F})\Phi = 0$ so that this equation has a nontrivial solution?
24a. What are the stationary solutions of the homogeneous off-critical balance equations?
  b. Are these solutions physically realizable?
25. What is commonly done in order to obtain information on the degree of off-criticality?
26. How can one achieve criticality if one is willing to alter the system (i.e., $\mathbf{M}$ or $\mathbf{F}$)? Describe three ways!
27. Why is a source addition not generally suitable to determine the degree of off-criticality?
28a. Give the precise definition of $k = k_{\text{eff}}$.
  b. Give two interpretations of $\lambda = 1/k$.
29. Express the static reactivity and the reactivity increment in terms of λ's and k's.
30. What are "higher modes"? Can they be physically realized alone (why or why not)? In combination with a fundamental mode?
31. Characterize a time-dependent neutronics problem in terms of criticality and physical realizability of the flux.
32a. Which factored expression has been used historically to estimate $k = k_{\text{eff}}$?
  b. Give the five-factor formula and explain the factors in words.
  c. Give the four factors of k_∞ in terms of ratios of reaction rates.
33. What is the difference between k_∞ and the infinite-medium k in

words and in terms of precise expressions of integrated reaction rates (no factoring of k_∞)?

34. Split the nonleakage probability, P, into two factors.

35a. In which cases or spatial areas is the flux, $\phi(\boldsymbol{r},E)$, separable in space and energy?

b. Give the formula for the separation, including the split of the dimensions.

36. Carry out the separation in space and energy for the energy-dependent diffusion equation.

37. Describe the solution procedure of the pair of equations that results from separation in E and $\boldsymbol{r}$ and discuss the physical realization of the solution:
   a. for given geometry and composition (bare, homogeneous)
   b. for an infinite homogeneous medium; give the formulas for the infinite-medium k and for k_∞
   c. for obtaining the fundamental spectrum
   d. for a critical (bare) core with a given composition
   e. for a critical (bare) core with a given geometry
   f. for a core with given λ.

38a. Give the mathematical definition of the material buckling.

b. How and where does the material buckling influence the flux distribution?

c. Derive a one-group model formula for B_m^2 from its general definition.

d. Derive B_m^2 for the one-group model and discuss its relation to k_∞.

39. Discuss the relationship of λ $(=1/k)$, B_m^2, and B_{geo}^2 for the one-group model.

40. What is the condition on the space dependence of a source $S(\boldsymbol{r},E)$ so that the inhomogeneous problem is separable in $\boldsymbol{r}$ and E?

41. Consider the one-group equation for a source problem:

$$\nabla^2\phi(\boldsymbol{r}) + B^2\phi(\boldsymbol{r}) = -S(\boldsymbol{r})/D \quad .$$

a. What is the condition on the $\boldsymbol{r}$ energy dependence of the source that leads to this type of one-group equation?

b. Is this energy dependence physically available?

c. What is the condition for the above equation to have a solution?

d. What is B^2 in the above equation? Why?

42a. Give the balance equation in operator form for the "external" source problem.

b. Give inhomogeneous boundary conditions for three simple cases: a spherical source at the center of a spherical geometry, a point source at $r = 0$, and a plane source.
c. Explain on the basis of the plane source boundary condition how a pronounced angular dependence of the source could influence the flux.
d. Give the typical form of the analytical solution for $\phi(x,E)$ with a plane source.
e. What is its asymptotic form? What is the effect of the transitory terms?

43. What is "asymptotic" reactor theory?

REFERENCES

1. S. Glasstone and M. C. Edlund, *The Elements of Nuclear Reactor Theory,* Chapter 5, Van Nostrand Co., Princeton, New Jersey (1952).
2. J. R. Lamarsh, *Introduction to Nuclear Reactor Theory,* Chapter 5, Addison-Wesley Publishing Co., Reading, Massachusetts (1966).
3. B. Davison, *Transport Theory of Neutrons,* Oxford Press, London (1956).
4. G. I. Bell and S. Glasstone, *Nuclear Reactor Theory,* Chapter 1, Van Nostrand Reinhold Co., New York (1970).
5. K. O. Ott and R. J. Neuhold, *Introductory Nuclear Reactor Dynamics,* American Nuclear Society, La Grange Park, Illinois (1985).
6. A. M. Weinberg and E. P. Wigner, *The Physical Theory of Neutron Chain Reactors,* p. 381, University of Chicago Press (1958).

Three

THE SPACE DEPENDENCE OF THE NEUTRON FLUX

3-1 Introduction

In the previous chapter, the general diffusion equation was derived, which approximately describes the production, absorption, and migration of neutrons throughout a system; basic types of solutions were discussed. The presentation concentrated on the basic concepts in order to make the physics and the structure of the problems transparent. In this chapter, the behavior of neutrons is explored in more detail by examining the space distributions of neutrons within various physical systems. Chapter 3 is followed by two chapters in which the energy dependence of the neutron flux is investigated. The combined space-energy dependence is then discussed in Chapter 7.

The space dependence of the neutron flux is derived within the context of diffusion theory, i.e., Fick's Law is assumed to describe the relation of current and flux. Furthermore, only steady-state behavior is considered. As with the traditional approach,[1,2] the spatial neutron flux relationships are derived for several different types of reactor configurations.

Only a description for "global" spatial variations of the neutron flux in simple configurations is sought in this chapter. Each of the various reactor models or geometries is assumed to have a uniform material distribution in each region, which consists of, or may be approximated by, a homogeneous composition; local flux depressions due to heterogeneous arrangement of materials are neglected. The "local" space variations (e.g., cell problems) are discussed in Chapter 6.

The basic starting point for this development of the space distribution of neutrons is the space-dependent component of the diffusion equation derived in the previous chapter, i.e., the wave equation or Helmholtz equation as it is sometimes called. To develop a conceptual understanding initially, the flux shape is first derived for simple one-dimensional media. More realism is then provided by next considering the general separated multidimensional case. Finally, multiregion reactor configurations are treated. This allows an assessment of the impact of reflectors in the case of thermal reactors and of blankets in the case of fast reactors.

In practical applications, the flux distributions along the main directions are normally calculated numerically. The analytical flux shapes, as evaluated in this chapter, are then applied to describe the flux variation along the simpler directions. For example, in the radial direction, a reactor core has a regional structure resulting from differences in the fuel enrichments and from the presence of a reflector. In the axial direction, the fresh fuel has no enrichment variation. Therefore, in scoping calculations, the axial direction can be treated approximately by applying the respective simple flux form (cosine) that is derived below for the axial dimension of a cylinder.

3-2 Basic Considerations

The steady-state neutron flux distribution is calculated in the diffusion approximation of Eq. (2.28). As derived and discussed in Chapter 2, the flux, $\phi(\boldsymbol{r},E)$, is separable in a bare homogeneous medium, i.e.,

$$\phi(\boldsymbol{r},E) = \phi(\boldsymbol{r}) \cdot \varphi(E) \quad .$$

Then, substitution of this separated expression for the flux into the diffusion equation separates Eq. (2.28) into two equations:

$$\nabla^2\phi(\boldsymbol{r}) + B^2\phi(\boldsymbol{r}) = 0 \tag{3.1}$$

and

$$[D(E)B^2 + \Sigma_t(E)]\varphi(E) - \int_{E'} \Sigma_s(E' \rightarrow E)\varphi(E')\, dE'$$

$$= \lambda\chi(E)\int_{E'} \nu\Sigma_f(E')\varphi(E')\, dE' \quad . \tag{3.2}$$

The solution for Eq. (3.1), subject to appropriate boundary conditions, provides the spatial variation of the flux in a bare reactor and is the subject of this chapter. Recall that the fundamental buckling eigenvalue, B^2, in Eq. (3.2) is called the "material buckling," B_m^2, if Eq. (3.2) is solved without the eigenvalue λ (see Sec. 2-7). If Eq. (3.1) is solved with given boundary conditions, the B^2 term is designated by B_{geo}^2 and called the "geometric buckling" (see Sec. 2-6). When a reactor is critical,

$B^2_{geo} = B^2_m$. Either the geometry or the composition may be varied to establish this equality. The same holds in any domain in which $\phi(\boldsymbol{r},E)$ is separable, where the B^2_m determines the geometrical flux curvature.

If a system is either too large or too small to be critical, i.e., if

$$B^2_{geo} < B^2_m \text{ or } B^2_{geo} > B^2_m \quad ,$$

no finite flux can be established; the two B^2 values in Eqs. (3.1) and (3.2) are different and the problem has only the trivial solution $\phi(\boldsymbol{r}) \equiv 0$.

If a nontrivial solution of Eqs. (3.1) and (3.2) is sought for a *given* geometry and *given* composition, Eq. (3.1) is solved first and the resulting B^2_{geo} is inserted in Eq. (3.2); Eq. (3.2) must then be solved as an eigenvalue problem in λ. This yields the multiplication constant $k = 1/\lambda$ of the off-critical system. The geometrical buckling, substituted for B^2 in Eq. (3.2), describes the corresponding geometry or size dependency of the neutron leakage.

As stated in Sec. 2-4E, the physically allowed solution of a (static) eigenvalue problem is always the fundamental mode, i.e., the eigenfunction belonging to the fundamental eigenvalue. This means that the space distribution of the flux in a bare core can be established by determining the smallest eigenvalue B^2 for the particular geometry from Eq. (3.1) and the corresponding eigenfunction. It also means that the physically allowed energy distribution in a bare core is the fundamental mode spectrum, which is found by solving Eq. (3.2) as an eigenvalue problem in either B^2_m (if $\lambda = 1$) or in λ (if $B^2 = B^2_{geo}$). But only the spectrum with $\lambda = 1$ can be physically realized (see Sec. 2-7).

In Sec. 3-4 flux distributions are calculated as they result from neutrons that are emitted from an independent source and diffuse through scattering and absorbing or even multiplying media. If the media, as such, contain no neutron sources (source-free media), i.e., particularly no fissionable material, one has source-sink problems, which obviously have a physically realistic solution independent of the geometrical size of the medium.

The balance equation for the flux distribution in the one-group treatment of the subcritical system with a source was derived in Sec. 2-6C:

$$\nabla^2\phi(\boldsymbol{r}) + B^2_m\phi(\boldsymbol{r}) = -\frac{1}{D}S(\boldsymbol{r}) \quad . \tag{3.3}$$

The source may also be localized in space, in which case the balance equation is homogeneous:

$$\nabla^2\phi(\boldsymbol{r}) + B^2_m\phi(\boldsymbol{r}) = 0 \quad . \tag{3.4}$$

The source appears in the boundary conditions, e.g., Eqs. (2.165), (2.166), and (2.167).

Note that source-problem balance equations contain the material buckling. Since source problems normally have a nontrivial solution, there is no need (and no possibility) for determining a geometrical buckling eigenvalue from the condition of a nontrivial spatial flux solution.

In Sec. 3-4E, simple two-group flux solutions are presented to illustrate the key features of the combined space-energy problem, which was introduced in Sec. 2-6D. There, the two-group material buckling example of Sec. 2-7 is completed by the space dependence in two simple cases.

Section 3-5 is devoted to the more realistic multiregion reactor problem, without an independent source. This problem is again an eigenvalue problem, but the space and energy distribution of the flux are not separable as they are for the bare core. Thus, the bare core relations of B^2_{geo}, B^2_m, and λ, as discussed above, do not apply. The treatment of the fully realistic problem requires the simultaneous solution of the space and energy dependence (multigroup theory, see Chapter 7). Since separate eigenvalue problems for the space and energy dependence do not appear in the simultaneous treatment, the concepts B^2_m and especially B^2_{geo} lose much of their usefulness. The λ eigenvalue problem then provides the information on the criticality of the system. The B^2 mode appears as an asymptotic spectrum, and B^2_m as an asymptotic curvature. But, geometrical buckling does not appear at all in multiregion problems (except for a possible description of the transverse leakage in one- and two-dimensional problems).

To gain insight into the space dependence of the flux in multiregion problems, the one-group model is applied to core/reflector configurations. Only the λ eigenvalue solution is calculated.

3-3 Flux Shapes in Homogeneous Single Regions

To understand the spatial nature of the neutron flux, it is instructive to initially consider rather simple idealized problems. Therefore, in this section, the spatial flux behavior is derived for several single-region homogeneous systems (e.g., "bare" cores in reactor terminology). Admittedly this class of systems is somewhat artificial as bare cores are never actually constructed except as special experimental facilities. However, these simple systems provide the background and basis for understanding the flux distribution in the more realistic multiregion reactor cores presented in later sections.

The simple flux distribution expressions are derived first for an infinite slab (of finite thickness) and a spherical bare core; they are followed by the calculation of the fluxes in an infinite cylinder, a finite-height cylinder, and a rectangular parallelepiped. In addition, the spatial flux shapes for several source-sink geometries are presented. In all cases, these treatments are subject to the boundary conditions that the flux $\phi(\boldsymbol{r})$ goes to zero at the extrapolated outer boundary (see Sec. 2-2E).

3-3A Infinite Slab Reactor

Figure 3-1 shows a bare core of finite thickness, H, which is composed of a homogeneous mixture of materials, including fissile isotopes. The core dimensions are assumed to be infinite in x-y directions (infinite slab reactor). The extrapolated thickness of the core has the extrapolated distance ($0.71\lambda_{tr}$) at each side:

$$H' = H + 1.42\lambda_{tr} \quad , \tag{3.5}$$

where λ_{tr} is the mean free transport path as defined by Eq. (2.68).

The slab system involves only the z dimension, and the wave equation, Eq. (3.1), can be expressed as:

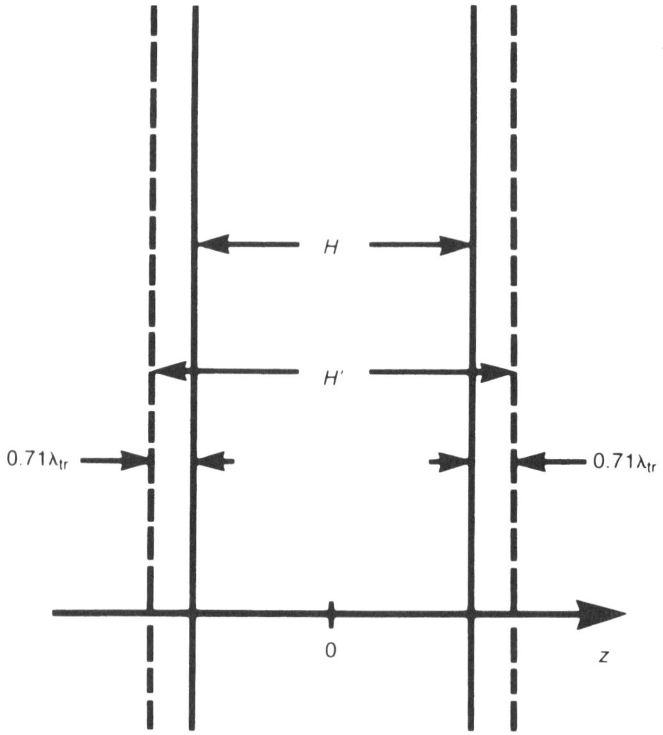

Fig. 3-1. Illustration of an infinite slab reactor.

$$\frac{d^2\phi(z)}{dz^2} + B^2\phi(z) = 0 \quad ; \tag{3.6a}$$

the Laplacian operator ∇^2 in Eq. (3.1), in one linear dimension, assumes the form d^2/dz^2. With the origin located at the center of the slab, the requirement that the flux vanish at the extrapolated distance can be expressed as:

$$\phi(z) = 0 \text{ at } z = \pm\frac{1}{2}H' \quad . \tag{3.6b}$$

These are the two boundary conditions required for a unique solution of the second-order differential equation. The problem is now completely specified.

The general solution to the second-order homogeneous differential equation, Eq. (3.6a), is different for positive and negative B^2. The positive B^2 solution is applicable to a slab with a surplus production of neutrons. This can be seen easily for the one-group equation, Eq. (2.80), where $DB^2 > 0$ if $\lambda\nu\Sigma_f - \Sigma_a > 0$. The general solution for positive B^2 is given by:

$$\phi(z) = C_1 \cos Bz + C_2 \sin Bz \quad . \tag{3.7}$$

Solutions for negative B^2 are discussed in Sec. 3-4D. Since B^2 is assumed to be positive, B is real in Eq. (3.7).

If the boundary conditions of Eq. (3.6b) are to be satisfied, B cannot have the same value in the two terms of Eq. (3.7). For a given B, only one of the two terms can satisfy the boundary conditions of Eq. (3.6b). It is the cosine term that can satisfy the boundary conditions with the smallest B^2 value. Thus, the fundamental eigenfunction, i.e., the one with the smallest B^2 eigenvalue, consists of the cosine term only [note that the sine term, as such, is not ruled out by the boundary condition Eq. (3.6b)[a]]:

$$\phi(z) = C_1 \cos Bz \quad . \tag{3.8}$$

Due to the boundary conditions, expressed by Eq. (3.6b), the flux must be zero at $z = \pm H'/2$; Eq. (3.8) then becomes

$$\phi\left(\pm\frac{H'}{2}\right) = C_1 \cos\frac{BH'}{2} = 0 \quad . \tag{3.9}$$

---

[a]An alternative and equivalent formulation is to consider the half-slab problem with the boundary conditions $\phi\left(\frac{1}{2}H'\right) = 0$ and $\frac{d\phi}{dz} = 0$ at $z = 0$. Then, the value of C_2 in Eq. (3.7) must be zero to satisfy the zero-gradient boundary condition.

Rejecting the trivial solution $C_1 = 0$, which would make $\phi(z)$ identical to zero, requires that

$$\cos \frac{BH'}{2} = 0 \quad . \tag{3.10a}$$

Then, only a discrete set of B^2 values is allowed: the eigenvalues. The acceptable B^2 values result from solving Eq. (3.10a). This gives

$$\frac{B_n H'}{2} = (n - 1/2)\pi \text{ for } n \geq 1 \quad , \tag{3.10b}$$

where n is an integer. For the various values of n, a set of eigenvalues is obtained. Since the geometric buckling, B^2_{geo}, is the smallest eigenvalue of the wave equation, $n = 1$ yields the geometric buckling:

$$B^2_{\text{geo}} = \left(\frac{\pi}{H'}\right)^2 \quad . \tag{3.11}$$

The spatial flux distribution for the bare infinite slab reactor of extrapolated thickness H' is obtained by simply substituting the expression for B as given by Eq. (3.11) into Eq. (3.8) and equating C_1 with the flux at $z = 0$:

$$\phi(z) = \phi_0 \cos\left(\frac{\pi z}{H'}\right) \quad . \tag{3.12}$$

The proportionality constant ϕ_0 is arbitrary; it cannot be calculated from the homogeneous equation, Eq. (3.6a). The proportionality constant is practically obtained by specifying the desired central reactor power density. The reactor power can be chosen arbitrarily within the acceptable range (the coolant flow must be able to transport the generated heat out of the core). The power level can be varied by making the reactor temporarily sub- or supercritical. The flux will then go through a transient and decrease or increase, respectively. Reestablishing criticality at the appropriate time leads to a new stationary state, with the flux level having the desired new value and the flux shape given by the solution of Eq. (3.6a).

As stated above, the mathematical reason for the arbitrariness of the flux level is the fact that the balance equation, together with the boundary conditions, represents a homogeneous problem. The physical reason for the homogeneity of the problem is the fact that the neutrons are produced through a chain reaction, which causes the neutron source to be proportional to the flux level. If the source is not dependent on the flux, then one has an inhomogeneous problem for which the flux level is uniquely determined by the solution of the problem (see Sec. 3-4D).

The eigenvalue problem represented by Eqs. (3.6) has more solutions than just the fundamental eigenfunction of Eq. (3.12). The higher eigenfunctions of the slab problem consist of cosine and sine functions [compare Eq. (3.7)] with buckling eigenvalues larger than the fundamental buckling, Eq. (3.11). The complete set of eigenfunctions of the slab problem is given by:

$$\cos B_n z \text{ and } \sin B_l z \quad , \tag{3.13}$$

with

$$B_n = \frac{2n-1}{H'}\pi \text{ and } B_l = \frac{2l\pi}{H'} \quad , \tag{3.14}$$

where n and l are positive integers.

Only the fundamental eigenfunction can appear as a physical solution since this is the only eigenfunction that is positive within the entire domain (compare Sec. 2-4A). The higher eigenfunctions of the simple problems of Eqs. (3.6), i.e., the functions of Eq. (3.13), are useful to construct the fundamental eigenfunction of the more complicated problems, such as multiregion problems (compare Fig. 2-8 and Secs. 3-4B and 3-5B). Since most reactor problems are now solved with high-speed computers, the practical importance of higher eigenfunctions is now considerably less than it was in the early times of reactor development.

3-3B Spherical Bare Core

Consider a bare spherical core with a uniform homogeneous composition. As is the solution for the slab, the fundamental solution is expected to be symmetrical, i.e., independent of the angles ϑ and α in standard spherical coordinate notation. The flux then depends only on the radius r. Using the angular independent Laplacian operator ∇^2 in spherical coordinates, given by (see Appendix A):

$$\nabla^2 = \frac{d^2}{dr^2} + \frac{2}{r}\frac{d}{dr} \quad , \tag{3.15a}$$

the wave equation, Eq. (3.1), then becomes:

$$\frac{d^2\phi(r)}{dr^2} + \frac{2}{r}\frac{d\phi(r)}{dr} + B^2\phi(r) = 0 \quad . \tag{3.15b}$$

The boundary conditions for the spherical reactor shape of physical radius R is that $\phi(r)$ is finite at $r = 0$, and is zero at the extrapolated boundary R':

$$\phi(r) \text{ is finite at } r = 0 \tag{3.16a}$$

and

$$\phi(r) = 0 \text{ at } r = R' \quad . \tag{3.16b}$$

The extrapolated radius, R', is expressed in the usual way by R and the mean free transport length λ_{tr} (see Sec. 2-2E):

$$R' = R + 0.71\lambda_{tr} \quad . \tag{3.16c}$$

The general solution to Eq. (3.15b), with B^2 again being positive, is obtained as (compare Sec. 3-4A for the technique to solve the wave equation in spherical geometry):

$$\phi(r) = \frac{C_1 \sin Br}{r} + \frac{C_2 \cos Br}{r} \quad . \tag{3.17}$$

Applying the condition Eq. (3.16a) that the flux $\phi(r)$ be finite at the origin ($r = 0$) requires that C_2 be zero. The only acceptable solution to Eq. (3.16b) is then:

$$\phi(r) = C_1 \frac{\sin Br}{r} \quad . \tag{3.18}$$

Applying the boundary condition Eq. (3.16b), namely, $\phi(r) = 0$ at $r = R'$ (the extrapolated boundary), gives:

$$C_1 \frac{\sin BR'}{R'} = 0 \quad . \tag{3.19}$$

Based on the physical requirement that R' is finite, and rejecting the trivial solution, $C_1 = 0$, the equation for finding the eigenvalues becomes simply:

$$\sin BR' = 0 \quad . \tag{3.20}$$

Thus, BR' must take on values of $n\pi$, where n is an integer:

$$B_n R' = n\pi \quad , \tag{3.21}$$

with $n = 1$ being the lowest eigenvalue ($n = 0$ leads to the trivial solution). Equation (3.21) yields the required expression for the geometrical buckling:

$$B^2_{\text{geo}} = \left(\frac{\pi}{R'}\right)^2 \quad . \tag{3.22}$$

The flux distribution in the sphere is then obtained by substituting the square root of the buckling expression, Eq. (3.22), into Eq. (3.18):

$$\phi(r) = \frac{C_1 \sin\left(\frac{\pi}{R'}r\right)}{r} \quad . \tag{3.23}$$

As in the previous problem for the slab reactor, C_1 can be obtained in terms of the maximum flux (i.e., at center of sphere). Again, letting ϕ_0 be the flux at $r = 0$ and applying l'Hôpital's rule to Eq. (3.23):

$$\phi_0 = \phi(r \to 0) = \lim_{r \to 0} C_1 \frac{\sin\left(\frac{\pi}{R'}r\right)}{r}$$

and

$$\phi_0 = C_1 \frac{\pi}{R'} \quad . \tag{3.24}$$

Hence, the expression for the flux distribution in a bare sphere, normalized to ϕ_0 at the center, becomes:

$$\phi(r) = \phi_0 \frac{\sin\left(\frac{\pi}{R'}r\right)}{(\pi r/R')} \quad . \tag{3.25}$$

3-3C Infinite Cylindrical Core

This last geometric example of a one-dimensional core assumes the reactor is again composed of a uniform and homogeneous composition containing fissile material. The core is assumed to be infinitely long in the axial direction, being bare in the radial direction. The wave equation, Eq. (3.1), in one-dimensional cylindrical coordinates then becomes (see Appendix A):

$$\frac{d^2\phi(r)}{dr^2} + \frac{1}{r}\frac{d\phi(r)}{dr} + B^2\phi(r) = 0 \quad . \tag{3.26}$$

The two boundary conditions are identical to those required for the spherical bare core, namely, that the flux be finite at the core centerline, dropping to zero at the extrapolated radial boundary R', i.e.:

$$\phi(r) \text{ is finite at } r = 0 \tag{3.27a}$$

and

$$\phi(r) = 0 \text{ at } r = R' \quad , \tag{3.27b}$$

where

$$R' = R + 0.71\lambda_{tr} \quad , \tag{3.27c}$$

with R being the actual physical outer core radius and λ_{tr} is the mean free transport length.

The general solution to the differential equation, Eq. (3.26), with positive B^2 consists of a linear combination of ordinary Bessel functions of the first $[J_0(r)]$ and second $[Y_0(r)]$ kind with constant coefficients, i.e.:

$$\phi(r) = C_1 J_0(Br) + C_2 Y_0(Br) \quad . \tag{3.28}$$

The zero subscript denotes the fact that Eq. (3.26) is a Bessel equation of zeroth order; i.e., the coefficient of ϕ is B^2 and does not contain the negative component $-m^2/r^2$, with m being the order of the Bessel function. Based on the first boundary condition, Eq. (3.27a), and the fact that $Y_0(Br)$ goes to $-\infty$ as r approaches zero (see Ref. 3 for Bessel function characteristics), the only permissible solution to Eq. (3.26) is with $C_2 = 0$, hence:

$$\phi(r) = C_1 J_0(Br) \quad . \tag{3.29}$$

To evaluate the specific conditions required so that Eq. (3.28) satisfies the reactor equation, the second boundary equation, Eq. (3.27b), is utilized, yielding:

$$\phi(R') = C_1 J_0(BR') = 0 \quad . \tag{3.30a}$$

Since C_1 must be nonzero to prevent the trivial situation of a zero flux occurring throughout the reactor, the only alternative is to require that:

$$J_0(BR') = 0 \quad . \tag{3.30b}$$

From the infinite number of roots of $J_0(\alpha)$, for which

$$J_0(\alpha_1) = J_0(\alpha_2) = J_0(\alpha_3) ... = 0 \quad , \tag{3.31}$$

the smallest value, $\alpha_n = \alpha_1$, is equal to 2.405. This value must be selected in order to obtain the positive fundamental mode solution. Hence BR' is equal to this 2.405 numerical value of the first root of the J_0 Bessel function and therefore:

$$B_{\text{geo}} = \frac{2.405}{R'} \text{ and } B^2_{\text{geo}} = \left(\frac{2.405}{R'}\right)^2 \quad . \tag{3.32}$$

Upon substitution of the expression of the square root of the buckling for the infinite cylindrical reactor as given by Eq. (3.32) into Eq. (3.29), the expression for the critical flux is simply:

$$\phi(r) = \phi_0 J_0\left(\frac{2.405}{R'}r\right) \quad . \tag{3.33}$$

Again, as was the case for the plane geometry, the proportionality constant C_1 is equal to the arbitrary value of the desired flux at the center [note that $J_0(0) = 1$].

3-3D Finite Cylindrical Core

The reactor geometries considered in previous sections were taken to be one-dimensional, either by design or by idealization. In this and the following section, more typical multidimensional reactor configurations are considered. As will be seen, the derivation of the flux shapes for multidimensional reactor geometries is quite similar to that for the single-dimensional systems, if the flux can be separated in the spatial variable [see, for example, Eq. (3.36)].

As an example of a typical two-dimensional reactor problem, consider a bare finite cylinder of extrapolated radius R' and extrapolated height H'. Expressing the wave equation in cylindrical coordinates including both the r and z dependencies, Eq. (3.1) becomes (see Appendix A):

$$\frac{1}{r}\frac{\partial}{\partial r}\left[r\frac{\partial\phi(r,z)}{\partial r}\right] + \frac{\partial^2\phi(r,z)}{\partial z^2} + B^2\phi(r,z) = 0 \quad . \tag{3.34}$$

The by-now-familiar boundary conditions require that the flux vanish at the extrapolated boundaries. With the origin of the coordinate system taken in the axial midplane at $r = 0$, the boundary conditions become:

$$\phi\left(r, \pm\frac{H'}{2}\right) = \phi(R',z) = 0 \tag{3.35a}$$

and

$$\phi(r,z) \text{ is finite at } r = 0 \quad . \tag{3.35b}$$

The solution to Eq. (3.34) under the boundary condition constraints of Eqs. (3.35) can be found by the method of separation of variables. For the uniform one-region core, the flux is separable into two directional components, i.e.:

$$\phi(r,z) = \phi(r)Z(z) \quad . \tag{3.36}$$

Let the radial factor $\phi(r)$ in Eq. (3.36) have the dimension of a flux; the axial component then has the character of a "form factor." This choice is arbitrary and irrelevant, as can be seen from Eq. (3.42). The same result is obtained if the flux dimension is assigned alternatively to the z-dependent factor. The substitution of Eq. (3.36) into Eq. (3.34) and division by $\phi(r,z)$ gives:

$$\frac{1}{\phi(r)}\frac{1}{r}\frac{d}{dr}\left[r\frac{d\phi(r)}{dr}\right]+\frac{1}{Z(z)}\frac{d^2Z(z)}{dz^2}+B^2=0 \quad . \tag{3.37}$$

Since the first and second terms in Eq. (3.37) depend only on r and z, respectively, they must each be equal to a constant. Therefore, letting $-B_r^2$ and $-B_z^2$ denote these constants, Eq. (3.37) becomes:

$$-B_r^2-B_z^2+B^2=0 \tag{3.38a}$$

or

$$B^2=B_r^2+B_z^2 \quad , \tag{3.38b}$$

where

$$\frac{1}{\phi(r)}\frac{1}{r}\frac{d}{dr}\left[r\frac{d\phi(r)}{dr}\right]=-B_r^2 \tag{3.39a}$$

and

$$\frac{1}{Z(z)}\frac{d^2Z(z)}{dz^2}=-B_z^2 \quad . \tag{3.39b}$$

The above two expressions, Eq. (3.39a) and (3.39b), for the two components of the separated flux are identical to those obtained earlier for the infinite cylinder (see Sec. 3-3C) and infinite slab (see Sec. 3-3A), respectively. Applying the boundary condition for the radial component of the flux, one obtains:

$$\phi(r)=\phi_0 J_0\left(\frac{2.405}{R'}r\right) \quad , \tag{3.40a}$$

where B_r is determined by the lowest root of the J_0 Bessel function, i.e.:

$$B_r^2=(2.405/R')^2 \quad , \tag{3.40b}$$

with B_r^2 being the "radial buckling." Similarly for the axial flux component, by using the boundary condition that the flux vanish at $z = \pm H'/2$, a solution to Eq. (3.39b) identical to that for the infinite slab reactor derived in Sec. 3-3A results:

$$Z(z)=\cos\left(\frac{\pi z}{H'}\right) \quad , \tag{3.41a}$$

where the constant in front of the cosine is equal to unity since the radial factor has been chosen to describe the flux at $r = z = 0$. The lowest eigenvalue of Eq. (3.39b) has been employed in Eq. (3.41a) so that one obtains the fundamental mode flux. This eigenvalue

$$B_z^2 = \left(\frac{\pi}{H'}\right)^2 , \qquad (3.41b)$$

is called the "axial buckling."

With the separated components of the neutron flux identified by Eqs. (3.40a) and (3.41a) for each of the two directions, the expression for the total flux in the finite cylindrical reactor is formed from Eq. (3.36):

$$\phi(r,z) = \phi_0 J_0\left(\frac{2.405}{R'}r\right)\cos\left(\frac{\pi z}{H'}\right) , \qquad (3.42)$$

where, again, ϕ_0 is merely a proportionality factor, which is equated to the maximum flux at the core center.

Similarly, the *total* buckling for this finite cylindrical reactor is obtained by substituting the individual bucklings obtained for the radial direction [Eq. (3.40b)] and axial direction [Eq. (3.41b)] into Eq. (3.38b), yielding:

$$B^2 = \left(\frac{2.405}{R'}\right)^2 + \left(\frac{\pi}{H'}\right)^2 . \qquad (3.43)$$

Thus, from this example, the total buckling, B^2, is the sum of the bucklings in each of the two directions considered in this finite cylindrical reactor. As is seen in the next section, this relationship between the total and individual bucklings holds for a three-dimensional reactor as well.

3-3E Rectangular Parallelepiped Reactor

For the final simple bare reactor shape, consider a three-dimensional rectangular parallelepiped reactor with extrapolated dimensions a, b, and c in the x, y, and z rectangular coordinates, respectively. The controlling wave equation, Eq. (3.1), in this rectangular system then becomes (see Appendix A):

$$\frac{\partial^2\phi(x,y,z)}{\partial x^2} + \frac{\partial^2\phi(x,y,z)}{\partial y^2} + \frac{\partial^2\phi(x,y,z)}{\partial z^2} + B^2\phi(x,y,z) = 0 . \qquad (3.44)$$

The required boundary conditions for the origin located at the core center express the fact that at each extrapolated core external surface, the flux must vanish, i.e.:

$$\phi\left(\pm\frac{a}{2},y,z\right) = \phi\left(x,\pm\frac{b}{2},z\right) = \phi\left(x,y,\pm\frac{c}{2}\right) = 0 \quad . \tag{3.45}$$

The similarity of Eqs. (3.44) and (3.45) with the wave equation and boundary conditions for the one-dimensional infinite slab reactor presented in Sec. 3-3A is quite evident and the solution, as expected, will be analogous. To demonstrate this, the method of separation of variables is again employed:

$$\phi(x,y,z) = \phi_0 X(x)Y(y)Z(z) \quad . \tag{3.46}$$

Upon substitution of Eq. (3.46) into Eq. (3.44) and with some rearrangement, the following equation results:

$$\frac{1}{X(x)}\frac{d^2X(x)}{dx^2} + \frac{1}{Y(y)}\frac{d^2Y(y)}{dy^2} + \frac{1}{Z(z)}\frac{d^2Z(z)}{dz^2} + B^2 = 0 \quad . \tag{3.47}$$

Notice that each of the first three terms in Eq. (3.47) is only a function of its respective coordinate with its sum equal to a constant ($-B^2$). For this to be true, each term must also be equal to a constant. Therefore, by defining these constants as $-B_x^2$, $-B_y^2$, and $-B_z^2$, the following three equations result:

$$\frac{d^2X(x)}{dx^2} + B_x^2X(x) = 0 \quad , \tag{3.48a}$$

$$\frac{d^2Y(y)}{dy^2} + B_y^2Y(y) = 0 \quad , \tag{3.48b}$$

and

$$\frac{d^2Z(z)}{dz^2} + B_z^2Z(z) = 0 \quad . \tag{3.48c}$$

From Eq. (3.47), it is evident that the total reactor buckling, B^2, is merely the sum of the bucklings in each of the three directions:

$$B^2 = B_x^2 + B_y^2 + B_z^2 \quad . \tag{3.49}$$

The solutions to Eqs. (3.48a), (3.48b), and (3.48c) can be written by inspection, as these equations and boundary conditions are identical to the slab derived in Sec. 3-3A; i.e., fluxes are proportional to the cosine term as the sine component is unacceptable since it requires a buckling value that is larger than the fundamental buckling. Considering the smallest eigenvalues gives the fundamental solutions for each of the three directions, which upon substitution back into the separated expression, Eq. (3.46), produces the desired relationship for the total flux:

$$\phi(x,y,z) = \phi_0 \cos\frac{\pi x}{a} \cdot \cos\frac{\pi y}{b} \cdot \cos\frac{\pi z}{c} \quad , \tag{3.50}$$

where again the normalization constant, ϕ_0, represents the flux at the reactor center.

The three constants B_x^2, B_y^2, and B_z^2 in Eqs. (3.48) can be interpreted as the bucklings in the x, y, and z directions, respectively. As derived in Sec. 3-3A for the slab, they take on the following corresponding values for the lowest indices of the infinite set of cosine solution functions, i.e.:

$$B_x^2 = \left(\frac{\pi}{a}\right)^2 , \; B_y^2 = \left(\frac{\pi}{b}\right)^2 , \; B_z^2 = \left(\frac{\pi}{c}\right)^2 \quad . \tag{3.51}$$

The total buckling for the rectangular parallelepiped reactor is then determined by substituting these individual bucklings expressed by Eq. (3.51) into Eq. (3.49):

$$B^2 = \left(\frac{\pi}{a}\right)^2 + \left(\frac{\pi}{b}\right)^2 + \left(\frac{\pi}{c}\right)^2 \quad . \tag{3.52}$$

3-3F Summary of One-Region-Reactor Flux Shapes

The solutions of the wave equation, Eq. (3.1), for simple bare core shapes using the appropriate coordinate system were derived in the previous sections. The solutions for the flux distributions and geometric buckling expressions derived above are tabulated in Table 3-I.

The spatial neutron flux shape is an important parameter in the analysis of a reactor since the energy release, and hence the heat generation rate in the core, is proportional to the neutron flux. The power limit or level of a reactor is largely determined by its maximum value with the energy output of a given core volume being proportional to the average power. Therefore, the ratio of the maximum energy generation to the average energy generation in the core is a very important quantity. Table 3-I also presents these ratios for several idealized reactor geometries. The derivation of these quantities is presented in the following section.

3-3G Power Peaking Factors and Optimum One-Region Geometries

From the spatial flux distributions, as derived in the previous sections, one can calculate important integral quantities that are needed for basic design considerations. Although practical designs employ more complicated geometries than treated in this section, the general types of integral considerations are the same in all cases; they can, therefore, be illustrated with simple geometries.

TABLE 3-I
Flux Distributions and Geometric Bucklings for Principle Core Geometries

Shape	Extrapolated Dimensions	Flux (ϕ/ϕ_{max})	Flux Peak to Average[a] ($\phi_{max}/\bar{\phi}$)	Geometric Buckling (B_{geo}^2)
Slab	Thickness H'	$\cos\left(\frac{\pi z}{H'}\right)$	1.570	$(\pi/H')^2$
Sphere	Radius R'	$\frac{\sin(\pi r/R')}{(\pi r/R')}$	3.290	$(\pi/R')^2$
Rectangular parallelepiped	Width a, length b, and height c	$\cos\left(\frac{\pi x}{a}\right)\cos\left(\frac{\pi y}{b}\right)\cos\left(\frac{\pi z}{c}\right)$	3.876	$\left(\frac{\pi}{a}\right)^2+\left(\frac{\pi}{b}\right)^2+\left(\frac{\pi}{c}\right)^2$
Cube	Side S'	$\cos\left(\frac{\pi x}{S'}\right)\cos\left(\frac{\pi y}{S'}\right)\cos\left(\frac{\pi z}{S'}\right)$	3.876	$3(\pi/S')^2$
Infinite cylinder	Radius R'	$J_0(\alpha_0 r/R')$[b]	2.315	$(\alpha_0/R')^2$
Finite cylinder	Radius R' and height H'	$J_0\left(\frac{\alpha_0 r}{R'}\right)\cos\left(\frac{\pi z}{H'}\right)$	3.636	$(\alpha_0/R')^2+(\pi/H')^2$

[a]Calculated from $\phi(\boldsymbol{r})$ within the extrapolated boundaries.
[b]The first root of the Bessel function $J_0(r)$ is at $\alpha_0 = 2.405$.

In the context of a one-energy-group model, the rate of energy release per unit volume at any point[b] in a reactor core is merely:

energy release rate:

$$\phi(\boldsymbol{r})\frac{\Sigma_f(\boldsymbol{r})}{c'} \quad \left[\mathrm{W/cm^3} = \frac{\mathrm{W \cdot s}}{\mathrm{cm^3 s}}\right] \quad , \tag{3.53}$$

where c' is the conversion factor from fissions to energy in a steady-state reactor; c' is $\approx 3.1 \times 10^{10}$ fission/W · s. Upon integrating over the entire reactor, the thermal power output is:

$$P_{tot} = \frac{1}{c'}\int_{core} \Sigma_f(\boldsymbol{r})\phi(\boldsymbol{r})\, dV \quad [\mathrm{W}] \quad . \tag{3.54}$$

By introducing $\overline{\phi}$, the spatially averaged neutron flux in the core, and $\overline{\Sigma}_f$, the core averaged Σ_f, one obtains:

$$P_{tot} = 3.2 \cdot 10^{-11}\overline{\Sigma}_f \cdot V_{core} \cdot \overline{\phi} \quad [\mathrm{W}] \quad , \tag{3.55}$$

where V_{core} is the total physical volume of the core and

$$\overline{\phi} = \frac{1}{V_{core}}\int_0^R \phi(\boldsymbol{r})\, dV \quad . \tag{3.56}$$

With a spherical core, the expression for the flux shape was presented in terms of the yet undetermined maximum flux, ϕ_0 [see Eq. (3.25)]. The average flux is then:

$$\overline{\phi} = \frac{1}{\frac{4}{3}\pi R^3}\int_0^R \phi_0 \frac{\sin\left(\frac{\pi}{R'}r\right)}{(\pi r/R')} \cdot 4\pi r^2\, dr \quad . \tag{3.57}$$

Simplifying Eq. (3.57) and performing the integration (assuming that $R' = R$ for simplicity, i.e., the extrapolation distance is neglected) gives:

$$\overline{\phi} = \phi_0 \frac{3}{\pi R^2}\int_0^R r \sin\left(\frac{\pi r}{R'}\right) dr \tag{3.58a}$$

[b]It is assumed here, for simplicity, that all energy is released at the location of the fission, i.e., the effect of energy dissipation through migration of neutrons and γ quanta is neglected (see Table 1-II).

or

$$\bar{\phi} = \phi_0 \frac{3}{\pi R^2} \left[\frac{\sin\left(\frac{\pi r}{R}\right)}{(\pi/R)^2} - \frac{r\cos\left(\frac{\pi}{R} r\right)}{(\pi/R)} \right]_0^R \quad . \tag{3.58b}$$

One then obtains

$$\bar{\phi}/\phi_0 = \frac{3}{\pi^2} \simeq 0.304 \quad . \tag{3.59}$$

Inverting this result, Eq. (3.59) yields the tabulated $\phi_0/\bar{\phi}$ value of 3.290 as presented in Table 3-I for the spherical core. Similar integrations are performed over other reactor core volumes under the assumption that the extrapolated distance is negligible. Form factors for these idealized core geometries are presented in Table 3-I.

A knowledge of the dimensions of a reactor for a given composition and value of B^2, which provides the minimum critical volume (and hence minimum fuel mass), is frequently desired. This problem can be easily determined for the simple geometries just presented. As an example, consider the case of the cylinder of volume $\pi R^2 H$ where the geometric buckling is given by:

$$B^2 = \left(\frac{\pi}{H}\right)^2 + \left(\frac{2.405}{R}\right)^2 \quad . \tag{3.60}$$

The actual and extrapolated dimensions have been assumed to be equal for simplicity. If the radius is eliminated by means of Eq. (3.60), the volume can then be expressed as:

$$V = \frac{H\pi(2.405)^2}{B^2 - (\pi/H)^2} \quad . \tag{3.61}$$

The volume assumes a minimum value if $\frac{dV}{dH}$ is zero or if:

$$H = \frac{\sqrt{3}\pi}{B} = \frac{5.441}{B} \quad . \tag{3.62}$$

Substituting this expression for H in Eq. (3.60) and solving for R results in:

$$R = \sqrt{3/2}\left(\frac{2.405}{B}\right) = \frac{2.945}{B} \quad . \tag{3.63}$$

The minimum volume (V_{min}) for the cylindrical reactor with a given B is then:

$$V_{min} = \pi R^2 H = \frac{148.2}{B^3} \quad , \tag{3.64}$$

and the "optimum" radius-to-height ratio is:

$$R/H = \frac{2.405}{\sqrt{2}\pi} = 0.541 \quad . \tag{3.65}$$

This result for the cylindrical reactor indicates that the minimum volume occurs when its diameter is ~ 10% larger than its height.

In Table 3-II the minimum volumes for a given material or geometrical buckling are presented for other reactor geometries. As might be expected for the rectangular parallelepiped case, this optimum condition occurs when the three sides are equal and hence the minimum volume is identical to that of the cube. Table 3-II shows also that the core with the minimum volume for a given B^2 (e.g., $B^2 = B_m^2$) is the sphere. This might be expected since the sphere, for a given volume, has the minimum surface area, which results in the smallest amount of neutron leakage per unit volume. However, the spherical reactor has severe design disadvantages and hence it is seldom used in practice.

Remember, the leakage can be approximately described by a DB^2 term, added to the absorption term in the neutron balance equation. Thus, Table 3-II indicates that the sphere has the smallest volume for a given leakage; or, if the volume is given, it has the smallest leakage and thus requires the smallest enrichment in an otherwise given composition. By the same reasoning, the cylinder is preferable to the cube. The corners of the cube, which are cut by going to a cylinder, have a relatively high contribution to the leakage.

TABLE 3-II
The Optimum Core Dimensions and Minimum Volumes for Various Reactor Geometries

Core Geometry	Optimum Dimensions	Minimum Volume
Cube	$S = \frac{\sqrt{3}\pi}{B}$	$161/B^3$
Finite cylinder	$R = \frac{2.405}{B}\sqrt{3/2},\ H = \frac{\sqrt{3}\pi}{B}$	$148/B^3$
Sphere	$R = \pi/B$	$130/B^3$

3-4 One-Group Flux Solutions in Source-Sink Problems and in Subcritical Reactors

In this section, the most rudimentary version of the diffusion theory, namely, one-group theory, is used to find the neutron flux shape in several source-free media. The neutron source is assumed to be localized (at a midplane or at a point). This class of problems is sometimes referred to as "source-sink problems." Reactor applications are, for example, the approximate treatment of the neutron flux in certain types of shielding problems (e.g., in determining the flux in the thermal shield or in other reactor structures). In addition, as will be seen, the solution of these types of problems provides a clearer physical understanding of the important diffusion length parameter that occurs so frequently in diffusion theory.

As in the previous sections, a steady-state situation of neutrons diffusing in a homogeneous medium is investigated with the source assumed to be independent of the neutron flux and constant in time. The neutron source, S, is taken to be zero except at the particular source location (i.e., point, line, or plane), with the diffusing region consisting only of a scattering and absorbing medium. Under these assumptions, the diffusion equation, Eq. (2.40), becomes simply:

$$D\nabla^2\phi(\boldsymbol{r}) - \Sigma_a\phi(\boldsymbol{r}) = 0 \quad , \tag{3.66}$$

or after dividing by D:

$$\nabla^2\phi(\boldsymbol{r}) - \frac{1}{L^2}\phi(\boldsymbol{r}) = 0 \quad , \tag{3.67}$$

with L being the diffusion length ($L = \sqrt{D/\Sigma_a}$).

The balance equation in a source problem contains the material buckling as it was shown in Sec. 2-6D [compare Eq. (2.168)]. The factor $-1/L^2$ is readily recognized as the limit of the material buckling for $\nu\Sigma_f \rightarrow 0$:

$$\lim_{\nu\Sigma_f \to 0} B_m^2 = \lim_{\nu\Sigma_f \to 0} \frac{k_\infty - 1}{L^2} = -\frac{1}{L^2} \quad . \tag{3.68}$$

The source is located at the boundary. It will have to be specified along with the appropriate boundary conditions in order to obtain the solution of the source-sink problem. Several typical solutions to Eq. (3.67) for common geometries are presented in this section.

Problems similar to source-sink problems are those in which the diffusing medium contains some fissionable material but not enough to sustain a chain reaction. Then an independent source is required to support a finite neutron flux. Problems of this kind are treated in Sec. 3-4D.

3-4A Point Neutron Source in an Infinite Region

As the first example, consider a point neutron source symmetrically emitting S_p monoenergetic neutrons per second into an infinite homogeneous diffusion medium. With the point source at the center of the spherical coordinate system, the terms in ∇^2, which differentiate with respect to angles, will be zero. The source-free diffusion equation, Eq. (3.67), then becomes (see Appendix A):

$$\frac{d^2\phi(r)}{dr^2} + \frac{2}{r}\frac{d\phi(r)}{dr} - \frac{\phi(r)}{L^2} = 0 \quad , \tag{3.69}$$

where r is the distance from the point source.

The boundary conditions for Eq. (3.69) as posed by this diffusion problem are:

1. The flux $\phi(r)$ is zero at the (extrapolated) outer boundary, which for simplicity is extended to infinity.
2. The net number of neutrons diffusing across a spherical surface must equal that emitted from the source as the radius r approaches zero.

The second condition requires that [compare Eq. (2.166)]:

$$\lim_{r \to 0} 4\pi r^2 J(r) = S_p \quad , \tag{3.70}$$

where $J(r)$ is the neutron current.

The solution of Eq. (3.69) is facilitated by substituting the function $\psi(r)/r$ for $\phi(r)$. Equation (3.69) then becomes:

$$\frac{d^2\psi(r)}{dr^2} - \frac{\psi(r)}{L^2} = 0 \quad . \tag{3.71}$$

Since the square of the diffusion length L^2 is always positive, Eq. (3.71) has the following general solution:

$$\psi(r) = C_1\exp(-r/L) + C_2\exp(r/L) \quad . \tag{3.72}$$

With $\psi(r) = r\phi(r)$, Eq. (3.72) gives for $\phi(r)$:

$$\phi(r) = \frac{1}{r}\left[C_1\exp(-r/L) + C_2\exp(r/L)\right] \quad . \tag{3.73}$$

To obtain the desired specific solution to Eq. (3.69), the boundary conditions presented above must be satisfied. The first condition, namely, that the flux vanishes at $r=\infty$, requires that C_2 in Eq. (3.73) must be zero.[c] The second boundary condition involves the limit of the neutron current as expressed by Fick's Law. In the framework of the diffusion equation, the current is given by [see Eq (2.27)].

$$J(r) = -D\frac{d\phi(r)}{dr} \quad . \tag{3.74}$$

Equation (3.73), with C_2 set equal to zero, is differentiated and yields:

$$\frac{d\phi(r)}{dr} = -C_1\exp(-r/L)\left(\frac{1}{r^2}+\frac{1}{rL}\right) \quad . \tag{3.75}$$

Applying the second boundary condition, Eq. (3.70), by first substituting the expression for $d\phi(r)/dr$ from Eq. (3.75) into Eq. (3.74), gives:

$$\lim_{r\to 0} 4\pi r^2 J(r) = \lim_{r\to 0} 4\pi DC_1\exp(-r/L)\left(\frac{r}{L}+1\right) = S_p \quad , \tag{3.76}$$

or in the limit as $r\to 0$:

$$4\pi DC_1 = S_p; \quad C_1 = S_p/4\pi D \quad . \tag{3.77}$$

The desired expression for the neutron flux is obtained by substituting zero for C_2 and the quantity for C_1 given by Eq. (3.77) into Eq. (3.73):

$$\phi(r) = \frac{S_p}{4\pi D}\frac{1}{r}\exp(-r/L) \quad . \tag{3.78}$$

As exhibited by Eq. (3.78), the shape of the neutron flux distribution emanating from a point source, S_p, in an infinite homogeneous medium decreases with the distance r from the source as $1/r$ times an exponential function.

The exponential function describes the attenuation governed by neutron absorption and scattering. If $\Sigma_a = 0$, $L=\infty$ and there is no absorption; only geometrical attenuation remains, which results in a diffusion theory flux that is proportional to $1/r$. Thus, diffusion results in a smaller decrease of the neutron population as in a vacuum ($D=\infty$), where the mere geometrical spread of the neutrons leads to a $1/r^2$ dependency of the flux.

At $r=0$, $\phi(r)$ per Eq. (3.78) becomes singular, although it satisfies the physical boundary condition of Eq. (3.76). This physically unrealistic singularity results from the mathematical artifice of a "point" source, i.e., a source with a density zero everywhere except in a single point

---

[c]For a finite medium, $C_2 \neq 0$ and the second term in Eq. (3.73) are obviously needed to satisfy the zero flux boundary condition at a finite radius.

(here $r=0$). Any extended source, e.g., a small sphere, would mathematically yield a finite flux with a vanishing gradient at $r=0$.

It is sometimes useful to apply the result given by Eq. (3.78) to incorporate the contribution from a number of point sources. Using the method of superposition, the flux at any location in a medium with several point sources can be obtained by simply summing the respective point source contributions given by Eq. (3.78).

3-4B Plane Neutron Source in an Infinite Medium

The plane source in an infinite geometry emitting neutrons at a rate of S neutrons per square centimeter per second is located at $x=0$, i.e., at the boundary of each half-space. The flux in this infinite system is independent of the y and z coordinates. The Laplacian operator is then simply d^2/dx^2 and the source-free diffusion equation, Eq. (3.67), is homogeneous:

$$\frac{d^2\phi(x)}{dx^2} - \frac{1}{L^2}\phi(x) = 0 \text{ for } x \neq 0 \quad . \tag{3.79}$$

The boundary conditions are inhomogeneous, since they contain the source (see Sec. 3-6 for a more refined boundary condition):

1. The flux is zero at the extrapolated outer boundary, which is again shifted to infinity for simplicity.
2. At $x \to 0$, the neutron current, $J(x)$, is one-half of the neutron source per square centimeter [compare Eq. (2.167)]:

$$\lim_{x\to 0} J(x) = S/2 \quad . \tag{3.80}$$

The second boundary condition follows from neutron conservation and from the symmetry of the problem; equal numbers of neutrons traveling in the positive and negative x direction do not contribute to the current. This boundary condition merely expresses the conservation of neutrons, which requires that at the surface, $x=0$, the net flow of neutrons must be equal to the source.

The general solution of Eq. (3.79) is:

$$\phi(x) = C_1\exp(-x/L) + C_2\exp(x/L) \quad , \tag{3.81}$$

where C_1 and C_2 are again constants to be determined from the boundary conditions. Applying the first boundary condition requires $C_2=0$ for the infinite medium (C_2 would be $\neq 0$ in a finite medium; compare Sec. 3-4D).

To determine the value of C_1, the second boundary condition (or source condition) is applied, again using Fick's Law. Differentiating Eq. (3.81) (with $C_2=0$) and taking the lim $x\to 0$ gives:

$$\lim_{x\to 0} J(x) = \lim_{x\to 0}\left[-D\frac{d\phi(x)}{dx}\right]$$

$$= \lim_{x\to 0}\left[\frac{DC_1\exp(-x/L)}{L}\right] = DC_1/L \quad . \tag{3.82a}$$

Thus, from the source condition, it follows that:

$$C_1 = \frac{SL}{2D} \quad . \tag{3.82b}$$

The desired expression for the flux, upon substitution of C_1 as given by Eq. (3.82b) into Eq. (3.81), becomes:

$$\phi(x) = \frac{SL}{2D}\exp(-x/L) \quad . \tag{3.83}$$

The result for the flux distribution given by Eq. (3.83) has several important features. Notice that, similar to Eq. (3.78), the flux is directly proportional to the source, which means that an increase in source intensity results in a similar increase in the diffusing neutron flux. Also, the diffusion length appears in the exponential term since it acts as a "relaxation length," i.e., flux decreases by a factor of e for every diffusion length distance from the source. Finally, it must be pointed out that since diffusion theory was applied Eq. (3.83) and the previous spherical expression Eq. (3.78) are somewhat inaccurate for distances less than about three diffusion lengths away from the source (compare Sec. 2-2C). But the inaccuracy of the diffusion theory flux in these source-sink problems is merely reflected in the flux shape. The flux "magnitude" and thus the flux integral are correct; they lead to the correct neutron conservation relation:

$$\int_0^\infty \Sigma_a\phi(x)\,dx = \frac{SL}{2D}\Sigma_a\int_0^\infty \exp(-x/L)\,dx = \frac{S}{2} \quad , \tag{3.84}$$

i.e., the number of neutrons absorbed in a half-space equals the neutrons emitted into that space.

3-4C Physical Interpretations of Diffusion Length

The flux solutions derived in the previous sections depend on cross sections in the combination D/Σ_a, the square of the diffusion length L. The solutions can therefore be employed to develop a physical interpretation and a measurement procedure for the diffusion length (see Sec. 3-4E). It is the flux distribution in the plane source-free medium that provides the simplest physical interpretation: The diffusion length L is the relaxation length of the flux in the plane source-free medium.

The flux distribution from a point source in an infinite medium allows a second interpretation. The solution derived in Sec. 3-4A is given by [see Eq. (3.78)]:

$$\phi(r) = \frac{S}{4\pi r D} \exp(-r/L) \quad . \tag{3.85}$$

The mean square distance, $\overline{r^2}$, traveled by neutrons from their point of origin to their final absorption can be expressed mathematically as:

$$\overline{r^2} = \frac{\int_0^\infty r^2 \Sigma_a \phi(r) 4\pi r^2 \, dr}{\int_0^\infty \Sigma_a \phi(r) 4\pi r^2 \, dr} = \frac{1}{S}\int_0^\infty \Sigma_a \phi(r) 4\pi r^4 \, dr \quad . \tag{3.86}$$

Upon substitution of the relationship for the flux from Eq. (3.78), Eq. (3.86) becomes:

$$\overline{r^2} = \frac{1}{L^2}\int_0^\infty \exp(-r/L) r^3 \, dr \quad . \tag{3.87}$$

The integral in Eq. (3.87) is equal to $6L^4$, hence:

$$\overline{r^2} = 6L^2 \tag{3.88a}$$

or

$$L^2 = \frac{1}{6}\overline{r^2} \quad . \tag{3.88b}$$

Therefore, the square of the diffusion length can also be interpreted as one-sixth of the mean square distance traveled by neutrons from their point of origin to the location where they are absorbed. Inherent in these interpretations is that the neutron diffusion takes place with constant neutron energy.

3-4D Flux Distributions in Subcritical Systems

In the previously treated source-sink problems, source and sink were spatially separated, e.g., by using a point or plane source. The medium outside of the source had only scattering and capture cross sections. A somewhat different problem arises when the scattering medium also contains fissionable material. Still, an independent source is required in addition to the fission source since the systems considered in this section are subcritical. As discussed in Chapter 2, in a subcritical reactor system, a steady-state neutron population cannot be sustained without the aid of neutrons from a supportive neutron source.

The infinite-medium multiplication rate, k_∞, in a subcritical system may be >1. The medium with $k_\infty > 1$ is "superproductive": It produces more neutrons than it absorbs. A system containing a superproductive medium is subcritical when the neutron loss through leakage more than compensates the surplus production of neutrons in the medium. Then, the system is smaller than a critical one of the same composition. If $k_\infty < 1$, i.e., if the medium is "subproductive," the system is subcritical for any size.

The material buckling values, B_m^2, of the sub- or superproductive media are negative or positive, respectively. Actually, it is the sign of the material buckling, the eigenvalue of Eq. (2.144), which initially decides whether a medium is sub- or superproductive: k_∞ is derived from the solution of this eigenvalue problem by forming the one-group constants in Eq. (2.145) and by then defining k_∞ through Eq. (2.146). Since k_∞ is defined by Eq. (2.146), the signs of B_m^2 and of $k_\infty - 1$ are the same and thus the sign of $k_\infty - 1$ can also be used to define sub- or superproductivity.

Several examples of source-sink problems with neutron multiplying media are discussed in the following. Plane geometry is used for simplicity. Other examples are formulated as homework problems.

The implication of negative material buckling on the spatial flux distribution is first investigated using a simple one-dimensional infinite slab. The geometry is similar to that selected for the critical case, namely, infinite in the x-y directions (see Sec. 3-3A). At $z = 0$, an independent neutron source is present (e.g., thermal column from a critical core or Pu-Be neutron source). The neutron balance equation for the space dependence of the neutron flux is given by Eq. (3.1). The buckling B^2 must be equal to the material buckling B_m^2 since the physically realizable flux distribution in a given medium is to be determined. This requires the factor λ in Eq. (2.138) to be equal to 1; then Eq. (2.138) yields B_m^2 as the eigenvalue, which is to be used as the buckling in Eq. (2.137) [see also Eq. (2.168)]:

$$\frac{d^2}{dz^2}\phi(z) + B_m^2\phi(z) = 0 \quad . \tag{3.89}$$

At one side of the slab, the source maintains a flux[d] $\phi(z = 0) = \phi_0$. At the

[d]The given flux boundary condition is used here primarily to introduce this form of an inhomogeneous boundary condition, which is different from specifying the current from a source.

other (extrapolated) boundary, $z = H$, the flux is assumed to vanish, in accordance with the diffusion theory boundary condition near a vacuum boundary:

$$\phi(0) = \phi_0$$

and

$$\phi(H) = 0 \quad . \tag{3.90}$$

The solution of Eq. (3.89) for negative B_m^2, is given by:

$$\phi(z) = C_1 \exp(|B_m|z) + C_2 \exp(-|B_m|z) \quad , \tag{3.91}$$

with

$$|B_m|^2 = -B_m^2 > 0 \quad .$$

From the boundary conditions expressed by Eqs. (3.90), it follows that:

$$C_1 + C_2 = \phi_0$$

and

$$C_1 \exp(|B_m|H) + C_2 \exp(-|B_m|H) = 0 \quad . \tag{3.92}$$

In the limit of infinitely large H, i.e., for a half-space, C_1 must be zero to eliminate the infinitely large exponential function. Then $C_2 = \phi_0$ and the solution becomes simply:

$$\phi(z) = \phi_0 \exp(-|B_m|z) \quad . \tag{3.93}$$

Thus, the flux in an infinitely large half-space of a subproductive medium decreases exponentially, with the inverse of $|B_m|$ as the relaxation length. This result is the generalization of Eq. (3.83), which holds for a nonmultiplying medium and has $1/L$ as a relaxation length. From Eq. (2.146) it follows that the limit of $|B_m|$ as the fission cross section tends to zero is:

$$|B_m| = \left(\left|\frac{\nu\Sigma_f - \Sigma_a}{D}\right|\right)^{1/2} \rightarrow \left(\left|\frac{\Sigma_a}{D}\right|\right)^{1/2} = \frac{1}{L} \quad . \tag{3.94}$$

If H is finite, the other exponential function is required in Eq. (3.91) to force the flux to zero at a finite height. The corresponding coefficients C_1 and C_2 can be readily obtained as the solutions of Eqs. (3.92):

$$C_1 = -\phi_0 \frac{\exp(-|B_m|H)}{\exp(|B_m|H) - \exp(-|B_m|H)} = -\frac{\phi_0 \exp(-|B_m|H)}{2\sinh(|B_m|H)} \qquad (3.95a)$$

and

$$C_2 = \phi_0 \frac{\exp(|B_m|H)}{\exp(|B_m|H) - \exp(-|B_m|H)} = \frac{\phi_0 \exp(|B_m|H)}{2\sinh(|B_m|H)} \quad . \qquad (3.95b)$$

Figure 3-2 shows a graphic representation of the flux distributions of Eqs. (3.93) and (3.91) with Eqs. (3.95) giving the C values.

If the medium is superproductive and thus $B_m^2>0$, the flux has a different shape. Since at each point in the medium there are more neutrons produced than absorbed, the current that leaves an infinitesimal slice of the system must be larger than the current that enters that slice from the side of the source. This requires an increasing negative slope as compared to a decreasing one in the case of a subproductive medium (compare the dashed-line tangents depicted in Figs. 3-2 and 3-3).

In most cases, one knows the source strength or the resulting current rather than the flux at the location of the source, as it has been assumed

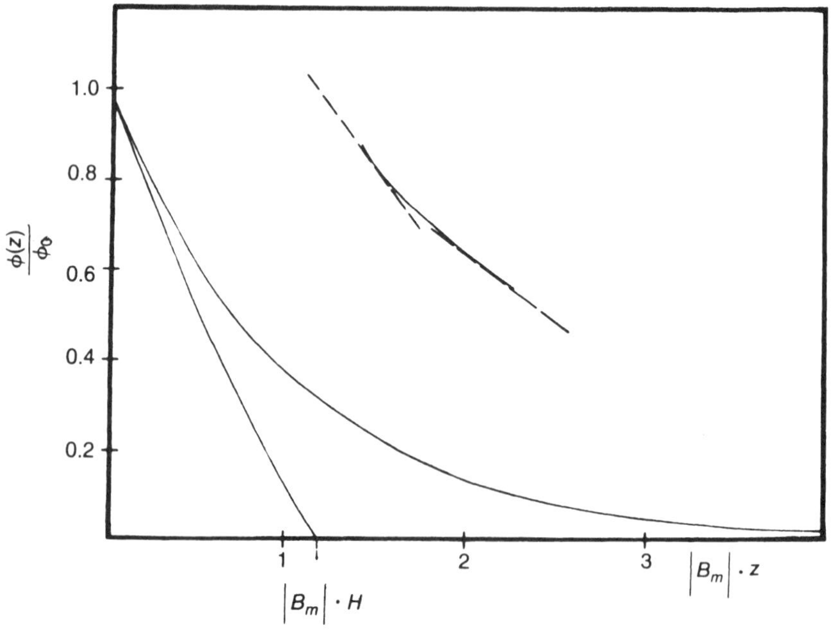

Fig. 3-2. Spatial flux distribution in a subcritical slab with $k_\infty < 1$, driven by a plane source at $z = 0$.

in previous examples. Therefore, the boundary conditions for the subcritical system with a superproductive medium are expressed by the current, as in Eq. (3.80). The source is assumed to be located at the center plane ($z=0$) of the superproductive slab. Thus, the boundary conditions are given by

$$-D\left[\frac{d\phi(z)}{dz}\right]_{z=\pm 0} = \pm J(0) = \pm J_0 = \frac{S_0}{2} \tag{3.96a}$$

and

$$\phi\left(\pm\frac{H}{2}\right) = 0 \quad . \tag{3.96b}$$

Equation (3.96a) expresses the boundary condition at the right and the left of the infinitesimally thin plane source. The source strength per square centimeter is equal to the sum of the two currents:

$$S_0 = 2J_0 \quad . \tag{3.97}$$

Since the slab region is split in the middle by the source, the solution of the differential equation, Eq. (3.89), consists of two branches, one for positive and one for negative z:

$$\phi^{\pm}(z) = C_c^{\pm}\cos(B_m z) + C_s^{\pm}\sin(B_m z) \quad . \tag{3.98}$$

The derivative of the cosine vanishes at $z=0$. Therefore, the boundary condition at the source plane determines the coefficient of the sine term:

$$C_s^{\pm} = \frac{\mp J_0}{DB_m} \quad . \tag{3.99}$$

From the second boundary condition, Eq. (3.96b), one obtains for the coefficient of the cosine term:

$$C_c^{+} = C_c^{-} = \frac{J_0}{DB_m}\tan\left(B_m\frac{H}{2}\right) \quad . \tag{3.100}$$

Thus, the flux distribution is given by:

$$\phi(z) = \frac{S/2}{DB_m}\left[\tan\left(B_m\frac{H}{2}\right)\cos(B_m z) - \sin(B_m z)\right], \; z \geqslant 0 \tag{3.101a}$$

and

$$\phi(z) = \frac{S/2}{DB_m}\left[\tan\left(B_m\frac{H}{2}\right)\cos(B_m z) + \sin(B_m z)\right], \; z \leqslant 0 \quad , \tag{3.101b}$$

where Eq. (3.101a) holds for positive and Eq. (3.101b) for negative z. Figure 3-3 shows a typical neutron flux in a subcritical superproductive medium calculated from Eqs. (3.101).

The coefficient of the cosine term in Eqs. (3.101) increases toward infinity if $B_m \cdot H/2$ approaches $\pi/2$, i.e., if the reactor approaches criticality:

$$\tan\left(B_m \frac{H}{2}\right) \rightarrow \infty, \text{ if } H \rightarrow \frac{\pi}{B_m} \quad . \tag{3.102}$$

Thus, if the height, H, of the slab reactor is increased and approaches the critical dimension, the flux shape approaches a cosine. The magnitude of the flux increases toward infinity, in accordance with the fact that no finite stationary flux can exist in a critical reactor in the presence of an independent source.

For any height below the critical value, the source neutrons are "multiplied" in the neutron multiplying medium. Suppose a neutron counter is placed near $z = 0$ where the sine term in Eqs. (3.101) vanishes. Then, the factor of the cosine gives the static "source multiplication factor," M_s, which describes the ratio of the fission source to the independent source S:

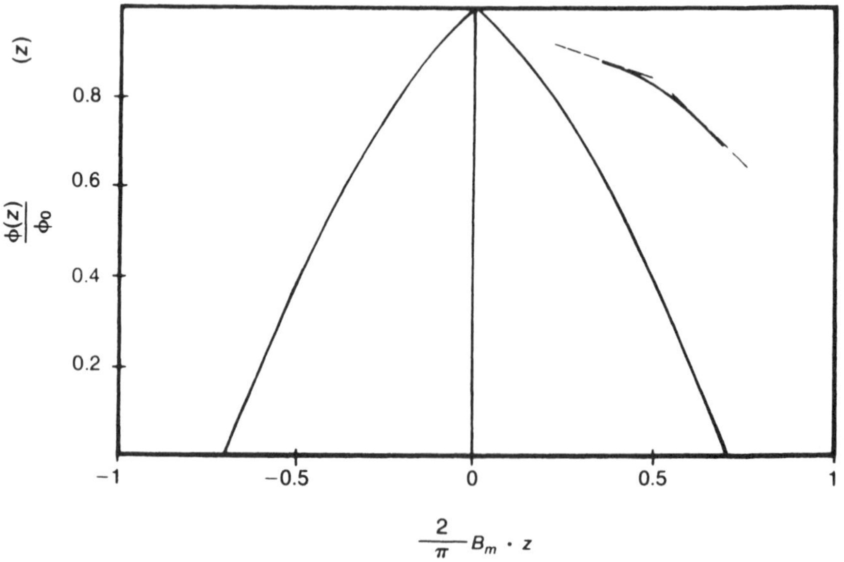

Fig. 3-3. Spatial flux distribution in a subcritical slab with $k_\infty > 1$, driven by a plane source in the slab midplane.

$$M_s = \frac{\nu\Sigma_f\phi(0)}{S} = \nu\Sigma_f \frac{\tan\left(B_m \frac{H}{2}\right)}{2DB_m} \quad . \tag{3.103}$$

The static source multiplication is often used during the loading of fuel into a reactor to obtain information on the degree of subcriticality and to extrapolate to the critical condition. One plots $1/M_s$, which can be extrapolated to zero to find the critical dimension. Figure 3-4 shows a $1/M_s$ diagram derived from Eq. (3.103) as a function of H/H_{crit}, with H_{crit} being the critical height. The dots may indicate M_s measurements as obtained after finite incremental increases of H (in practical situations, the abscissa may be plotted as an "effective" radius or the number of

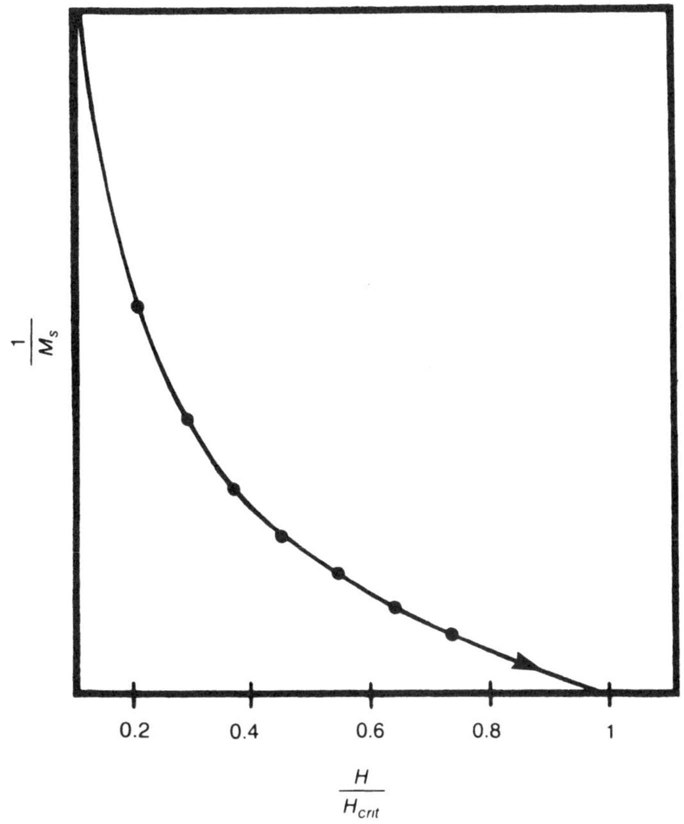

Fig. 3-4. Dependence of the inverse source multiplication factor ($1/M_s$) as a function of the height of a slab reactor.

fuel subassemblies, instead of H). Note that the extrapolation to the critical dimension must use the shape of the $1/M_s$ dependence.

Finally, it should be noted that the flux shapes in the subcritical medium slab system derived above are only approximate solutions. The inaccuracy occurs near the neutron source due to the neglect of transition phenomena in the energy and angular distribution of the source neutrons (see the discussion at the end of Sec. 2-6D).

3-4E The Exponential Pile

An example of a device in which the flux distribution in a subcritical medium is physically realized is the so-called "exponential pile." This experimental facility is analogous to that alluded to in the previous sections; it is applied in order to measure the diffusion length and material buckling. It consists of an assemblage of fissionable and/or nonfissionable reactor material. The purpose of the experiment is to obtain the material buckling from the steady-state measurement of the thermal flux distribution by employing a thermal neutron source and using the same composition and lattice structure as the intended reactor design. From the results obtained from this subcritical system, which is considerably smaller and less expensive than a critical system, the critical configuration of the reactor can be inferred. Exponential piles are no longer in much use today, having been replaced by other means (e.g., critical facility or sophisticated computational techniques) for obtaining integral reactor data.

A typical exponential pile is illustrated in Fig. 3-5. It consists of a long rectangular block of fissionable material with a neutron source at one end. Figure 3-5 depicts a point source [e.g., a Ra-Be source where the neutrons are produced by the (α,n) reaction] embedded in a large slab of graphite, which creates a distributed source of thermal neutrons at the face of the exponential pile. An alternate method of creating the source of thermal neutrons could be achieved by placing the pile in a "thermal column," which is an extension of the reflector of a critical reactor. Some exponential piles have a cylindrical geometry with a neutron source at the center axis, for example, the Fast Breeder Blanket Facility at Purdue University.[4,5] Sometimes, "exponential" cylindrical regions were embedded in a ring of higher neutron multiplication, and the whole system was critical.

The diffusion equation describing the fuel moderator system of rectangular geometry of the exponential pile depicted in Fig. 3-5 is given by Eq. (3.44):

$$\frac{\partial^2\phi(x,y,z)}{\partial x^2} + \frac{\partial^2\phi(x,y,z)}{\partial y^2} + \frac{\partial^2\phi(x,y,z)}{\partial z^2} + B_m^2\phi(x,y,z) = 0 \quad , \qquad (3.104)$$

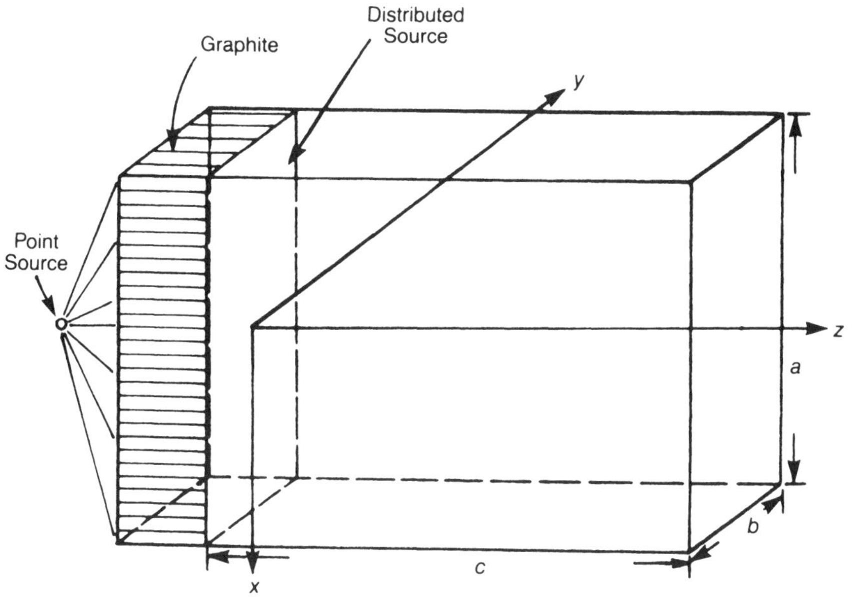

Fig. 3-5. Illustration of an exponential pile.

where B_m^2 is the material buckling to be determined from the exponential pile experiment. The approximate boundary conditions at the extrapolated boundaries are:

$$\phi(x,y,z) = \phi_0(x,y) \text{ at the plane } z = 0 \quad , \tag{3.105a}$$

and

$$\phi(x,y,z) = 0 \text{ at the planes } x = \pm a/2, y = \pm b/2, \text{ and } z = c \quad . \tag{3.105b}$$

The solution to the problem posed by Eq. (3.104) makes use of the approximate separability of the flux in the bare system:

$$\phi(x,y,z) = \phi_0 X(x) Y(y) Z(z) \quad . \tag{3.106}$$

Substitution of Eq. (3.106) into Eq. (3.104) and division by $\phi(x,y,z)$ results in:

$$\frac{1}{X}\frac{d^2X}{dx^2} + \frac{1}{Y}\frac{d^2Y}{dy^2} + \frac{1}{Z}\frac{d^2Z}{dz^2} + B_m^2 = 0 \quad , \tag{3.107}$$

which requires that each term be a constant. Since the exponential pile investigated here is unreflected in the x and y dimensions, the x and y terms in Eq. (3.107) must be equal to the respective geometrical bucklings. On one side of the z dimension, the source neutrons enter the system. Thus, the z component in Eq. (3.107) is not given by B_z^2; it is set equal to a constant, γ^2, which is yet to be determined. It turns out that this constant is positive:

$$\frac{1}{X}\frac{d^2X}{dx^2} = -B_x^2, \frac{1}{Y}\frac{d^2Y}{dy^2} = -B_y^2, \frac{1}{Z}\frac{d^2Z}{dz^2} = \gamma^2 \quad . \tag{3.108}$$

Inserting Eq. (3.108) into Eq. (3.107) gives the positive constant for γ^2:

$$\gamma^2 = B_x^2 + B_y^2 - B_m^2 \quad . \tag{3.109}$$

If B_m^2 is negative, then γ^2 is clearly positive. But for a positive B_m^2, one also obtains a positive γ^2, since all exponential piles are so small in the x and y direction that the transverse buckling (here, $B_x^2 + B_y^2$) is larger than the material buckling. Note that in the hypothetical example of an infinite slab, which was treated above, B_x^2 and B_y^2 were zero and:

$$\gamma^2 = -B_m^2 \text{ for a slab infinite in } x \text{ and } y \quad . \tag{3.110}$$

The fundamental solution to the three separated equations (Eq. 3.108) with the boundary conditions, Eqs. (3.105a) and (3.105b), is symmetrical about the z axis and is zero at $x = \pm a/2$, $y = \pm b/2$, and $z = c$:

$$X(x) = \cos B_x x = \cos\frac{l\pi x}{a} \quad ,$$

$$Y(y) = \cos B_y y = \cos\frac{n\pi y}{b} \quad , \tag{3.111}$$

and

$$Z(z) = \sinh[\gamma_{ln}(c - z)] \quad ,$$

with

$$\gamma_{ln}^2 = \left(\frac{l\pi}{a}\right)^2 + \left(\frac{n\pi}{b}\right)^2 - B_m^2 \text{ and } l,n = 1, 3, 5, \ldots \quad . \tag{3.112}$$

The solution of Eq. (3.104) then is the combination of all terms of Eq. (3.111), i.e.:

$$\phi(x,y,z) = \sum_{l=1}^{\infty}\sum_{n=1}^{\infty} C_{ln}\cos\frac{l\pi x}{a}\cos\frac{n\pi y}{b}\sinh[\gamma_{ln}(c - z)] \quad . \tag{3.113}$$

The constants C_{ln} can be obtained by applying the boundary condition, Eq. (3.105) [that is, $\phi(x,y,z) = \phi_0(x,y)$], at $z = 0$, then multiplying Eq. (3.113) by $\cos\frac{l\pi x}{a}\cos\frac{n\pi y}{b}$ and integrating over the x and y space (i.e., x from

$-a/2$ to $+a/2$ and y from $-b/2$ to $+b/2$. For $\phi_0(x,y) = \phi_0$ (independent of x and y), one obtains:

$$C_{ln} = \left(\frac{4}{\pi}\right)^2 \frac{\phi_0}{ln\sinh\gamma_{ln}c}(-1)^{\frac{l+n}{2}-1} \quad . \tag{3.114}$$

Far away from the source plane at $z = 0$, the higher harmonics die away as γ_{ln} increases with l and n. Therefore, in some distance from the interfaces, the neutron flux can be approximated by the term with $l = 1$ and $n = 1$ ($\gamma_{11} = \gamma$):

$$\phi(x,y,z) \simeq 2C\cos\frac{\pi x}{a}\cos\frac{\pi y}{b}\sinh[\gamma(c - z)] \tag{3.115a}$$

and

$$\phi(x,y,z) \simeq C\cos\frac{\pi x}{a}\cos\frac{\pi y}{b}\exp[-\gamma(z - c)] \quad , \tag{3.115b}$$

where

$$\gamma^2 = \left(\frac{\pi}{a}\right)^2 + \left(\frac{\pi}{b}\right)^2 - B_m^2 \tag{3.116a}$$

and

$$C = \left(\frac{4}{\pi}\right)^2 \frac{\phi_0}{\sinh(\gamma c)} \cdot \frac{1}{2} \quad . \tag{3.116b}$$

This simple solution is also obtained when the flux at the source interface is approximated by a product of cosine functions:

$$\phi_0(x,y) = \phi_0\cos\frac{\pi}{a}x\cos\frac{\pi}{b}y \quad . \tag{3.117}$$

In the application of the above exponential pile theory, small foils or counters are used to obtain the neutron flux distribution along the z axis. The flux data can then be applied to Eq. (3.115), from which a value for γ is obtained by fitting the experimental values to $\exp[-\gamma(z - c)]$. Using Eq. (3.116a) and the physical dimensions at the pile (i.e., a and b), the material buckling (B_m^2) of the configuration can be determined.

3-5 Multiregion Flux Shape Problems

In the previous sections, only bare systems were considered, i.e., it was assumed that neutrons that had diffused out of the system were not scattered back. This geometry is very inefficient in terms of neutron economy and, hence, is not used in actual power reactors. Actual reactors are surrounded by reflectors or blankets and by structural and shielding material. Therefore, many neutrons are actually scattered back into the core. Thermal reactor cores are generally surrounded by a reflector; fast reactor cores are surrounded by blankets in which a large part of the breeding of fissile material takes place.

The study of multiregion cores is also closer to reality since in most reactors the core consists of several regions with distinctly different fuel enrichment in order to flatten the power distribution.

The approximate flux shape distributions in multiregion reactors are discussed in this section. The derivation concentrates on reflected systems only, again within the context of the one-group theory. Although, in principle, these flux separation methods could be applied to the fast reactor core-blanket regions; the one-group approximation, however, is too inaccurate in fast reactors. In fact, even in thermal reflected systems, at least two groups are required in practice for a first-order description of the effect of a reflector on the neutron behavior (compare Sec. 7-4).

3-5A General Features of Reflected Cores

It is general reactor design practice to surround a core by a reflector region whose primary function is to scatter a large fraction of the neutrons back into the core. This function requires that the reflector material possess a large scattering cross section with an attendant low capture cross section. Furthermore, fast neutrons leaking into the reflector are moderated to lower energies before they are returned to the core section, which is a distinct benefit in thermal reactors. It follows that a good moderator is well suited as a reflector.

The presence of the reflector in a thermal core also tends to flatten the neutron flux distribution in the core region. This flattening makes the power distribution more uniform and has the advantage of evening out the fuel burnup distribution across the core as well as the radial coolant temperature distribution.

To illustrate the multiregion theory, a single core region surrounded by a reflector is considered with the neutrons again taken to be in one energy group. The analytical methods are similar to those employed for the geometry shapes presented in the sections on bare cores (see Secs. 3-3 and 3-4). The main difference is due to the fact that macroscopic cross sections are discontinuous at the regional boundaries, i.e., the coefficients of the diffusion equation are not differentiable at the regional boundaries. Therefore, separate diffusion equations are required for each region. To describe the neutron shape behavior in the whole reactor, the regional solutions are coupled at the interfacing boundaries. The core region parameters are denoted here by the subscript c [i.e., D_c, $\Sigma_{a,c}$, $\Sigma_{f,c}$, $\phi_c(\boldsymbol{r})$] and the reflector by the subscript r [i.e., D_r, $\Sigma_{a,r}$, $\phi_r(\boldsymbol{r})$].

As indicated in Sec. 3-2, the actual calculation of the flux in a multiregion problem requires the simultaneous treatment of space and energy dependence, which then yields the (smallest) eigenvalues, $\lambda(=1/k)$, and fundamental λ mode, $\phi_\lambda(\boldsymbol{r},E)$. To illustrate this procedure, the one-group model solutions are derived in this section basically in the same way as in the general problem.

The one-group diffusion equation for the λ eigenvalue problem is given by:

$$-\nabla\cdot D\nabla\phi(\boldsymbol{r})+\Sigma_a\phi(\boldsymbol{r})=\lambda\nu\Sigma_f\phi(\boldsymbol{r}) \quad . \tag{3.118}$$

Since the cross sections in Eq. (3.118) and particularly D are not differentiable at the regional interfaces, Eq. (3.118) is rewritten into a set of two equations, one for each region. For the core/reflector problem (indices c and r, respectively), one obtains:

$$-D_c\nabla^2\phi_c(\boldsymbol{r})+\Sigma_{a,c}\phi_c(\boldsymbol{r})=\lambda\nu\Sigma_{f,c}\phi_c(\boldsymbol{r}) \tag{3.119a}$$

and

$$-D_r\nabla^2\phi_r(\boldsymbol{r})+\Sigma_{a,r}\phi_r(\boldsymbol{r})=0 \quad . \tag{3.119b}$$

The reflector is assumed to contain no fissionable material (source-free medium). Since the diffusion constant is assumed to be constant in each of the regions, the D values can be taken out of the respective derivatives. Equations (3.119) can then be brought into the standard form of the wave equation:

$$\nabla^2\phi_c(\boldsymbol{r})+B_\lambda^2\phi_c(\boldsymbol{r})=0 \tag{3.120a}$$

and

$$\nabla^2\phi_r(\boldsymbol{r})-\kappa^2\phi_r(\boldsymbol{r})=0 \quad , \tag{3.120b}$$

with

$$B_\lambda^2=\frac{\lambda k_\infty-1}{L_c^2} \tag{3.121a}$$

and

$$\kappa^2=\frac{1}{L_r^2}=\frac{\Sigma_{a,r}}{D_r}=-B_{m,r}^2 \quad . \tag{3.121b}$$

The eigenvalue λ is an unknown quantity in Eq. (3.121a); it is to be determined such that the entire problem has a nontrivial solution. The quantity λk_∞ is assumed to be >1 so that the inner regions may qualify as core. The B_λ^2 is positive. The material buckling of the reflector is negative since $k_\infty = 0$ due to the absence of fissionable material.

The quantity B_λ^2 is called the "λ-modified material buckling." It is to be calculated, in general, as a B^2 eigenvalue from Eq. (2.138) for a given value of λ; i.e., B_λ^2 is calculated from the same equation that gives B_m^2 in case of $\lambda = 1$. For $\lambda = 1$, one has:

$$B_\lambda^2(\lambda = 1) = B_m^2 \quad . \tag{3.122}$$

Thus, B_λ^2 for $\lambda \neq 1$ describes the B_m^2 analog of the off-critical system.

The solutions of the set of Eqs. (3.120) are determined by the system geometry, which enters the problem through interface and boundary conditions. It should be noted that for the core region the governing equation, Eq. (3.120a), is formally identical to the one for the bare core, which was solved and discussed in Sec. 3-3. The reflector equation, Eq. (3.120b), on the other hand, is identical to that for the source-free medium problems of Sec. 3-4. In addition to the outer boundary conditions (which are the same as for the bare core, i.e., the flux being zero at the extrapolated distance), interface conditions are required to link the regional flux branches. The interface conditions require the flux and neutron current at the core/reflector interface to be continuous (compare Sec. 2-2E):

$$\phi_c(\boldsymbol{r}_b) = \phi_r(\boldsymbol{r}_b) \tag{3.123a}$$

and

$$D_c \nabla\phi_c(\boldsymbol{r}_b) = D_r \nabla\phi_r(\boldsymbol{r}_b) \quad , \tag{3.123b}$$

with $\boldsymbol{r}_b$ describing core boundary, i.e., the core/reflector interface.The outer boundary condition is:

$$\phi_r(\boldsymbol{r}_v) = 0 \quad , \tag{3.123c}$$

with $\boldsymbol{r}_v$ describing the extrapolated outer boundary, i.e., the interface with the vacuum. The solution procedure is illustrated in the next section using an infinite plane slab reactor as an example.

3-5B Reflected Infinite Slab Core

The solution procedure for the flux distribution in a reflected core is illustrated for the simple case of a core slab, infinite in x-y, of half-thickness $H/2$ surrounded on both sides by a reflector of thickness δH:

$$H' = H + 2\delta H \quad .$$

For simplicity, δH shall include the extrapolation length. The geometry is symmetrical about $z = 0$, so that only positive values of z need to be considered. As a variation of the example presented in Sec. 3-3A, the boundary conditions

$$\phi\left(\frac{H'}{2}\right) = 0 \tag{3.124a}$$

and

$$\left.\frac{d\phi(z)}{dz}\right|_{z=0} = 0 \tag{3.124b}$$

are used. The same treatment can also be applied to the bare slab. Equations (3.124a) and (3.124b) replace the equivalent boundary condition, $\phi(\pm H'/2) = 0$, in the bare slab case.

The flux distribution in the core region, obtained from the solution of Eq. (3.120a), is identical to that previously derived for the bare slab, i.e.:

$$\phi_c(z) = C_1 \cos B_\lambda z \quad . \tag{3.125}$$

The solution for the flux shape in the reflector region involves solving Eq. (3.120b) in the Cartesian system (i.e., $\nabla^2 = d^2/dz^2$). This general reflector solution ($\kappa^2 > 0$) can be written in the form of exponential or of hyperbolic functions:

$$\phi_r(z) = C_2 \cosh \kappa_r z + C_3 \sinh \kappa_r z \quad . \tag{3.126}$$

Applying the outside surface boundary condition of Eq. (3.123c), which requires that the flux must vanish at the extrapolated boundary, $z = H'/2$, results in:

$$\phi\left(\frac{H'}{2}\right) = C_2 \cosh \kappa_r \frac{H'}{2} + C_3 \sinh \kappa_r \frac{H'}{2} = 0 \quad , \tag{3.127}$$

from which C_2 can be obtained:

$$C_2 = -C_3 \tanh \kappa_r \frac{H'}{2} \quad . \tag{3.128}$$

Inserting this expression for C_2 into Eq. (3.126), the desired flux distribution in the reflector region is obtained after some simple arithmetic:

$$\phi_r(z) = C_4 \sinh \kappa_r \left(\frac{H'}{2} - z\right) \quad , \tag{3.129}$$

where C_4 is another constant. It can readily be seen that Eq. (3.129) solves the balance equation and satisfies the outer boundary condition.

To establish the two constants, C_1 in the core flux expression, Eq. (3.125), and C_4 in the reflector flux expression, Eq. (3.129), the interface boundary conditions are required, i.e., continuity of the flux and the current at the core/reflector interface (i.e., at $z = H/2$). Expressed mathematically, these two conditions are:

$$\phi_c(H/2) = \phi_r(H/2) \tag{3.130a}$$

and

$$D_c \frac{d\phi_c(H/2)}{dz} = D_r \frac{d\phi_r(H/2)}{dz} \quad . \tag{3.130b}$$

From the first condition here and the flux expressions, Eqs. (3.125) and (3.129), the following relationship is obtained:

$$C_1 \cos\left(B_\lambda \frac{H}{2}\right) = C_4 \sinh(\kappa_r \delta H) \quad . \tag{3.131}$$

Applying the second boundary condition to the same two flux branches results in:

$$-C_1 D_c B_\lambda \sin\left(B_\lambda \frac{H}{2}\right) = -C_4 D_r \kappa_r \cosh(\kappa_r \delta H) \quad . \tag{3.132}$$

Dividing Eq. (3.132) by Eq. (3.131) eliminates the constants and yields the condition for a nontrivial solution of the infinite reflected slab reactor in the one-group approximation:

$$D_c B_\lambda \tan\left(B_\lambda \frac{H}{2}\right) = D_r \kappa_r \coth(\kappa_r \delta H) \quad . \tag{3.133a}$$

For $\lambda = 1$, Eq. (3.133a) becomes the criticality condition:

$$D_c B_m \tan\left(B_m \frac{H}{2}\right) = D_r \kappa_r \coth(\kappa_r \delta H) \quad . \tag{3.133b}$$

Equation (3.133a) is a transcendental equation with given parameters (i.e., D_c, D_r, κ_r, and δH); it can be solved for B_λ or H by an appropriate iterative method. Here only a graphic solution method and approximate analytic solutions are discussed.

The graphic solution is facilitated by rewriting Eq. (3.133a) in such a way that the right side is a constant. If H is given, and B_λ is to be determined, one can write:

$$\left(B_\lambda \frac{H}{2}\right) \tan\left(B_\lambda \frac{H}{2}\right) = \frac{H}{2} \frac{D_r \kappa_r}{D_c} \coth(\kappa_r \delta H) \quad . \tag{3.133c}$$

Figure 3-6 illustrates the graphic solution of Eq. (3.133c): the function on the left side equals the value of the right side at a certain value of $B_\lambda \frac{H}{2}$. It is the first branch of the tangent that yields the desired smallest eigenvalue corresponding to the fundamental solution. The corresponding value of B_λ can be established by variation of the composition, or of λ, in the two usual ways: composition or λ eigenvalue search. The result for $\lambda = 1$ gives the critical core half-thickness for a known reflector thickness δH. Alternately, the critical enrichment of the core can be determined from Eq. (3.133b) by varying B_m, knowing the dimensions and other properties of the reflected system.

Equation (3.133a) is the generalization of the simple criticality equation, which is obtained from Eqs. (3.133) for $\delta H \to 0$. If δH is reduced to zero, the right side of Eq. (3.133a) approaches an infinite value. The

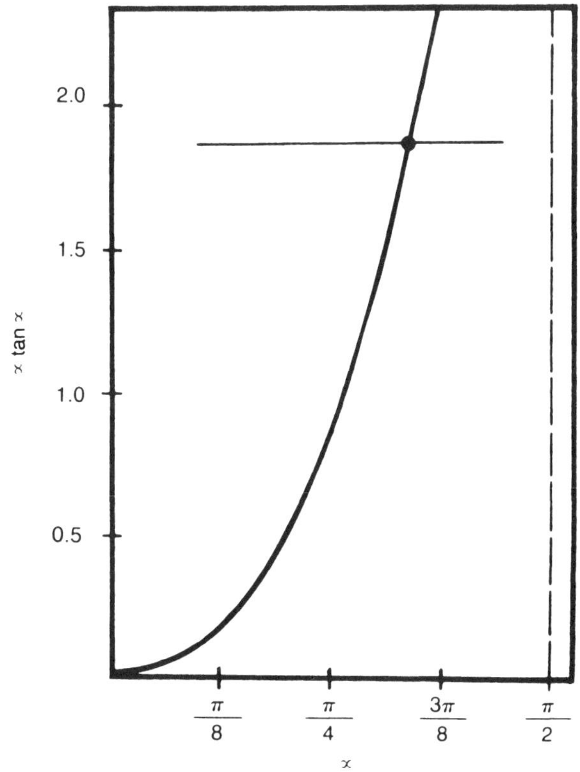

Fig. 3-6. Illustration of the graphic solution of the transcendental equation for determining the B_λ eigenvalue.

resulting solution, $B_\lambda H/2 = \frac{\pi}{2}$, then agrees with the solution for a bare slab.

As was the case with the flux shapes obtained for the bare cores, the magnitude of the flux distribution is unattainable since the mathematical problem is homogeneous. Thus, the flux always contains a free constant; e.g., ϕ_0. Equating C_1 with ϕ_0, and by solving Eq. (3.131) for C_4:

$$C_4 = \phi_0 \frac{\cos B_\lambda H/2}{\sinh \kappa_r \delta H} \quad . \tag{3.134}$$

The flux in the core is given by Eq. (3.125) with $C_1 = \phi_0$. The flux distribution in the reflector, after inserting the expression for C_4 [as given by Eq. (3.134)] into Eq. (3.129), then becomes:

$$\phi_r(z) = \phi_0 \frac{\cos(B_\lambda H/2) \sinh \kappa_r \left(\frac{H'}{2} - z \right)}{\sinh \kappa_r \delta H} \quad . \tag{3.135}$$

The core flux distribution, which has been normalized to the maximum flux, ϕ_0, at the slab center, is again integrated over the slab thickness to obtain the average flux in the core region. The maximum-to-average flux ratio in the core is then:

$$\frac{\phi_0}{\bar{\phi}} = \frac{B_\lambda H/2}{\sin(B_\lambda H/2)} \quad . \tag{3.136}$$

In addition to the graphic solution of Eq. (3.133c) described above, the formal analytical solution is also of interest:

$$\frac{B_\lambda H}{2} = \arctan \left[\frac{D_r \kappa_r}{D_c B_\lambda} \coth(\kappa_r \delta H) \right] \quad . \tag{3.137}$$

As expected, for a bare core, B_λ is equal to π/H and $\phi_0/\bar{\phi}$ is ~1.57 (see Table 3-I). For a very thick reflector, δH can be assumed to approach infinity. Then, Eq. (3.137) becomes simply:

$$\frac{B_\lambda H}{2} = \arctan \left(\frac{D_r \kappa_r}{D_c B_\lambda} \right) \quad , \tag{3.138}$$

as $\coth(\kappa \delta H)$ approaches one. Equation (3.138) and Eq. (3.136) give the flux peaking factor for the limiting case of a large reflector.

3-5C Reflector Savings

A major benefit of the presence of a reflector surrounding a core is the reduction in the core size. The term describing the savings in core thickness due to the presence of the reflector is called "reflector savings." For the infinite slab described in the previous section, the reflector savings is defined as:

reflector savings = { half-thickness of *bare* critical core minus half-thickness of *reflected* critical core with the same composition.

Denoting the reflector savings by s and neglecting the extrapolation distance, the half-thickness of a bare slab of modified material buckling B_λ^2 is just $\pi/2B_\lambda$. The reflector savings is then given by:

$$s = \frac{\pi}{2B_\lambda} - \frac{H}{2} \quad . \tag{3.139}$$

The reflected half-height is given by its unreflected value minus the reflector savings:

$$H/2 = \frac{\pi}{2B_\lambda} - s \quad . \tag{3.140}$$

Substituting this expression for $H/2$ into the infinite reflected slab eigenvalue equation, Eq. (3.137), gives:

$$s = \frac{1}{B_\lambda}\left\{\frac{\pi}{2} - \arctan\left[\frac{D_r\kappa_r}{D_c B_\lambda}\coth(\kappa_r\delta H)\right]\right\} \quad , \tag{3.141a}$$

which can be simplified to (note: $\pi/2 - \arctan\alpha = \arctan 1/\alpha$):

$$s = \frac{1}{B_\lambda}\arctan\frac{D_c B_\lambda}{D_r\kappa_r}\tanh(\kappa_r\delta H) \quad . \tag{3.141b}$$

Equation (3.141b) can be solved for the reflector savings as it applies to various reflector thickness values δH.

If the core is assumed to be large, such that B_λ is small, Eq. (3.141b) reduces to its value for large cores (note: $\arctan\alpha \simeq \alpha$ for very small α):

$$s \simeq \frac{D_c}{D_r}L_r\tanh\frac{\delta H}{L_r} \quad , \tag{3.142}$$

with κ_r denoted by $1/L_r$. For the case where the diffusion length of the reflector, L_r, is large compared to the reflector thickness, δH, one can further approximate:

$$\tanh(\delta H/L_r) \cong \delta H/L_r \quad . \tag{3.143a}$$

Equation (3.142) then gives the reflector savings for a large reactor with a thin reflector as approximately equal to the reflector thickness, i.e.:

$$s \cong \frac{D_c}{D_r}\delta H \quad . \tag{3.143b}$$

For a very thick reflector (i.e., $\delta H > L_r$), on the other hand, one can apply the approximation:

$$\tanh(\delta H/L_r) \cong 1 \quad . \tag{3.143c}$$

Equation (3.137) then gives the *maximum* reflector savings:

$$s \simeq \frac{D_c}{D_r}L_r \quad , \tag{3.144a}$$

or if $D_c \simeq D_r$, simply:

$$s = L_r \quad . \tag{3.144b}$$

This result suggests that the reflector savings, s, reaches a limiting value equal to the diffusion length, L_r, as the reflector's thickness, δH, increases. Figure 3-7 illustrates this effect of variation in reflector savings with reflector thickness. As can be seen from Fig. 3-7, the maximum reflector savings is practically attained when the reflector thickness is approximately twice the diffusion length (i.e., $\tanh 2 = 0.964$). There is, therefore, no incentive from the standpoint of reflector savings for mak-

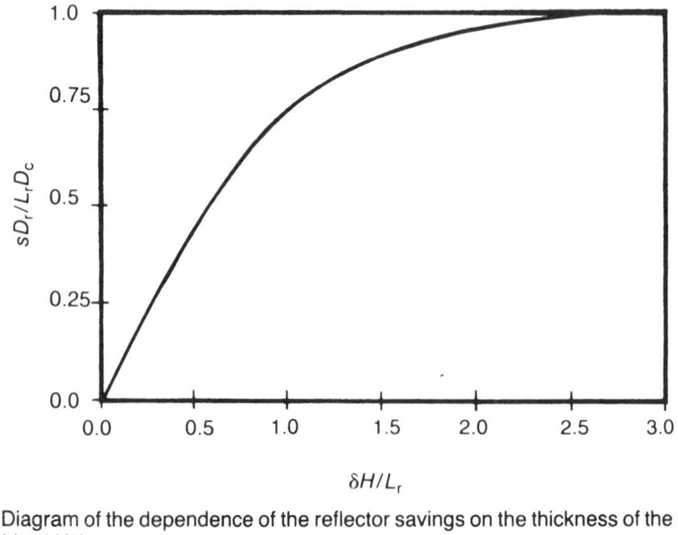

Fig. 3-7. Diagram of the dependence of the reflector savings on the thickness of the reflector, measured by $\delta H/L_r$.

ing the reflector thickness more than about twice the diffusion length of the reflector.

The concept of reflector savings can also be explained and derived using a somewhat different, more intuitive, approach. Consider the simplified one-group thermal flux distribution of neutrons in the core and reflector of a slab geometry system. As discussed below, the reflector savings, s, can be interpreted approximately as the distance between the core/reflector interface, z_c, and the location where the dotted line of Fig. 3-8 crossed the z axis. The dotted line is merely the linear extrapolation of the core neutron flux shape as established by the tangent to the flux shape curve at z_c. It should be noted that for the slab, the core flux in Fig. 3-8 is a cosine shape, whereas the reflector shape is an exponential function with a negative exponent. In one-group theory, the characteristic inflection point occurs near the core/reflector interface location z_c. From Fig. 3-8, it is evident that the effect of the reflector on the flux is equivalent to an increase of the actual core size (denoted by z_c) by an amount equal to the reflector savings s. For some practical purposes, then, the presence of the reflector can be neglected in subsequent reactor analysis with the core dimension taken to be the new pseudo-height z_c' (i.e., actual core height z_c plus reflector savings s).

The above interpretation for reflector savings is employed to derive

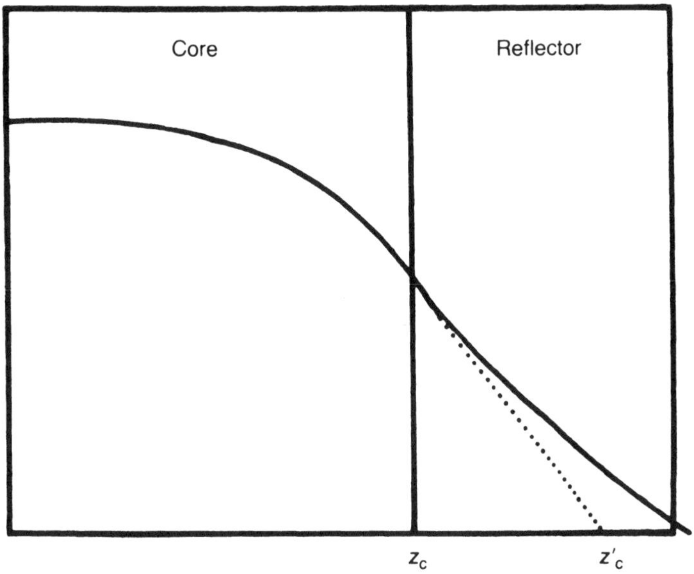

Fig. 3-8. Illustration of the one-group flux dependence in a reflected assembly.

the previous expression, Eq. (3.144a), for the reflector savings by considering the tangent of the curve at $z = z_c$ of Fig. 3-8:

$$\frac{d\phi_c(z_c)}{dz} = \frac{\phi_r(z_c') - \phi_c(z_c)}{(z_c' - z_c)} = -\frac{\phi_c(z_c)}{s} \quad . \tag{3.145}$$

Using Eq. (3.145) and the continuity relationships for the neutron flux and current at the core/reflector interface, i.e.:

$$\phi_c(z_c) = \phi_r(z_c)$$

and

$$-D_c \frac{d\phi_c(z_c)}{dz} = -D_r \frac{d\phi_r(z_c)}{dz} \quad , \tag{3.146}$$

the expression for s becomes:

$$s = \frac{-\dfrac{D_c}{D_r}\phi_r(z_c)}{\left[\dfrac{d\phi_r(z_c)}{dz}\right]} \quad , \tag{3.147}$$

in terms of reflector flux quantities. To obtain an expression for the derivative in Eq. (3.147), the exponential solution for a thick reflector,

$$\phi_r(z) = \left[\phi_r(z_c)\exp\left(-\frac{z - z_c}{L_r}\right)\right] \quad , \tag{3.148}$$

is employed.

Differentiating Eq. (3.148) and solving for L_r results in the desired flux expression:

$$\frac{\phi_r(z_c)}{\left[\dfrac{d\phi_r(z)}{dz}\right]} = -L_r \quad . \tag{3.149}$$

Then substituting this flux expression into Eq. (3.147) yields:

$$s = \frac{D_c}{D_r} L_r \quad , \tag{3.150}$$

which is identical to Eq. (3.144a).

The concept of reflector savings can be applied to other reactor shapes as well. The curvature of the core surface is small in most cases so that the reflector savings for a slab can be applied. If the one-group model is not accurate enough, the reflector savings must be determined numerically from multigroup calculations. The basic condition for determining the reflector savings value is the preservation of k_{eff}:

$$k_{eff} \text{ (bare system, dimensions enlarged by } s)$$
$$= k_{eff} \text{ (reflected system).}$$

Numerical multigroup determinations of reflector savings are recommended as homework problems.

3-6 Analytical Two-Group Flux Solutions in Subcritical Systems

The concepts evolving from the separation of the space-energy dependent diffusion equation, i.e., the material buckling, B_m^2, the fundamental spectrum, φ_m, and the corresponding higher eigenvalues and eigenspectra, were illustrated in Sec. 2-7 for a two-group problem. In this section, this illustration is completed by showing how these concepts affect the spatial flux distribution and its curvature.

Two examples are presented, using the same material data as in Sec. 2-7B and thus the same material buckling eigenvalues and eigenspectra.

Case 1 represents a superproductive medium, a core-type material. Two subcritical slab configurations and the critical slab configurations are compared. A flux in the subcritical configuration is established by a surface source in the high-energy group; the source could be a spontaneous fission source or an (α, n) source. Both sources emit neutrons only into group 1.

Case 2 represents a subproductive medium. Its B_m^2 value is negative; the higher eigenvalue must then be more negative. The components of the flux solutions for the semi-infinite medium are both exponential functions, the higher one with a shorter relaxation length than the fundamental one.

3-6A Two-Group Fluxes in a Superproductive Medium

The boundary condition is determined by the isotropic high-energy surface source. For the critical slab, the source needs to be removed or—to fit better into the sequence of slabs of increasing thickness—the source strength is reduced to zero.

The surface source provides the neutrons in the form of an incoming current, J_g^{Sin}, with $g = 1$ and 2. Let S be the source strength on each side, i.e., neutrons emitted per square centimetre and second. Applying the definition of the current, Eq. (1.10), to the isotropic source gives (for the right and left surface of the slab; $x = \pm a/2$):

$$J_1^{Sin}\left(\frac{a}{2}\right) = \frac{S}{4\pi}\int_0^{2\pi}\int_{-1}^{0} \mu \, d\mu \, d\alpha = -\frac{S}{4}$$

and

$$J_1^{Sin}\left(-\frac{a}{2}\right) = \frac{S}{4\pi}\int_0^{2\pi}\int_0^{1} \mu \, d\mu \, d\alpha = \frac{S}{4} \quad . \tag{3.151}$$

The minus sign in front of $S/4$ in the first of these equations indicates that the incoming neutrons are emitted against the positive μ direction. The sum of both components, the total net current, is zero as expected for isotropic sources or fluxes. The total number of neutrons emitted into the slab is S, with $S/2$ entering each side.

The calculation of the incoming neutron currents as determined from the linear approximation of the angular flux was presented in Sec. 2-2E. Application to the slab yields

$$J_g^{in}\left(\frac{a}{2}\right) = \int_{-1}^{0} \mu\phi\left(\frac{a}{2},\mu\right) d\mu$$

$$= -\frac{1}{4}\phi_g\left(\frac{a}{2}\right) + \frac{1}{2}J_g\left(\frac{a}{2}\right)$$

and

$$J_g^{in}\left(-\frac{a}{2}\right) = \int_0^{1} \mu\phi\left(-\frac{a}{2},\mu\right) d\mu$$

$$= \frac{1}{4}\phi_g\left(-\frac{a}{2}\right) + \frac{1}{2}J_g\left(-\frac{a}{2}\right) \quad . \tag{3.152}$$

Equations (3.151) and (3.152) give the source boundary conditions for the two groups:

$$J_1^{in}\left(\frac{a}{2}\right) = -\frac{1}{4}\phi_1\left(\frac{a}{2}\right) + \frac{1}{2}J_1\left(\frac{a}{2}\right) = -\frac{S}{4}$$

$$J_2^{in}\left(\frac{a}{2}\right) = -\frac{1}{4}\phi_2\left(\frac{a}{2}\right) + \frac{1}{2}J_2\left(\frac{a}{2}\right) = 0$$

$$J_1^{in}\left(-\frac{a}{2}\right) = +\frac{1}{4}\phi_1\left(-\frac{a}{2}\right) + \frac{1}{2}J_1\left(-\frac{a}{2}\right) = \frac{S}{4}$$

$$J_2^{in}\left(-\frac{a}{2}\right) = +\frac{1}{4}\phi_2\left(-\frac{a}{2}\right) + \frac{1}{2}J_2\left(-\frac{a}{2}\right) = 0 \quad . \tag{3.153}$$

These boundary conditions are used to determine the two free constants in the analytical expression for the flux, which is composed of the spectra calculated in Sec. 2-7B and the spatial components given in Sec. 2-7C:

$$\begin{pmatrix} \varphi_1(x) \\ \varphi_2(x) \end{pmatrix} = C_m \begin{pmatrix} \varphi_1 \\ \varphi_2 \end{pmatrix}_m \cos B_m x + C_2 \begin{pmatrix} \varphi_1 \\ \varphi_2 \end{pmatrix}_{(2)} \cosh|B_2|x \quad . \qquad (3.154)$$

The calculation of the two coefficients is suggested as a homework problem. Here, only the results are shown in the form of graphs as follows.

On a linear scale, Fig. 3-9 shows the two components of the flux of Eq. (3.154) for both groups. It becomes apparent that the cosh term, with its $|B|$ value being much larger than B_m, provides a correction only near the interface. The flux in the larger inner area is governed by the cos term with the material buckling determining the curvature.

To better illustrate the curvature, the fluxes are depicted in Figs. 3-10 on a log scale. All inner flux distributions, for both energy groups as well as for all thickness values of the slab, including the critical slab, are proportional to $\cos B_m x$. This can be seen very well on the logarithmic scale, where a proportionality factor appears as additive displacement.

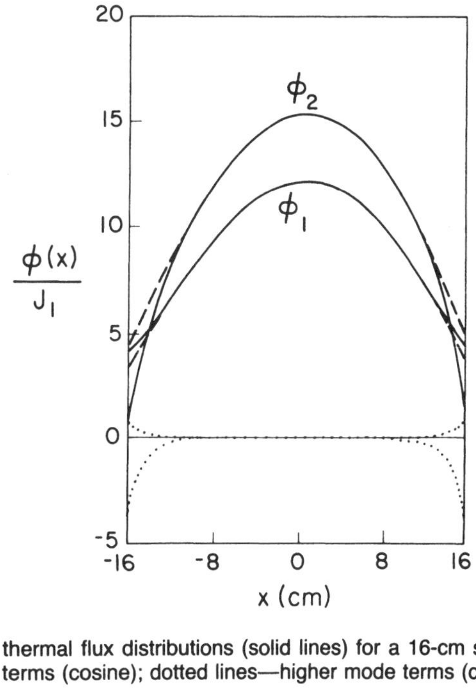

Fig. 3-9. Fast and thermal flux distributions (solid lines) for a 16-cm slab; dashed lines—fundamental mode terms (cosine); dotted lines—higher mode terms (cosh).

The normalization factor of the critical flux and the source strengths for the two subcritical slabs have been chosen such that the three curves each of Figs. 3-10 are fairly close together so that the proportionality and thus the identity of the relative curvatures can be seen clearly.

The solid lines in Figs. 3-10 represent the group fluxes, whereas the dashed lines depict the $\cos B_m x$ contribution alone. Several centimetres away from the interface, the "asymptotic" flux distribution is established. This can be understood in terms of the transport mfp:

$$\lambda_{tr} = \frac{1}{\Sigma_{tr}} = 3D \quad . \tag{3.155}$$

With the data of Table 2-I, one obtains for the two groups

$$\lambda_{tr1} = 3.9 \text{ cm}; \ \lambda_{tr2} = 0.75 \text{ cm} \quad . \tag{3.156}$$

Although λ_{tr} is much smaller for the thermal than for the fast group, the adjustment near the interfaces reaches farther into the medium for the thermal group than for the fast group since no thermal neutrons

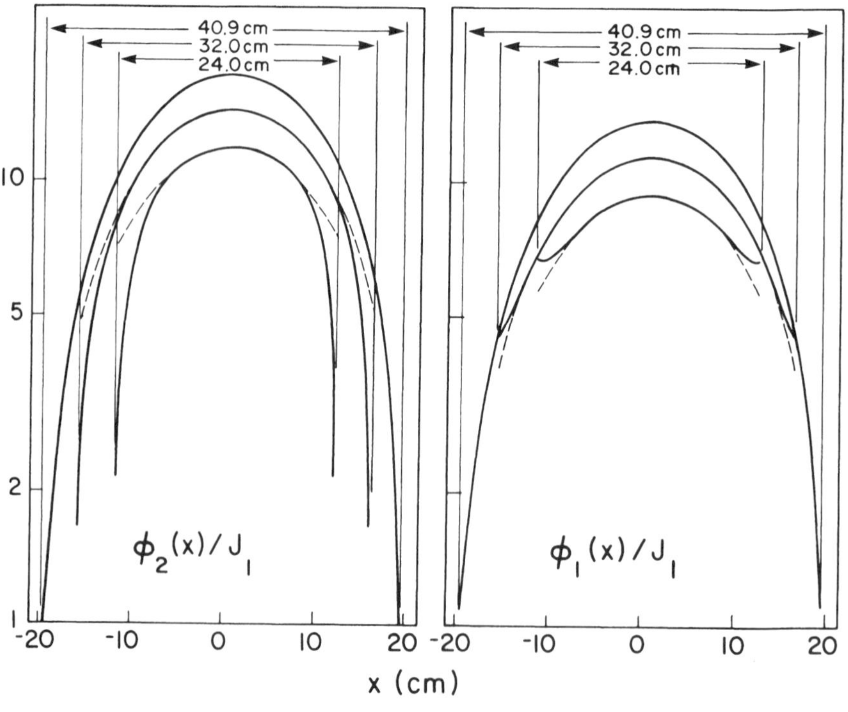

Fig. 3-10. The fast and thermal flux distributions for three different slab thicknesses, as indicated; critical slab: 40.9 cm (solid lines); dashed lines—fundamental mode terms (cosine).

come from the source and thus a more involved adjustment is required.

Equation (3.154) with $|B_2| >> B_m$ indicates mathematically why the cosh term fades away quickly, leaving the cos term dominating the entire inner area. The physical understanding of the fact that the flux shape and thus the curvature is the same for both groups in the large inner area of the slab comes from considering the *coupling* of the fast and thermal group fluxes:

1. The fast neutron flux yields the source of thermal neutrons through slowing down.
2. The thermal neutron flux feeds the fast group with fission neutrons.

In doing so, the two groups mutually impose their spatial shape on each other, which then results in a common flux distribution. The same physical explanation applies to the shape of the multigroup flux.

3-6B Two-Group Flux Distributions in a Subproductive Medium

A semi-infinite medium is considered in this case with a material as described by the group constants of Table 2-I. This two-group solution is an extension of the one-group solution derived in Sec. 3-4D.

The boundary condition is formed in analogy to Eq. (3.153); the incoming current from the source is set equal to the incoming current of the flux, again considering the source neutrons to be in group 1. The derivation of the analytical solution is suggested as a homework problem. Only the graphical representations of the solutions are discussed.

Recall that in the one-group case the equation for the intrinsic neutron balance has only one solution, B_m^2; there are no higher eigenvalues. Thus, the flux distribution in a semi-infinite medium is just a single exponential function,

$$[\phi(x)]_{1\text{-group}} = \phi_0 \exp(-|B_m|x) \quad , \tag{3.157}$$

as given by Eq. (3.93). In the two-group case, one has two B^2 eigenvalues and two eigenspectra. Again, the combination of these two spectra with the spatial solutions for the two B^2 eigenvalues gives the analytical solution:

$$\begin{pmatrix} \phi_1(x) \\ \phi_2(x) \end{pmatrix} = C_m \begin{pmatrix} \varphi_1 \\ \varphi_2 \end{pmatrix}_m e^{-|B_m|x} + C_2 \begin{pmatrix} \varphi_1 \\ \varphi_2 \end{pmatrix}_{(2)} e^{-|B_2|x} \quad . \tag{3.158}$$

The two spectra for the material considered in this example (case 2) have been calculated in Sec. 2-7. Figures 3.11 show the components of the two flux distributions. As no thermal neutrons enter the system, the thermal flux goes through a buildup phase before its eventual expo-

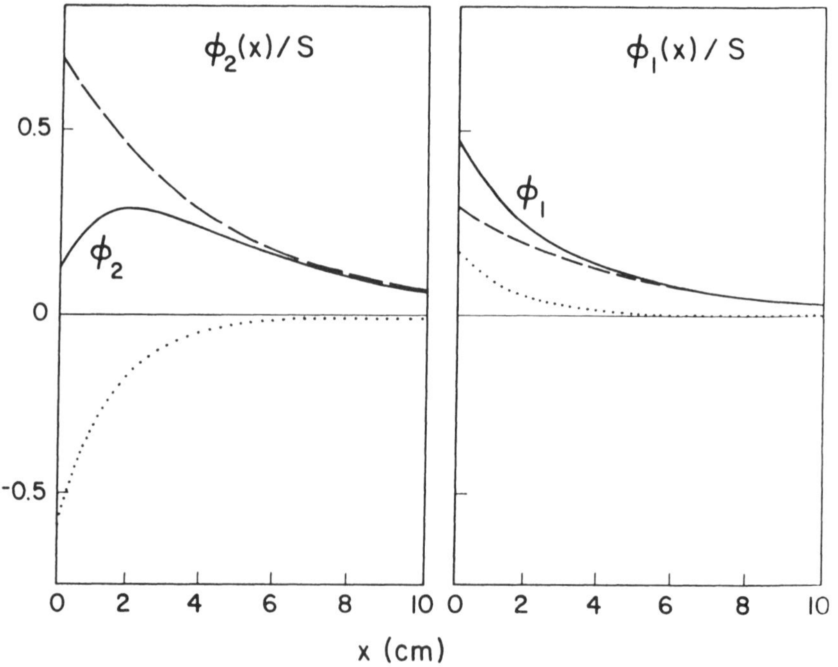

Fig. 3-11. The fast and thermal flux distributions for a semi-infinite medium (solid lines); dashed lines—fundamental mode terms; dotted lines—higher mode terms.

nential attenuation. This behavior displays the characteristics of the flux entering a reflector as indicated in Sec. 7-4D, Fig. 7-2. The analytical treatment of a semi-infinite medium is of course much simpler than the finite two-region problem depicted in Fig. 7-2. It therefore allows a clear understanding of this buildup feature in terms of the analytical two-group solution.

3-7 Summarizing Remarks

In this chapter, the spatial nature of the neutron flux has been explored. In part of the derivations, the neutron flux was assumed to be separable in space and energy. This basic assumption allowed the space-dependent component of the flux to be broken up for separate examination of the space- and energy-dependent components. Although these spatial neutron distributions are valid only for a separable reactor physics model, they do approximate the neutron flux shape in nonseparable reactor situations, especially within larger regions.

It was shown that, within the context of the separability assumption and also assuming a homogeneous material distribution, analytical expressions for the neutron flux shape can be derived for several simple reactor cores and neutron source-sink configurations. It should be pointed out, however, and stressed again, that these results are based on highly idealized conditions, which are never attained in real reactors. First, the actual neutron flux shape is distorted due to various effects such as burnup, control rods, or temperature effects both in the moderator as well as in the fuel. In some reactors (i.e., boiling water) there also exists a great difference in the density of the moderator due to a phase change itself (depicted by a variation in void fraction axially as well as radially across the core). Control rod patterns and uneven fission product buildup (e.g., xenon) affects the spatial distribution of the flux in subtle ways that are not described in these simple reactor models. Furthermore, the buildup of fission product poisons and burnup of the fuel changes the physical properties unevenly within the core. Therefore, the reactor no longer consists of a single homogeneous material. Similarly, the effects of the reflector or blanket on the neutron behavior have only been crudely treated with the single-group model. The above effects can only be treated accurately with multigroup reactor theory, which is presented in Chapter 7.

In addition to the overall spatial behavior of the neutron flux, several reactor physics concepts were explored in more depth in this chapter. The role that the concept of buckling (presented initially in Chapter 2) plays in establishing reactor criticality was made more concrete with the determination of the geometric buckling for various simple reactor geometries. Also provided with a physical meaning were other important reactor parameters such as the diffusion length and the reflector savings. As was the case with the flux shapes, the analyses were all based on reactor models or geometries composed of uniform and homogeneous materials. The "local" space variations within reactor fuel cells (i.e., fuel rod unit cells) are discussed in Chapter 6.

In the next two chapters the second component of the total neutron flux, namely, the energy dependency, is discussed.

Homework Problems

1. Consider a nonmultiplying slab that is four diffusion lengths thick (infinite in the other two directions) with a uniform infinite plane neutron source S neutron/cm^{2}s located one diffusion length from one side and three diffusion lengths from the other side. Develop an expression for the flux distribution across the slab.
2. If the slab in problem 1 is H_2O ($D = 0.16$ cm, $\Sigma_a = 0.02$ cm^{-1}),

what fraction of the source neutrons escapes from the sides? Does this ratio change if the source is placed on the centerline of the slab?

3. Derive the Table 3-I peak-to-average flux ratios for bare critical reactors in the shape of:
   a. a finite cylinder of minimum volume
   b. a cube.
4. Show that the fraction of neutrons that leaks from the surface of a rectangular parallelepiped reactor to those lost due to leakage *and* absorption is

$$L^2B^2/(1+L^2B^2) \quad .$$

5. Consider a finite slab of a nonmultiplying medium of thickness H (infinite in the other two directions) containing a uniform plane source of neutrons S neutron/cm^{2}s located on its left side at $x = 0$.
   a. Give the boundary conditions for this asymmetrical problem.
   b. Show that the flux is given by:

$$\phi(x) = \frac{SL}{2D} \frac{\sinh\left(\dfrac{H-x}{L}\right)}{\cosh(H/L)} \quad .$$

   c. Compare the flux expression obtained in part (a) with that for an infinite slab (Eq. 3.83) by plotting the relative flux versus the distance from the source measured in diffusion lengths. Consider the effect for several slab thicknesses expressed in diffusion lengths.
   d. What can one conclude concerning the representation of a finite slab flux distribution with the infinite slab equation?
6. Develop an expression for the smallest radius of a cyclindrical reactor that will just sustain criticality for a given fuel-to-moderator mole ratio (thus B_m^2).
7. Prove that the dimensions of the rectangular parallelpiped for a given value of B^2 are those of a cube when the minimum critical volume is achieved. What is the total power of a cubical reactor 4 m on a side containing 80 kg of uniformly distributed ^{235}U if the maximum thermal neutron flux is 5×10^{13} neutron/cm^{2}s? Use $\sigma_f = 575$ b for ^{235}U.
8. Determine the reduction in the critical dimensions of a slab reactor consisting of a uniform mixture of uranium and graphite if a 0.5-m-thick reflector is placed on both faces. The material buckling of the bare core is 2×10^{-4}cm^{-2}. For both the core and the reflector, use the diffusion coefficient for graphite ($D_G = 0.85$ cm).

9. Show that the criticality relationship for a reflected spherical reactor (core radius R and reflector thickness T) is given by:

$$\cot RB_\lambda = \frac{1}{RB_\lambda}\left(1 - \frac{D_r}{D_c}\right) - \frac{D_r\kappa_r}{D_cB_\lambda}\coth \kappa\delta R \quad .$$

10. Using the criticality expression developed in problem 9, estimate the effect on the critical size of a spherical core due to the presence of a 0.5-m-thick reflector ($k_{\text{eff}} = 1$ with reflector). Use the material and buckling conditions specified in problem 8 for this reactor comparison.
11. For the reflected core of problem 10, determine the peak-to-average flux ratio and compare it to that for a bare spherical reactor.

Review Questions

Note: The questions for this chapter are concerned merely with the space-dependent problem as it results either from separation or from integration (over E).

1. Give the boundary conditions at the extrapolated boundaries for one-dimensional slab, sphere, and cylindrical geometries.

2a. Give the balance equation (in its wave equation form) for one-dimensional planar, cylindrical, and spherical geometries.

b. Give the general form of the ∇^2 operator as used in computer programs for these three geometries.

3a. Give the fundamental solutions (flux and geometrical) buckling for the one-dimensional, one-region slab, cylinder, and sphere.

b. Discuss in particular the elimination of the respective second solution.

4. Give the wave equation for the finite cylinder and the parallelepiped.

5. Derive the total flux solutions for the finite cylinder and the parallelepiped.

6a. Order the three geometries, cube, sphere, and finite cylinder, with respect to decreasing volume for given leakage; discuss the result.

b. What is the approximate value of H/D for the finite cylinder with the highest criticality for a given composition; explain the result qualitatively.

7a. Give the balance equation that determines the material buckling for a medium without fissionable material (in energy-dependent diffusion theory).

b. Derive the one-group solution of that equation.

8a. Give the balance equation of the total flux and the boundary conditions in an infinite medium without fissionable material with a point source.
b. Give the solution of that problem.
c. Discuss the behavior of the solution near zero and asymptotically; compare it with the solution in a vacuum.
d. Check the overall neutron balance.
e. Discuss how the analytical solution would differ for the finite sphere.

9a. Give the balance equation of the total flux and the boundary conditions for a plane source in an infinite medium without fissionable material.
b. Give the solution of that problem.
c. Discuss the solution near the plane; why is it not singular as is the solution near the point source?
d. Discuss the asymptotic solution; compare it with the solution in a vacuum.
e. Check the overall neutron balance.
f. Discuss the difference of the solution for infinite and finite media.

10. Same as question 9 for a medium with fissile material but $B_m^2 < 0$.

11a. Give the balance equation of the total flux in a finite slab with a center plane source for a medium with $B_m^2 > 0$.
b. Give the maximum size of the medium in which you can establish a physical static solution (B_m^2 = given).
c. Give the general analytical form of the solution, including the typical coefficients.
d. Discuss the behavior of the two coefficients if the size of the slab is increased toward criticality.

12. Describe the use of the source multiplication factor during an initial reactor loading operation.

13. Describe, semiquantitatively, the procedure to measure B_m^2 in a parallelepiped exponential pile. Disregard boundary effects at the boundary opposite the source, for simplicity, but include the discussion of the flux separation.

14. What are the advantages of placing a reflector around an otherwise bare one-region core?

15a. What are the effects of a reflector for enrichment or core size?
b. In what form are the core-size effects expressed conceptually?

16a. Describe the procedure to find the one-group flux in a reflected system.
b. What type of equation does one get for the eigenvalue λ?

17. What is the basic condition for determining the reflector savings in a multigroup case?

18. Suppose for a one-dimensional slab problem that k_{eff} is calculated to be 1.01. Remove a symmetrical half and replace it by an ideal neutron-reflecting mirror (i.e., $d\phi/dz = 0$ at the cut). Estimate qualitatively the resulting change in k_{eff}. Justify your answer.
19. Describe the procedure for the analytical solution of a two-group source problem in slab geometry. Consider both $B_m^2>0$ and $B_m^2<0$.

REFERENCES

1. J. R. Lamarsh, *Introduction to Nuclear Reactor Theory*, Chapter 9, Addison-Wesley Publishing Co., Reading, Massachusetts (1966).
2. S. Glasstone and M. C. Edlund, *The Elements of Nuclear Reactor Theory*, Chapter 7, D. Van Nostrand Co., Princeton, New Jersey (1952).
3. M. Abramowitz and I. A. Stegun, Eds., *Handbook of Mathematical Functions*, Applied Math Series 55, National Bureau of Standards (1970).
4. F. G. Krauss, K. O. Ott, and F. M. Clikeman, "The Conceptual Design of a Fast Subcritical Blanket Facility," *Nucl. Technol.*, **25,** 429 (1975).
5. F. M. Clikeman, Ed., "Fast Breeder Blanket Facility," NP-1657, Electric Power Research Institute (Jan. 1981).

Four

NEUTRON ENERGY DEPENDENCY DURING SLOWING DOWN

4-1 Introduction

Neutron energy generally changes when a neutron collides with a nucleus. Therefore, the neutron flux depends on space *and* on energy. The general balance equations that govern the simultaneous dependence of the flux on space and energy were derived and discussed in Chapter 2. In large regions, the space and energy dependence of the flux can be approximately separated; the separation holds exactly in bare cores as well as asymptotically in very large regions. Employing this special situation, the balance relationship for the separated flux consists of two equations, one for the spatial and one for the energy dependence, Eqs. (2.137) and (2.138).

The pure space dependence of the flux was treated in Chapter 3. In simple geometries, analytic solutions of the spatial balance equation were obtained. Even if the flux is not separable, the results of Chapter 3 give a semiquantitative indication of the one-group flux, i.e., of the flux integral over all energies or only over the thermal energy range.

This chapter is devoted to the treatment of the energy dependence of the flux, separated from its space dependence. Analytic solutions for the neutron energy dependence, i.e., for the spectrum, can be obtained if the neutron cross sections are sufficiently simplified. In this chapter, the neutron cross sections are assumed to be smooth functions of energy. The resonances that are superimposed on the smooth cross sections are separately treated in Chapter 5.

The results of Chapters 4 and 5 describe the spectrum accurately for areas in which the flux is separable in space and energy; but they hold approximately if the flux is not exactly separable, in full analogy to the space dependencies derived in Chapter 3. However, in contrast to the separated space dependencies, which are not practically applied if the separability is violated, the separated energy spectra are generally applied to generate the average group cross sections for the simultaneous treatment of space and energy dependence (see Chapter 7). This emphasizes the importance of understanding neutron spectra and their practical calculation.

The basis for the calculation of neutron spectra is the nuclear reactions in which the neutrons lose or gain energy, i.e., the scattering processes (see Sec. 4-2). As long as neutrons only *lose* energy, they are "slowed down." Neutrons will lose energy if their energy is much larger than the thermal motion energy of the scattering nuclei. One therefore calls the energy range above the thermal energies the "slowing down energy range," and the neutron spectrum as it results from energy losses the "slowing down spectrum."

The simple *numerical* method by which the slowing down spectrum may be calculated directly is described first (Sec. 4-3). No analytic representation of the scattering cross sections is needed for the numerical solution procedure.

The high energy end of the neutron spectrum in reactors is of importance for "fast fission," particularly in ^{238}U and ^{232}Th. Inelastic scattering is the dominating mechanism for energy loss in this high-energy range (Sec. 4-4).

At medium energies, elastic scattering dominates. The simple analytical form of the elastic cross section (compare Sec. 4-2A) allows an instructive analytical treatment of energy dependence in this domain (compare Secs. 4-5 and 4-6). To gain an in-depth understanding of the energy dependency of the neutron flux, the neutron spectra are calculated for a sequence of models with increasing complexity. In this way, the effect of individual physical properties of the system becomes fully transparent.

The decreasing neutron energy eventually becomes comparable to the energy of the thermal motion of the nuclei in the scattering medium. Then, neutrons may gain as well as lose energy in collisions. This largely terminates further slowing down. The result is a neutron spectrum that is primarily determined by the energy distribution of the thermal motion (see Sec. 4-7).

Figure 4-1 illustrates the sequence of phenomena and types of scattering processes along the energy axis. Also indicated is the basic structure of capture and fission cross sections as it prevails in the various energy ranges. At low energies, these two cross sections have an inverse velocity dependence ($1/v$ cross section); at higher energies, one has resonance cross sections, which have been identified in Fig. 4-1 although their treatment is deferred until Chapter 5.

Energy	1 eV	1 keV	1 MeV
Energy Range Character-ization	Thermalization	Slowing down	Source
Type of Reaction Scattering	Up and down scattering	Down scattering CMS isotropic elastic scattering	Inelastic scattering; elastic scattering with forward bias
Capture	$1/v$	Resonances	smooth cross sections
Fission (fissile)	$1/v$	Resonances	smooth cross sections
Fission (fissionable)	0	0	Threshold cross section

Fig. 4-1. Illustration of the major physical phenomena.

4-2 Scattering Processes

4-2A Elastic Scattering at Epithermal and High-Neutron Energies

Elastic collisions between neutrons and nuclei result merely in an exchange of energy and momentum. Collisions in which the nuclei themselves absorb energy are inelastic events, for example, inelastic scattering as explained in Sec. 4-2C. Inelastic scattering is possible only above a certain threshold energy, which corresponds to the excitation energy of the first excited state in the target nucleus (e.g., 45 keV in ^{238}U). Thus, in the major part of the logarithmic energy range, only elastic scattering needs to be considered.

Thermal reactors are designed to slow neutrons down into the thermal energy range. This requires emphasis on neutron slowing down through elastic scattering, which is practically achieved by putting a proper fraction of light materials with small capture cross sections (moderators) into the core. Typical moderators are H_2O, D_2O, and graphite. The neutron physical properties of these moderators are discussed later in this section.

The elastic exchange of energy and momentum is governed by the conservation laws of energy and momentum: *Since no inelastic "deformations" of either scattering partner are involved, the total kinetic energy must be the same before and after the collision. Furthermore, the total momentum is conserved in the collision.* In the center-of-mass coordinate system (CMS), one has by definition zero total momentum (before and after the collision). For detailed treatment of elastic collisions, the reader is referred to standard physics textbooks[1,2] or advanced neutron slowing down theory treatises.[3] Here only brief derivations are presented, and the results are reviewed, illustrated, and discussed.

Figure 4-2 illustrates an elastic collision in the CMS, as well as in the laboratory system (LS). All quantities in the CMS are distinguished

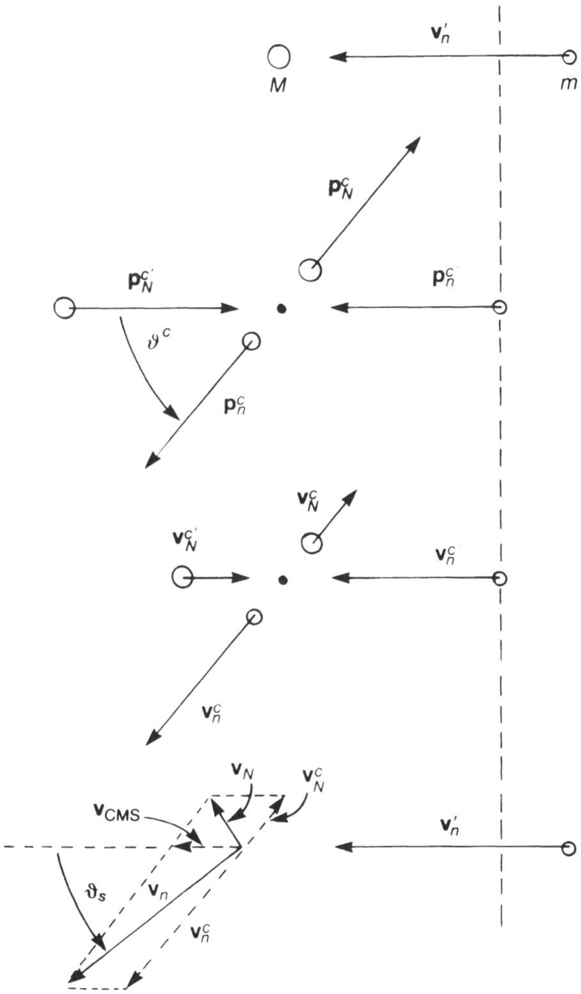

Fig. 4-2. Vector diagrams of elastic scattering in the LS and CMS.

from the corresponding quantities in the LS by a superscript c. The initial quantities are indicated by a prime. The nucleus (mass M) is initially at rest in the LS; the neutron (mass m) is moving with the velocity $\boldsymbol{v}_n'$ toward the nucleus (first line in Fig. 4-2). The momentum vectors in the CMS, before and after the collision, are shown in the second diagram of Fig. 4-2. Note that the momentum vectors of the neutron, $\boldsymbol{p}_n^c$, and the nucleus, $\boldsymbol{p}_N^c$, are of equal length and of opposite direction in the CMS before and after the collision, i.e.,

$$\boldsymbol{p}_n^{c'} = -\boldsymbol{p}_N^{c'} \text{ and } \boldsymbol{p}_n^c = -\boldsymbol{p}_N^c \text{ in the CMS} \quad . \tag{4.1a}$$

The angle ϑ^c gives the angular deflection of the neutron in the CMS. For energies below ~50 keV, the angular distribution in the CMS is essentially isotropic, i.e., neutrons are scattered with the same probability to any value of ϑ^c. This is called "s-wave scattering."[a]

The conservation of the energy requires that the absolute values of the momenta also remain unchanged in an elastic collision in the CMS, i.e.,

$$p_n^{c'} = p_n^c = p_N^{c'} = p_N^c \text{ in the CMS} \quad . \tag{4.1b}$$

(In inelastic scattering, energy and momenta would be reduced.)

The third part of Fig. 4-2 shows the corresponding velocity vectors in the CMS; the velocity vector of the CMS itself, $\boldsymbol{v}_{\text{CMS}}$, equals $-\boldsymbol{v}_N^{c'}$. Since only the neutron was in motion prior to the collision, its momentum $p_n' = mv_n'$ determines the momentum of the CMS and v_{CMS} as the average velocity:

$$M_{\text{CMS}}v_{\text{CMS}} = (A+1)mv_{\text{CMS}} = mv_n' \quad ; \tag{4.2a}$$

the mass of the nucleus, M, is commonly approximated by Am in elastic scattering kinematics, with A being its atomic weight. Thus,

$$v_{\text{CMS}} = \frac{1}{A+1}v_n' \quad . \tag{4.2b}$$

The magnitude of the nucleus velocity in the CMS, $v_N^{c'}$, must be equal to v_{CMS}, since the transformation back to the LS (performed by adding $\boldsymbol{v}_{\text{CMS}}$) must yield a nucleus at rest:

$$\boldsymbol{v}_N^{c'} + \boldsymbol{v}_{\text{CMS}} = 0 \quad . \tag{4.3a}$$

Therefore,

---

[a]See Sec. 5-2 and Appendix B.

$$v_{\mathrm{CMS}} = v_N^{c'} = \frac{1}{A+1} v_n' \quad . \tag{4.3b}$$

The magnitudes of the neutron and nucleus velocities in the CMS are unchanged in the scattering process due to the conservation of the momentum, as expressed in Eq. (4.1b):

$$v_n^c = v_n^{c'} \text{ and } v_N^c = v_N^{c'} \quad ; \tag{4.4a}$$

or due to taking the squares of the respective velocities:

$$(v_n^c)^2 = (v_n^{c'})^2 \text{ and } (v_N^c)^2 = (v_N^{c'})^2 \quad . \tag{4.4b}$$

For the velocities in the CMS before and after the collision, Eqs. (4.1b), (4.2b), and (4.3a) yield the following relations:

$$v_n^c = \frac{A}{A+1} v_n' = v_n^{c'} \tag{4.5a}$$

and

$$v_N^c = \frac{1}{A+1} v_n' = v_N^{c'} \quad . \tag{4.5b}$$

The vector addition of the CMS velocity, $\boldsymbol{v}_{\mathrm{CMS}}$, to the velocities of the particles in the CMS gives the corresponding velocities in the LS system. This is depicted in the lower part of Fig. 4-2. The transformation from the velocity diagram in the CMS into the corresponding LS diagram is indicated by the vector parallelogram for the addition of $\boldsymbol{v}_{\mathrm{CMS}}$ (dotted auxiliary lines). The neutron velocity is smaller than the initial velocity since, after the collision, the velocity vector $\boldsymbol{v}_{\mathrm{CMS}}$ is added at an angle ϑ^c, whereas before the collision, it is added parallel to $\boldsymbol{v}_n^{c'}$. The nucleus has taken over part of the energy as "recoil" energy. The angular deflection of the neutron in the LS is denoted by ϑ_s, and the cosine of the scattering angle is denoted by μ_s:

$$\mu_s = \cos\vartheta_s = \text{cosine of the scattering angle, the angular deflection in the LS.} \tag{4.6}$$

The completion of the graphic transformation in the lower part of Fig. 4-2 for the nucleus shows qualitatively that the nucleus is always scattered into the forward part of the solid angle (as seen from the initial direction of the neutron). The neutron, however, may also be scattered backward in the LS: This occurs when the neutron is scattered so far backward in the CMS system that the transformation from the CMS into the LS leaves the neutron velocity vector in the domain $\vartheta_s > 90$ deg.

If the scattering nuclei are very heavy ($v_{\mathrm{CMS}} << v_n^{c'}$), an isotropic angular distribution in the CMS is reflected as a nearly isotropic angular distribution in the LS. For light nuclei, however, forward scattering is strongly pronounced compared to backward scattering, since v_{CMS} is more comparable with v_n^c. Then, the angular distribution in the LS has a significant forward bias. For scattering of neutrons on protons, one has $v_N^c = v_n^c$ and the backward scattering disappears completely. Thus, the scattering of neutrons on hydrogen nuclei always has, in the LS, a strong forward bias (no neutrons can be scattered backwards), although the scattering in the CMS is isotropic (s-wave scattering).

The derivation described above is reviewed in Table 4-I. It begins with the description of the incoming particles and proceeds to the transformation into the CMS, *subject to the conservation of the momentum.* The third column describes the particles as they leave the scattering process (out), where the *conservation of energy* in this elastic process determines the velocities and thus the momenta (the relative direction of the respective vectors is still determined by the conservation of momentum). The final column indicates the transformation back into the LS, yielding the relations between angles and energies, which are further discussed in the remainder of this section.

The mathematical investigation of the elastic scattering, which consists essentially of the transformation of the simple scattering in the CMS into the LS, yields the important relationships between energies and scattering angles as well as the angular distribution of the scattered

TABLE 4-I
Review of the Elastic Collision Derivations

LS	CMS		LS
In	In[a]	Out[b]	Out[a]
v_n'	$v_n^{c'} = \frac{A}{A+1} v_n'$	$v_n^c = v_n^{c'}$	$\boldsymbol{v}_n = \boldsymbol{v}_n^c + \boldsymbol{v}_{CMS}$
$v_N' = 0$	$v_N^{c'} = v_{CMS} = \frac{1}{A+1} v_n'$	$v_N^c = v_N^{c'}$	$\boldsymbol{v}_N = \boldsymbol{v}_N^c + \boldsymbol{v}_{CMS}$
$p_n' = mv_n'$	$p_n^{c'} = p_N^{c'} = mv_n^{c'}$	$p_n^c = p_n^{c'}$	$\boldsymbol{p}_n = m\boldsymbol{v}_n$
$p_N' = mAv_N' = 0$	$p_N^{c'} = Amv_N^{c'} = \frac{A}{A+1} mv_n'$	$p_N^c = p_N^{c'}$	$\boldsymbol{p}_N = mA\boldsymbol{v}_N$
$v_{CMS}' = \frac{1}{A+1} v_n'$	$v_{CMS}^{c'} = 0$	$v_{CMS}^c = 0$	$v_{CMS} = v_{CMS}'$

[a]From conservation of momentum.
[b]From conservation of energy.

neutrons. The vector addition depicted in the lower parts of Fig. 4-2 yields the desired relations:

$$\boldsymbol{v}_n = \boldsymbol{v}_n^c + \boldsymbol{v}_{\text{CMS}} \quad . \tag{4.7}$$

The square of Eq. (4.7) gives a relation between the angle ϑ^c and the velocities or energies:

$$v_n^2 = (v_n^c)^2 + 2v_n^c v_{\text{CMS}} \cos\vartheta^c + v_{\text{CMS}}^2 \quad . \tag{4.8}$$

By dividing Eq. (4.8) with $(v_n')^2$, expressing v_n^c by v_n' as in Eq. (4.5a), and v_{CMS} by v_n' as in Eq. (4.2b), and then converting the velocities to energies, one obtains for E/E':

$$\frac{E}{E'} = \frac{A^2 + 2A\mu^c + 1}{(A+1)^2} \quad . \tag{4.9}$$

Solving Eq. (4.9) for $\mu^c = \cos\vartheta^c$ gives

$$\mu^c = \frac{1}{2A}\left[(A+1)^2\frac{E}{E'} - A^2 - 1\right] \quad . \tag{4.10}$$

Solving Eq. (4.7) for $\boldsymbol{v}_n^c$ and then taking the square gives the corresponding relations with $\mu_s = \cos\vartheta_s$, since ϑ_s is the angle between $\boldsymbol{v}_n$ and $\boldsymbol{v}_{\text{CMS}}$:

$$\boldsymbol{v}_n^c = \boldsymbol{v}_n - \boldsymbol{v}_{\text{CMS}} \quad , \tag{4.11}$$

with the square

$$(v_n^c)^2 = v_n^2 - 2v_n v_{\text{CMS}}\mu_s + v_{\text{CMS}}^2 \quad . \tag{4.12}$$

Again, application of Eqs. (4.3) and (4.5), and solving for μ_s in terms of E/E' yields:

$$\mu_s = \cos\vartheta_s = \frac{1}{2}\left[(A+1)\sqrt{\frac{E}{E'}} - (A-1)\sqrt{\frac{E'}{E}}\right] \tag{4.13}$$

and

$$\sqrt{\frac{E}{E'}} = \frac{(A^2 - 1 + \mu_s^2)^{1/2} + \mu_s}{A+1} \quad . \tag{4.14}$$

Inserting Eq. (4.14) into Eq. (4.9) allows one to relate the cosine of the two scattering angles:

$$\mu^c = \frac{1}{A}[\mu_s^2 + \mu_s(\mu_s^2 + A^2 - 1)^{1/2} - 1] \quad . \tag{4.15}$$

The angular distribution of the scattered neutrons can also be obtained by the same approach, namely, by first finding the simple angular distribution in the CMS, $p^c(\mu^c)$, and transforming it into the LS. The term $p^c(\mu^c)$ is constant, reflecting the isotropy of the s-wave scattering (see Sec. 5-2 and Appendix B):

$$p^c(\mu^c) = \frac{1}{2} \quad , \tag{4.16}$$

with

$$\int_{-1}^{1} p^c(\mu^c)\, d\mu^c = 1 \quad . \tag{4.17}$$

The transformation into the LS must conserve the integral:

$$\int_{-1}^{1} p^c(\mu^c)\, d\mu^c = \int_{-1}^{1} p^c[\mu^c(\mu_s)] \frac{d\mu^c}{d\mu_s}\, d\mu_s$$
$$= \int_{-1}^{1} p(\mu_s)\, d\mu_s \quad . \tag{4.18}$$

Thus,

$$p(\mu_s) = p^c[\mu^c(\mu_s)] \frac{d\mu^c}{d\mu_s} = \frac{1}{2} \frac{d\mu^c}{d\mu_s} \quad . \tag{4.19}$$

Differentiating Eq. (4.15) gives the angular distribution of elastic scattering in the LS:

$$p(\mu_s) = \frac{1}{2A}\left[\frac{A^2 - 1 + 2\mu_s^2}{(A^2 - 1 + \mu_s^2)^{1/2}} + 2\mu_s\right] \quad . \tag{4.20}$$

Figure 4-3 shows the angular distribution of elastically scattered neutrons for various A in the form of a polar diagram. The distance between the scattering center and a point on the curves represents the probability for the corresponding angular deflection. It appears that the scattering on heavy isotopes (e.g., ^{238}U) is practically isotropic in the LS; the very small velocity, v_{CMS}, hardly changes the angular distribution in the transformation from the CMS to the LS. For light nuclei however, the addition of v_{CMS} to the isotropically distributed vectors, $v_n^{c'}$, tends to concentrate the velocity vectors in the forward direction, increasingly so for lighter nuclei. The resulting forward bias in the angular distribution is very pronounced. For scattering on protons, the backward scattering is eliminated completely (as in the collision of two billiard balls).

Figure 4-4 depicts the same angular distributions as in Fig. 4-3; in this case, in a regular diagram for $p(\mu_s)$ as functions of μ_s.

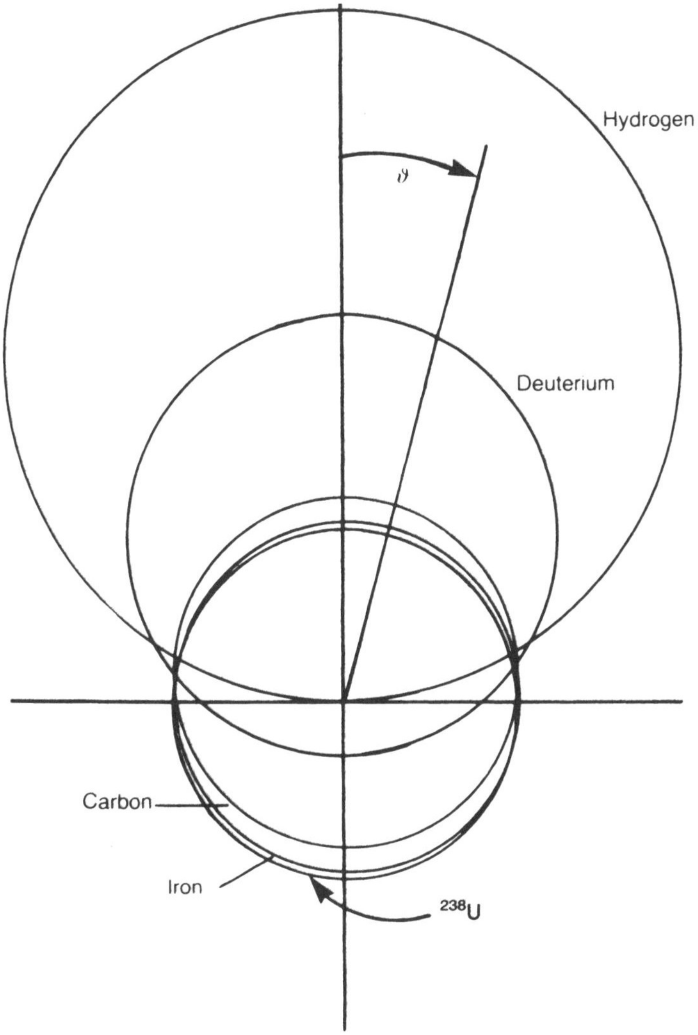

Fig. 4-3. Polar diagram of the angular distribution in CMS isotropic scattering for hydrogen, deuterium, carbon, iron, and ^{238}U.

At energies above ~100 keV, the angular distribution of elastic scattering already shows a forward bias in the CMS (*p*-scattering), which increases with increasing energy. This leads to a further forward bias in the angular distribution in the LS, which is superimposed on the forward bias depicted in Figs. 4-3 and 4-4 (see Appendix B for an explanation of the terminology *s*- and *p*-scattering).

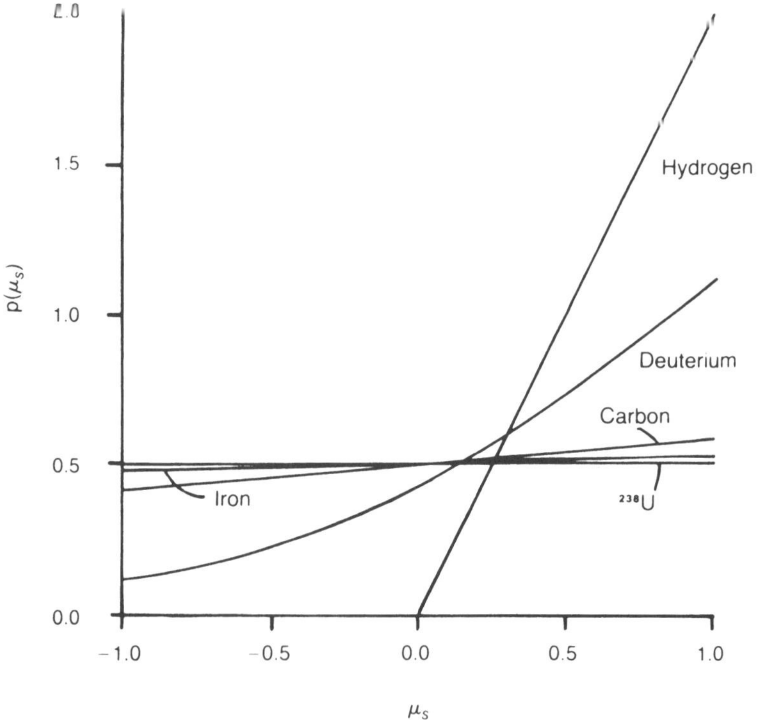

Fig. 4-4. The angular distribution in CMS isotropic scattering for hydrogen, deuterium, carbon, iron, and ^{238}U.

Based on the angular distribution, Eq. (4.20), one can readily calculate the important average cosine, $\overline{\mu}_s$:

$$\overline{\mu}_s = \int_{-1}^{1} \mu_s p(\mu_s)\, d\mu_s \quad . \tag{4.21}$$

Since all terms in Eq. (4.20) except the last one in brackets are symmetrical about $\mu_s = 0$, their contribution to the integral Eq. (4.21) cancels. The last term in Eq. (4.20) gives the important result:

$$\overline{\mu}_s = \frac{1}{A}\int_{-1}^{1} \mu_s^2\, d\mu_s = \frac{2}{3A} \tag{4.22}$$

for CMS isotropic elastic scattering.

The angular deflection is directly related to the velocities and, thus, the energies before, E', and after, E, the collision are as expressed by Eq. (4.13), which yields:

$$\frac{E}{E'} = \frac{[(A^2 - 1 + \mu_s^2)^{1/2} + \mu_s]^2}{(A + 1)^2} \tag{4.23}$$

for E/E' as a function of μ_s.

The minimum energy loss, which is obviously zero ($E = E'$), is associated with zero angular deflection ($\mu_s = 1$). The maximum energy loss, i.e., the minimum E/E' ratio, occurs for a central collision in which a neutron is scattered back ($\mu_s = -1$). Equation (4.23) gives

$$\left(\frac{E}{E'}\right)_{\min} = \left(\frac{A-1}{A+1}\right)^2 = \alpha \quad . \tag{4.24}$$

The ratio in Eq. (4.24) is commonly denoted by α. Thus, the post-scattering energy E appears somewhere in the interval between E' and $\alpha E'$:

$$E_{\min} = \alpha E' \leq E \leq E' = E_{\max} \quad . \tag{4.25}$$

For $A = 1$, true backscattering cannot occur even for $\mu_s = -1$.

The probability for the neutron to assume a special energy E in this interval is called the "elastic scattering kernel," $K(E' \rightarrow E)$. As a probability distribution, $K(E' \rightarrow E)$ is normalized to 1:

$$\int_{E_{\min}}^{E'} K(E' \rightarrow E)\, dE = 1 \quad . \tag{4.26}$$

If the elastic scattering is isotropic in the CMS, then the neutrons appear with equal probability at any energy between E' and $E_{\min}$; i.e., $K(E' \rightarrow E)$ is independent of E; it can be calculated from the isotropic angular distribution, Eq. (4.16), in the same way as $p(\mu_s)$ in Eq. (4.19):

$$\int_{\alpha E'}^{E'} K(E' \rightarrow E)\, dE = \int_{-1}^{1} p^c[\mu^c(E',E)] \frac{d\mu^c}{dE}\, dE \quad . \tag{4.27}$$

By differentiating Eq. (4.10), one finds that a CMS isotropic angular distribution leads to the following scattering kernel:

$$K(E' \rightarrow E) = \begin{cases} \dfrac{1}{(1-\alpha)E'} & \text{for } \alpha E' \leq E \leq E' \\ 0 & \text{for } \begin{cases} E > E' \\ E < \alpha E' \end{cases} \end{cases} \tag{4.28}$$

The integral of $K(E' \rightarrow E)$ with respect to E is equal to one, as required by Eq. (4.26).

The differential scattering cross section $\sigma_s(E' \rightarrow E)$ is related to $\sigma_s(E')$ and the scattering kernel by

$$\sigma_s(E' \rightarrow E) = \sigma_s(E')K(E' \rightarrow E) \quad . \tag{4.29}$$

The scattering kernel can be employed to find the average energy after the collision or the average energy loss:

$$\overline{(E' - E)} = \int_{\alpha E'}^{E'} \frac{(E' - E)\, dE}{(1 - \alpha)E'} = \frac{1}{2}(1 - \alpha)E' \quad ; \tag{4.30}$$

i.e., the average energy loss is half of the maximum loss. Two important dependencies become evident from Eq. (4.30):

1. The average energy loss increases linearly with increasing initial energy. For example, if a neutron loses 2 eV in an average collision at 10 eV, it loses 2000 eV at 10 keV. The fractional energy change, however, is constant:

$$\frac{\overline{(E' - E)}}{E'} = \frac{1 - \alpha}{2}, \text{ independent of } E' \quad . \tag{4.31}$$

2. The average energy loss decreases with increasing atomic weight, A, since it is proportional to $1 - \alpha$:

$$\overline{(E' - E)} = \frac{2A}{(A + 1)^2} E' \quad . \tag{4.32}$$

   For example, the average loss is 50% of the original energy for $A = 1$, but only ~1% for $A = 238$.

Both observations have methodological implications. The smallness of the energy loss in elastic scattering on heavy elements can be used to greatly simplify the calculation of the neutron spectrum in the presence of heavy nuclei. The first observation, the equality of the energy losses in terms of relative energies, suggests the use of some relative energy in the analytical treatment of neutron slowing down.

4-2B The Lethargy

The relative average energy loss during CMS isotropic elastic scattering is independent of the initial energy [see Eq. (4.31)]. This suggests the use of some form of relative energy in the theoretical treatment of neutron slowing down. The so-called "lethargy," u, has been introduced for this purpose:

$$\text{lethargy} \equiv u = \ln \frac{E_0}{E} \quad , \tag{4.33}$$

with E_0 being an arbitrary "base" energy. In fission reactors, E_0 is usually chosen as 10 MeV, since the number of neutrons above 10 MeV is negligibly small [$\chi(E)$ is then normalized to unity below E_0]. One then

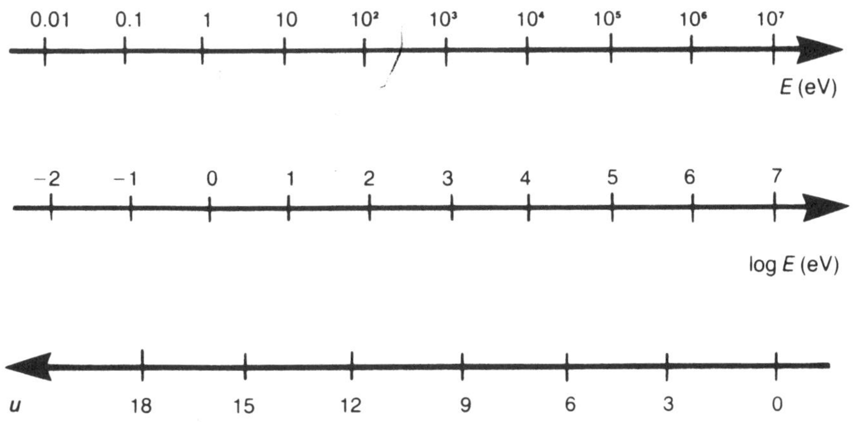

Fig. 4-5. Comparison of energy and lethargy scales.

has a positive lethargy over the entire energy range of interest; the neutron lethargy u is zero at $E = E_0$ and increases as the neutrons are slowed down. Near thermal energies, u is ~18; it would increase toward infinity if E would approach zero. The lethargy scale and the common logarithmic energy scale are depicted in Fig. 4-5.

Inversion of Eq. (4.33) gives E as a function of u:

$$E = E_0 e^{-u} \quad . \tag{4.34}$$

Instead of the average change in energy, the average change in lethargy, denoted by ξ, is now considered. One expects the average lethargy change to be constant (i.e., independent of E') since it is a function of E'/E:

$$\xi = \int_{\alpha E'}^{E'} \left(\ln\frac{E_0}{E} - \ln\frac{E_0}{E'} \right) K(E' \to E)\, dE \quad , \tag{4.35a}$$

$$\xi = \int_{\alpha E'}^{E'} \ln\frac{E'}{E} K(E' \to E)\, dE \quad , \tag{4.35b}$$

and

$$\xi = 1 + \frac{\alpha}{1 - \alpha} \ln\alpha = 1 - \frac{\alpha}{1 - \alpha} \ln\frac{1}{\alpha} \quad . \tag{4.36}$$

By using lethargy as the variable, the slowing down process is more evenly spread out over the domain of the independent variable than when energy is used as the variable. The slowing down process occurs as a sequence of jumps along the lethargy scale; their average is independent of energy.

Generally, the average energy change that corresponds to ξ is larger than the average relative energy loss. This is due to the presence of $\ln(E'/E)$ in the integrand of Eq. (4.35b), which places more weight on E values farther away from E'. But, it is ξ, rather than the average energy loss, which plays an important role in slowing down theory (see Sec. 4-6).

From Eqs. (4.34) and (4.36), it follows that $\xi = 1$ for hydrogen, which gives a larger average loss than Eq. (4.32). The difference is smaller for larger A than for $A = 1$. In the limit of $A \gg 1$, one can readily derive approximate formulas for both quantities:

$$\xi = \frac{2}{A + \frac{2}{3}} \quad \text{for } A \gg 1 \tag{4.37}$$

and

$$\frac{\overline{(E' - E)}}{E'} = \frac{2}{A + 2} \quad \text{for } A \gg 1 \quad . \tag{4.38}$$

Table 4-II gives values of α and ξ for some reactor materials.

It is common to use the same symbols for both the functions of energy and the corresponding functions of lethargy even though the functional dependencies are different. In the derivation of the relation of these two functions, a different notation is introduced temporarily: The functions of lethargy are marked by a tilde, ˜.

TABLE 4-II

The Values of α, ξ, and $\ln \frac{1}{\alpha}$ for Some Reactor Materials

Nuclei	A	α	ξ	$a = \ln \frac{1}{\alpha}$
Hydrogen	1	0	1.000	∞
Deuterium	2	0.111	0.725	2.198
Helium	4	0.360	0.425	1.022
Beryllium	9	0.640	0.209	0.446
Carbon	12	0.716	0.158	0.334
Oxygen	16	0.779	0.120	0.250
Sodium	23	0.840	0.083	0.174
Iron	56	0.931	0.036	0.072
Uranium	238	0.983	0.0084	0.017

Introducing the new independent variable u requires that all functions be transformed from E to u dependencies. In doing this, two different conservation prescriptions must be applied—conservation of functional values and conservation of integrals—depending on the type of quantity to be transformed.

The substitution in cross sections is to be carried out directly, i.e., only the functional values must be conserved; e.g.:

$$\tilde{\sigma}_t(u) = \sigma_t[E(u)] \quad . \tag{4.39a}$$

As a specific example, we consider the substitution of u for E in a $1/v$ cross section:

$$\sigma(E) = \frac{\sigma_0}{[E(\mathrm{eV})]^{1/2}} \quad , \tag{4.39b}$$

with $[\sigma_0 = \sigma(1\ \mathrm{eV})]$. This yields:

$$\sigma[E(u)] = \tilde{\sigma}(u) = \frac{\sigma_0}{[E_0(\mathrm{eV})]^{1/2}} e^{u/2} = \tilde{\sigma}[u(E)] \quad . \tag{4.39c}$$

Apparently, in the transformation from E to u and back, the values of σ are conserved, but the functional form is quite different.

The substitution in the scattering kernel is more complicated since it requires the conservation of the normalization integral, Eq. (4.26):

$$\int_{u_{\min}}^{u_{\max}} \tilde{K}(u' \rightarrow u)\, du = \int_{\alpha E'}^{E'} K(E' \rightarrow E)\, dE = 1 \quad . \tag{4.40}$$

Introducing u' and u into the energy integral, converting it into an integral with respect to u, and interchanging the limits of integration gives:

$$\int_{u_{\min}}^{u_{\max}} -K[E'(u') \rightarrow E(u)] \frac{dE}{du}\, du = 1 \quad , \tag{4.41}$$

with

$$u_{\min} = u(E') \text{ and } u_{\max} = u(\alpha E') \quad . \tag{4.42}$$

By comparing Eqs. (4.41) and (4.40), one obtains

$$\tilde{K}(u' \rightarrow u) = -K[E'(u') \rightarrow E(u)] \frac{dE}{du} \quad , \tag{4.43}$$

with dE/du expressed in terms of u. From Eq. (4.34), it follows that

$$\frac{dE}{du} = -E_0 e^{-u} = -E(u) \quad . \tag{4.44}$$

Thus,

$$\tilde{K}(u' \rightarrow u) = K[E'(u') \rightarrow E(u)]E(u) \quad . \tag{4.45}$$

From the general substitution of Eq. (4.45) into the special scattering kernel of Eq. (4.28), it follows that

$$K(u' \rightarrow u) = \frac{\exp[-(u-u')]}{1-\alpha}, \text{ for } u' \leq u \leq u' + \ln\frac{1}{\alpha} \quad . \tag{4.46}$$

The kernel for CMS isotropic elastic scattering depends only on the difference of the lethargies (K is expressed as a function of the positive lethargy difference $u - u'$):

$$K(u' \rightarrow u) = K(u - u') \quad . \tag{4.47}$$

This fact that the difference of $u - u'$ appears in the kernel plays an important role in slowing down theory.

By means of Eq. (4.45), the definition of ξ can be rewritten as:

$$\xi = \int_{u'}^{u' + \ln\frac{1}{\alpha}} (u - u')K(u - u')\, du \quad . \tag{4.48}$$

The lower energy limit, E', corresponds to the upper lethargy limit:

$$\ln\frac{E_0}{E'\alpha} = u' + \ln\frac{1}{\alpha} \quad . \tag{4.49}$$

The positive quantity $\ln\frac{1}{\alpha}$ appears in the following, frequently in the integration limit. A special abbreviating notation is therefore advised:

$$\ln\frac{1}{\alpha} = \mathrm{a} \quad . \tag{4.50}$$

Denoting $u - u'$ by w brings Eq. (4.48) to the form

$$\xi = \int_0^a wK(w)\, dw \quad . \tag{4.51}$$

In transforming the flux from energy into lethargy, one must also conserve the integral in order to conserve reaction rates:

$$\int_{u_{\min}}^{u_{\max}} \tilde{\Sigma}(u)\tilde{\varphi}(u)\, du = \int_{E_{\min}}^{E_{\max}} \Sigma(E)\varphi(E)\, dE \quad , \tag{4.52}$$

This leads, in the same way as for the scattering kernel, to the following transformation formulas:

$$\tilde{\varphi}(u) = E(u)\varphi[E(u)] \quad ,$$

or in short (with the tilde being omitted):

$$\varphi(u) = E\varphi(E) \quad . \tag{4.53}$$

The same holds for the fission spectrum:

$$\tilde{\chi}(u) = E(u)\chi[E(u)] \quad , \tag{4.54}$$

or in short:

$$\chi(u) = E\chi(E) \quad . \tag{4.55}$$

From these transformations, it follows that a cross section that is constant in energy is also constant in lethargy, but a spectrum $\varphi(u)$, which is constant in lethargy, corresponds to a $1/E$ spectrum in energy.

The basic neutron balance equation from which neutron spectra are calculated is given by Eq. (2.138). It is often applied after its transformation into lethargies:

$$[D(u)B^2 + \Sigma_t(u)]\varphi(u) - \int_0^\infty \Sigma_s(u' \to u)\varphi(u')\, du'$$

$$= \lambda\chi(u)\int_0^\infty \nu\Sigma_f(u')\varphi(u')\, du' \quad . \tag{4.56}$$

In plotting a neutron spectrum, one obtains the most informative diagram, if equal areas represent an equal number of reactions in case of a constant cross section. This is achieved if $\varphi(E)$ is plotted over a linear energy scale or if $\varphi(u)$ is shown as a function of u. But, if $\varphi(E)$ is plotted over a logarithmic energy scale, equal areas are not at all representative of the number of reactions. One therefore normally plots $E\varphi(E)$ over a $\log E$ scale where the equal areas represent equal reactions in the same way as $\varphi(u)$ plotted over u.

4-2C Elastic Scattering Cross Sections

The energy and angular distribution of elastically scattered neutrons have been discussed in the previous section. Conservation of energy and momentum determines the energy-angular relations and the scattering kernel. The nuclear physics of the isotropic elastic scattering process in the CMS determines the magnitude, i.e., the scattering cross section, $\sigma_s(E)$.

The elastic scattering process as such can occur in two different ways: Either the neutron is scattered by the entire nucleus, or the neutron is first absorbed by the nucleus and then reemitted.

The first kind of elastic scattering is called "potential scattering" for the following reason: The force (f) that one particle (located at $r = 0$) exerts on another particle at point r is represented by a vector field, $f(r)$. Most "force fields" can be represented by the gradient of a scalar field, $V(r)$, the so-called "potential": $f(r) = -\text{grad}\,V(r)$. Scattering on the joint potential of all nucleons in the nucleus is called "potential scattering," (σ_p).

In the second kind of elastic scattering, a neutron and nucleus temporarily form a so-called "compound" nucleus. Formation of a compound nucleus predominantly occurs around "resonance energies." The subsequent "decay" of the compound nucleus yields the various nuclear reactions (capture, fission, resonance elastic scattering, and inelastic scattering). If the neutron energy is smaller than the excitation energy for the first level of the nucleus, then the nucleus will always be in its ground state after the neutron emission. For example, the lowest nuclear level in ^{238}U is at 45 keV. Thus, scattering in ^{238}U below 45 keV must always be elastic scattering, i.e., neither of the two scattering partners absorbs any energy in the collision. Both scattering partners are only elastically "deformed" in the collision. Elastic scattering via interim formation of a compound nucleus leads to "resonance elastic scattering," which is added to σ_p to make up $\sigma_s(E)$.

The potential cross sections are responsible for most of the slowing down in thermal reactors. The analytical treatment of neutron slowing down, therefore, usually employs the assumption of scattering by the potential cross section only, which is often also treated as independent of energy. It is important to know how accurate this assumption is and, in particular, in which energy range one can make this assumption. Table 4-III, therefore, gives the "constant" potential scattering cross-section values and the energy range in which $\sigma_p(E)$ does not deviate by >5% from this value. It can be seen from Table 4-III that the potential cross sections for most moderators are fairly constant over a wide range of

TABLE 4-III

Energy-Independent Scattering Cross Sections in Light Nuclei; Potential Scattering Cross Sections

Nucleus	σ_p (b)	Energy Range with <5% Variation in σ_p
Hydrogen	20	1 eV to 10 keV
Deuterium	3.3	1 eV to 600 keV
Beryllium	6.0	0.1 eV to 100 keV
Carbon	4.7	0.1 eV to 100 keV
Oxygen	3.8	0.1 eV to 100 keV

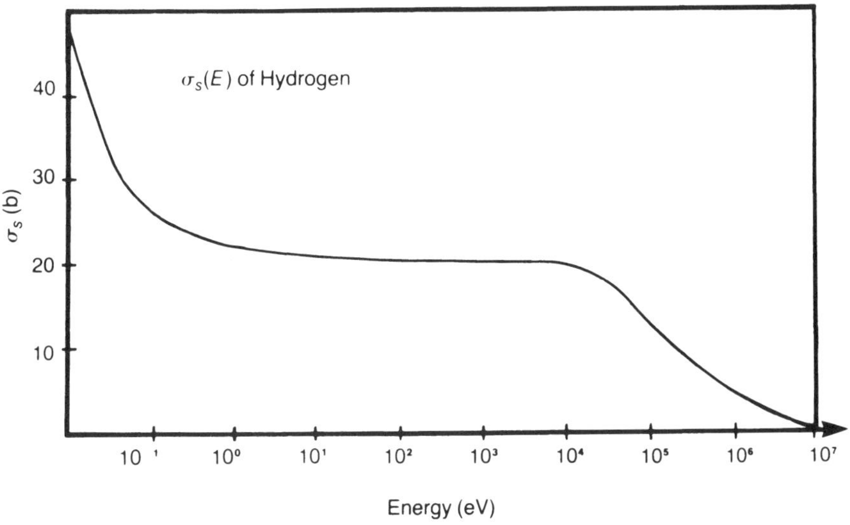

Fig. 4-6. The $\sigma_s(E)$ of hydrogen (from Ref. 1).

energy: over the range indicated, the cross sections vary <5%. Above this range, the potential cross section generally decreases; below this energy range, the scattering cross section in general increases or shows a rather complex behavior due to crystal binding effects (e.g., in beryllium).

Table 4-III also shows that the value of the potential cross section, σ_p, may vary irregularly with the atomic weight. The proton, i.e., the lightest nucleus, has the largest potential cross section of all nuclei. The deuteron, the second lightest nucleus, has—at 3.3 b—one of the smallest scattering cross sections. Superimposed on irregular variations is a general increase of σ_p with the size of the nucleus.

Figure 4-6 gives as an example the scattering cross section of hydrogen, $\sigma_s^H(E)$. The decrease toward high energies is clearly reflected in the neutron spectrum in light water reactors (see Sec. 4-5D).

Most intermediate-weight and all heavy nuclei have resonance cross sections superimposed on the potential cross sections in the entire energy range (see for example, Fig. 5-1). These nuclei do not exhibit a large range in which their total scattering cross sections are constant, although the potential cross sections are constant in ranges similar to those indicated in Table 4-III.

4-2D Inelastic Scattering

For neutron energies above the first excited level of the scattering nucleus, the nucleus may, after reemission of the neutron, be left in an excited state. De-excitation of the nucleus will occur by subsequent emission of a γ-quantum (e.g., in ^{238}U with $E_\gamma \simeq 45$ keV). Obviously, the neutron is emitted from the nucleus with an energy reduced by E_γ. In this scattering process, the nucleus appears inelastically "deformed" immediately after the neutron emission, i.e., the neutron has transferred energy to the nucleus beyond the elastic exchange of energy and momentum.

If the nucleus has several excited states below the neutron energy, the nucleus can be left in any one of these excited states. The excitation energy is then emitted in the form of one or several γ-quanta.

In Table 4-IV, the first and second excited energy levels, E^*, are tabulated for some nuclei of interest in reactors. The corresponding inelastic threshold energies are also shown. The threshold energy is larger than E^*, since only the kinetic energy of the neutron in the CMS is available in the compound nucleus to form an excited state. From Eqs. (4.5), it follows that:

$$E_n^c + E_N^c = \frac{A^2 + A}{(A+1)^2} E = \frac{A}{A+1} E \quad .$$

Based on this relationship, for an inelastic scattering event to occur, the

TABLE 4-IV
First and Second Excited Levels in Some Reactor-Material Nuclei*

	First Level		Second Level	
Nucleus	E_1^* (keV)	Inelastic Threshold Energy E_1^s (keV)	E_2^* (keV)	Inelastic Threshold Energy E_2^s (keV)
Carbon	4430	4800	7650	8290
Oxygen	6070	6440	6140	6520
Sodium	44	46	2080	2170
^{56}Fe	84	86	2080	2120
^{235}U	13	13	46	46
^{238}U	45	45	148	148
^{239}Pu	8	8	57	57

*The threshold energy of the first level gives the threshold for inelastic scattering.

neutrons are required to have a laboratory kinetic energy in excess of $\left(\frac{A+1}{A}\right)$ times that of the excited energy level:

$$E_{\text{threshold}} = E^s = \frac{A+1}{A} E^* \quad .$$

This dependency on mass number effectively increases the threshold energy in lighter nuclei (see Table 4-IV). Evident in Table 4-IV is the general decrease in threshold energy with increasing mass number. Thus, inelastic scattering is more important in heavy materials than in light elements where very energetic neutrons are required for an inelastic event to occur.

An elastic exchange of energy and momentum is always superimposed on an inelastic scattering event. At first, the nucleus is set into motion by collision with the neutron, forming the compound nucleus. This motion is subsequently changed by the recoil energy resulting from the neutron reemission. However, the kinetic energy of the neutron in the CMS is smaller after the collision than before; this leads to an exchange of energy and momentum, which is modified compared to elastic scattering where both kinetic energies are the same. A quantitative description of the "modified" exchange of energy and momentum is practically unimportant, at least for thermal reactors. It is introduced here only for conceptual clarity. Let K' describe the scattering kernel of the modified "elastic" component of the inelastic scattering. The kernel for inelastic scattering in which the nucleus is excited in its first level (energy E_1^*) is then given approximately by:

$$K_{\text{in}}(E' \rightarrow E) = K'(E', E' - E_1^* \rightarrow E) = K'(E' \rightarrow E) \quad . \tag{4.57}$$

The modified elastic scattering kernel depends on E' and $E' - E_1^*$ due to the difference of the neutron kinetic energy before and after the collision in the CMS. The corresponding inelastic energy transfer cross section is given by

$$\sigma_{\text{in}}(E' \rightarrow E) = \sigma_{\text{in}}(E') K_{\text{in}}(E' \rightarrow E) \quad . \tag{4.58}$$

At high-neutron energies, excitation of higher nuclear levels contributes to the inelastic scattering process. Then, the scattering kernel is given by superposition of several terms of the form of Eq. (4.58):

$$K_{\text{in}}(E' \rightarrow E) = \frac{1}{\sigma_{\text{in}}(E')} \sum_k \sigma_{\text{in}}^{(k)}(E') K'(E', E' - E_k^* \rightarrow E) \quad , \tag{4.59}$$

with E_k^* being the excitation energy of level k and $\sigma_{\text{in}}^{(k)}(E')$ the inelastic cross section for excitation in level k.

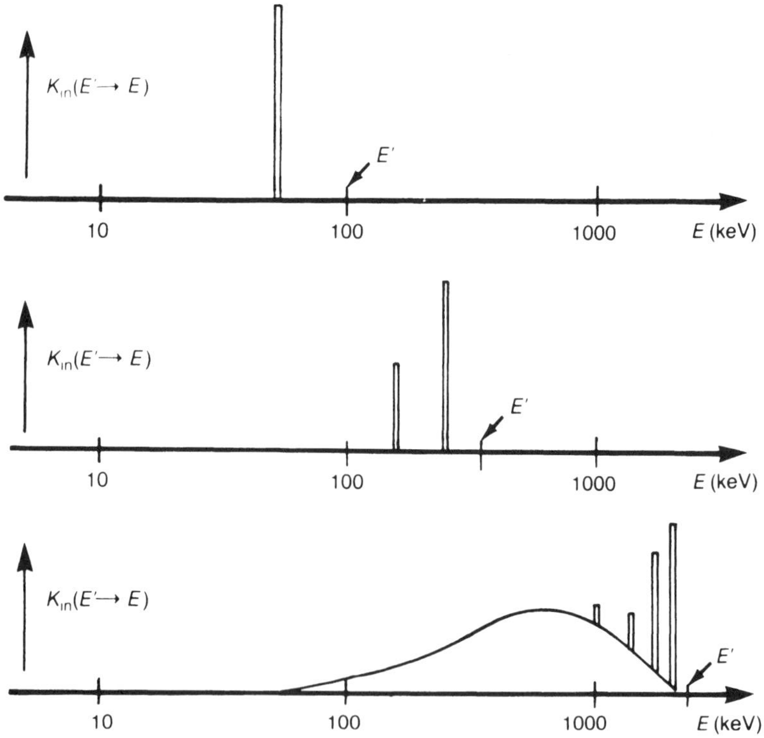

Fig. 4-7. Illustration of the inelastic scattering kernel of ^{238}U, $K_{in}(E' \rightarrow E)$, for three energies, $E' = 100$ keV, 300 keV, and 2 MeV.

Figure 4-7 shows qualitatively inelastic scattering kernels for ^{238}U for a sequence of initial energy values. For $E' = 100$ keV, only the first level can be excited ($E \gtrapprox E' - 45$ keV); at $E' = 300$ keV, the second level, occurring at 148 keV, can also be excited and the scattering kernel consists of two parts. At 2 MeV, inelastic scattering involves many levels. Several of these levels are not resolved; the kernel for unresolved levels appears to be continuous.

As seen in Fig. 4-7, the width of the "elastic" kernels, K', is quite narrow for heavy isotopes. For light nuclei, the width of K' is significant; but the excited levels of light nuclei appear only at very large energies (compare Table 4-IV) so that inelastic scattering on light nuclei plays virtually no role at all. Therefore, the width of the scattering kernel, K', in Eq. (4.59) is largely neglected and the inelastic scattering kernel is written approximately in the form:

$$K_{\text{in}}(E' \rightarrow E) \simeq \frac{1}{\sigma_{\text{in}}(E')} \sum_k \sigma_{\text{in}}^{(k)}(E') \delta[(E' - E_k^*) - E] \quad , \tag{4.60}$$

in which the kernels K' are replaced by δ functions.[b]

Fissile isotopes, such as ^{235}U and ^{239}Pu, generally have much lower excitation energies than nonfissile heavy isotopes. This is indicated in Table 4-IV by the fact that the first two levels of ^{235}U and ^{239}Pu are at comparatively low energies. Then, a fairly large number of levels can be excited in inelastic scattering with megaelectron volt neutrons. Most individual levels are not resolved experimentally; thus, the information [E_k^* and $\sigma_{\text{in}}^{(k)}(E')$] is not available to express the scattering kernels explicitly as sums in the form of Eqs. (4.59) or (4.60). The kernel, based on available experimental information on higher levels, appears to be a continuous function rather than a sum of δ functions. Only the contribution of some resolved lower levels stands out (see Fig. 4-7).

Thermodynamic considerations can be applied to find a theoretical expression for the inelastic scattering kernel (evaporation model; compare textbooks on nuclear physics, for example, Ref. 4). The excitation energy of the nucleus resulting from the temporary absorption of a neutron is treated as a "temperature" increase of the nucleons in the nucleus. Subsequently, there can be an emission of an "evaporation" neutron. The energy of this neutron can be obtained from the assumed Maxwellian temperature distribution of the nucleons in the nucleus.

Inelastic scattering is of minor importance in thermal reactors but of very high importance for the much harder neutron spectrum of fast reactors or in fusion systems. Inelastic scattering dominates elastic scattering in fast reactors over a large part of the lethargy range. In the energy domain of the fission neutron source, inelastic scattering is also the major slowing down mechanism in thermal reactors (see Sec. 4-4).

The basic information on inelastic scattering for reactor calculations is condensed into "group constant sets" (see Chapter 7). The scattering kernel appears as a scattering matrix in a group constant set. The more detailed information needed for nuclear reactor calculations is part of the Evaluated Nuclear Data File[5] (ENDF), which is widely available in the United States. Other countries have similar cross-section documentation.

[b]Some basic information of δ functions is provided in Appendix C.

4-3 Numerical Calculation of Slowing Down Spectra

4-3A Preliminary Form of the Multigroup Equations

For the discussion of the numerical solution procedure for slowing down spectra, a preliminary form of the multigroup equations is introduced. As in the rest of Chapter 4, the energy dependency is treated in isolation; the space dependence is included in Chapter 7. The major difference between this preliminary formulation and the practically used formulation of Chapter 7 is the neglect of the within-group weighting spectra. While this neglect does not affect the formal solution procedures, it restricts the practical applicability of the so-defined group constants to cases with very many groups where the within-group weighting becomes unimportant (except for resonance absorption as discussed in Chapter 5).

The slowing down of neutrons can be solved as a source-sink problem in which all sources are combined in a total source, S_t, which may also contain an eigenvalue in front of the fission source [see Eq. (2.94)]:

$$\Sigma_{tl}(E)\varphi(E) - \int_E^{\infty} \Sigma_s(E' \rightarrow E)\varphi(E')\, dE' = S_t(E) \quad . \tag{4.61}$$

Equation (4.61) results from a separation of the neutron flux in a large medium into its space and energy dependencies (see Sec. 2-6A). The neutron losses through leakage then appear to be added to the total cross section. The sum is abbreviated here with $\Sigma_{tl}(E)$ (the index, tl, indicates a combination of total and leakage-loss cross section):

$$\Sigma_{tl}(E) = \Sigma_t(E) + D(E)B^2 \quad , \tag{4.62}$$

with B^2 being the material or the geometrical buckling. In a critical multiregion system, B_m^2 should be used in Eq. (4.62).

For the numerical solution, Eq. (4.61) is written in lethargies instead of energies. By using the same functional symbols as in Eq. (4.61) [compare the discussion in Sec. 4-2B and Eq. (4.56)], one obtains:

$$\Sigma_{tl}(u)\varphi(u) - \int_0^{u} \Sigma_s(u' \rightarrow u)\varphi(u')\, du' = S_t(u) \quad . \tag{4.63}$$

The upper integration limit in the scattering integral is u [corresponding to E in Eq. (4.61)], which reflects the absence of upscattering. The lower limit 0 is an approximation, in which neutrons with energies $E>E_0$ are not treated explicitly. To conserve the neutron balance, $\chi(E)$ is then normalized to unity in the energy range below E_0.

For a numerical solution, the lethargy scale must be subdivided into group intervals. One often chooses the intervals to be equal:

$$\delta u_\gamma = u_\gamma - u_{\gamma-1} = \delta u \quad . \tag{4.64}$$

Based on the discrete lethargy scale, discrete values must be introduced for the continuous functions that appear in Eq. (4.63). For very small intervals, this could be done, for example, by using specific values of the continuous function corresponding to *specific points* on the abscissa, such as in the middle or at the end of the intervals. However, in treating the energy dependence of the neutron flux, it is common to use the spectrum *integral* (group flux) and the cross-section *averages* (group constants) as discrete representations of the continuous functions. The advantage of this approach is that it formally conserves the reaction rates, which are integrals over the lethargy (or energy).

Integrating Eq. (4.63) over each of the lethargy intervals obviously provides as many equations as one has unknown φ_γ values:

$$\int_{\delta u_\gamma} \Sigma_{tl}(u)\varphi(u)\, du - \int_{\delta u_\gamma} \int_0^u \Sigma_s(u' \rightarrow u)\varphi(u')\, du'\, du$$

$$= \int_{\delta u_\gamma} S_t(u)\, du \quad . \tag{4.65}$$

As discrete spectrum values, one uses the integrals (representing a many-group representation of the continuous spectrum):

$$\varphi_\gamma = \int_{u_\gamma}^{u_{\gamma-1}} \varphi(u)\, du \quad . \tag{4.66}$$

If the intervals δu are small,[c] the spectrum and the cross sections will have only minor variations within all δu_γ's. Therefore, the cross sections can be replaced in the u and u' integrations by their average values in the respective intervals:

$$\int_{\delta u_\gamma} \Sigma_{tl}(u)\varphi(u)\, du \simeq \Sigma_{tl\gamma} \int_{\delta u_\gamma} \varphi(u)\, du = \Sigma_{tl\gamma}\varphi_\gamma \quad , \tag{4.67}$$

with

$$\Sigma_{tl\gamma} = \frac{1}{\delta u_\gamma} \int_{\delta u_\gamma} \Sigma_{tl}(u)\, du \tag{4.68}$$

being the average cross section in the γ intervals.

---

[c]The extension to larger intervals leads to the "multigroup" concept of Chapter 7.

The average of the scattering kernel is to be taken in both intervals, δu_γ and $\delta u_{\gamma'}$. This converts the kernel into a scattering matrix, with each element $\sigma_{s\gamma'\gamma}$ describing the scattering from interval γ' to γ:

$$\sigma_s(\gamma' \to \gamma) = \sigma_{s\gamma'\gamma} \quad . \tag{4.69}$$

Since downscattering always leads to an increase in lethargy, γ is always larger than or equal to γ'. Then, the scattering matrix is lower triangular; it does not have any terms above the diagonal. The transfer $\gamma'\gamma$ is commonly reflected in the order of the indices $\gamma'\gamma$, which is reversed compared to the standard order of row and column indices:

$$\Sigma_s(\gamma' \to \gamma) = \Sigma_{s\gamma'\gamma} = \begin{pmatrix} \Sigma_{s11} & 0 & 0 & \cdots \\ \Sigma_{s12} & \Sigma_{s22} & 0 & \cdots \\ \Sigma_{s13} & \Sigma_{s23} & \Sigma_{s33} & \cdots \\ \cdot & \cdot & \cdot & \cdots \\ \cdot & \cdot & \cdot & \cdots \\ \cdot & \cdot & \cdot & \cdots \end{pmatrix} \quad . \tag{4.70}$$

The integration of the right side of Eq. (4.65) yields for the source "value" the corresponding source integral in the interval δu_γ:

$$S_{t\gamma} = \int_{\delta u_\gamma} S_t(u)\, du \quad . \tag{4.71}$$

Equations (4.68), (4.70), and (4.71) define the discrete interval version of Eq. (4.63) and represent a preliminary formulation of the multigroup theory:

$$\Sigma_{tl\gamma}\varphi_\gamma - \sum_{\gamma' \leq \gamma} \Sigma_{s\gamma'\gamma}\varphi_{\gamma'} = S_{t\gamma}, \quad \gamma = 1,\ldots,\Gamma \quad , \tag{4.72}$$

or in explicit form:

$$\begin{aligned} &(\Sigma_{tl1} - \Sigma_{s11})\varphi_1 + \quad 0 \quad + \quad 0 \quad + 0 + \ldots = S_{t1} \\ &- \Sigma_{s12}\varphi_1 + (\Sigma_{tl2} - \Sigma_{s22})\varphi_2 + \quad 0 \quad + 0 + \ldots = S_{t2} \\ &- \Sigma_{s13}\varphi_1 \quad - \Sigma_{s23}\varphi_2 + (\Sigma_{tl3} - \Sigma_{s33})\varphi_3 + 0 + \ldots = S_{t3} \end{aligned} \tag{4.73}$$

4-3B Solution of the Source-Sink Problem

For the solution of neutron slowing down as a source-sink problem, S_t is formally treated as an independent source.

Due to the triangular nature of the matrix on the left side of Eqs. (4.72) and (4.73), the solution can then be readily found by solving first for φ_1, inserting φ_1 into the second equation and finding φ_2, inserting φ_1 and φ_2 into the third equation and solving for φ_3, etc.:

$$\varphi_1 = \frac{S_{t1}}{\Sigma_{rl1}} \quad ,$$

$$\varphi_2 = \frac{S_{t2} + \Sigma_{s12}\varphi_1}{\Sigma_{rl2}} \quad ,$$

and

$$\varphi_3 = \frac{S_{t3} + \Sigma_{s13}\varphi_1 + \Sigma_{s23}\varphi_2}{\Sigma_{rl3}}$$

$$\vdots \qquad\qquad\qquad\qquad . \tag{4.74}$$

The numerator of each flux value in Eq. (4.74) gives the entire neutron source as it appears in the respective interval, coming either directly from the source or from higher energy intervals through scattering. The denominator contains the difference of the total cross section and the within-group scattering cross section. This difference describes the total removal out of the interval γ (also see Chapter 7):

$$\Sigma_{rl\gamma} = \begin{cases} \text{average removal and leakage-loss} \\ \text{cross section in interval } \gamma \quad . \end{cases} \tag{4.75}$$

Since the numerical solution of the slowing down spectrum can be readily found without an iteration, a large number of intervals can be used to obtain an accurate solution. Even early computer codes for slowing down in thermal reactors (for example, MUFT of Ref. 6) use ~60 intervals or groups. Codes for spectra calculation in fast reactors usually use up to 2000 groups (for example, MC^2 of Ref. 7). Some special codes treat tens or hundreds of thousands of groups.

4-3C Solution of the Eigenvalue Problems

If the neutron slowing down equation is treated as an inhomogeneous problem, it always has a nontrivial solution, even if $S_t = S_f$, the fission source emitting a neutron spectrum $\chi(u)$. If, however, the fission source is explicitly included [as $\chi(u)$ times the respective integral over $\nu\Sigma_f\phi$], then one has a homogeneous problem. It is still a pure slowing down problem, though a homogeneous one, if the entire thermal energy domain is represented by a single group or—as in the case of a fast reactor—neutrons arriving in the thermal domain can be neglected.

In this homogeneous problem, an eigenvalue needs to be included.

If both the composition and geometrical buckling are given in Eq. (4.63), an eigenvalue is needed in front of the fission source on the right side in order to obtain a nontrivial solution of the homogeneous problem. Thus, $S_t(u)$ is replaced by:

$$S_f'(u) = \lambda\chi(u)\int_0^\infty \nu\Sigma_f(u')\varphi(u')\,du' = \chi(u)S_f' \tag{4.76a}$$

or in finite interval form (S_f is the total fission source):

$$S_{f\gamma}' = \lambda\chi_\gamma\sum_{\gamma'}\nu\Sigma_{f\gamma'}\varphi_{\gamma'} = \chi_\gamma S_f' \quad . \tag{4.76b}$$

The prime on S_f indicates the modification of S_f by a λ eigenvalue.

The solution of the inhomogeneous problem, Eq. (4.73), can be found in one "sweep," i.e., without iteration. Eigenvalue problems, however, usually require an iterative solution. Suppose the right side of Eq. (4.73) were to depend on the spectrum in a general form. Then, the right side could not be precalculated since the spectrum is not yet known. A practical solution procedure would then require an iterative scheme that starts with a first guess for the right side. In each step of the iteration, the inhomogeneous problem would be solved as described above. The solution would be an improved spectrum, which would be used to find an improved right side, etc. The converged result would yield the eigenvalue and the eigenfunction. (Clearly, more sophisticated methods are employed in most practical calculations, using rather elegant numerical techniques. These specialized mathematical approaches, which greatly improve the speed and accuracy of the computer solution, are outside the scope of this book.)

However, for a right side of the form of Eq. (4.76a), one does not have to iterate to solve the eigenvalue problem. The reason is that the spectrum appears only under a definite integral, with $\chi(u)$ as a factor in front of the integral. The eigenvalue problem can then be solved in the following way: One picks an arbitrary value for S_f', i.e., for λ times the total fission neutron source in Eq. (4.76b), say $S_f'^0$. One then calculates $\varphi = \varphi^0$, which corresponds to the chosen magnitude of the source. From φ^0, a fission source can be calculated: S_f^0 (without prime), Eq. (4.77b). The ratio of the arbitrarily chosen value, $S_f'^0$, and the subsequently calculated one, S_f^0, gives the eigenvalue:

$$\lambda = \frac{S_f'^0}{S_f^0} = \frac{1}{k} \quad , \tag{4.77a}$$

with

$$S_f^0 = \int_0^\infty \nu\Sigma_f(u')\varphi^0(u')\,du' \quad . \tag{4.77b}$$

For example, if 100 ($= S_f'^{0}$) neutrons start from the arbitrarily assumed initial fission source and only 99 ($= S_f^0$) new fission neutrons are produced by the spectrum φ^0, then the reactor is subcritical with $k = 0.99 = 1/\lambda$.

For calculation of the asymptotic spectrum, B^2 in Eq. (4.62) is not known. Furthermore, no λ appears in front of the fission source and no independent source is present. The resulting balance equation is

$$B_m^2 D_\gamma \varphi_\gamma + \Sigma_{t\gamma}\varphi_\gamma - \sum_{\gamma' \leq \gamma} \Sigma_{s\gamma'\gamma}\varphi_{\gamma'} = \chi_\gamma \sum_{\gamma'} \nu\Sigma_{f\gamma'}\varphi_{\gamma'} \quad . \tag{4.78}$$

Equation (4.78) is to be solved as an eigenvalue problem that yields the material buckling, the asymptotic spectrum. By using a general iterative code for solving the matrix eigenvalue problems, the higher B_m^2 eigenvalues and eigenfunctions are also obtained. If only the fundamental mode is to be found, the same solution procedure used for the λ eigenvalue problem can be used, but B^2 needs to be varied such that λ eventually becomes equal to unity. The resulting B^2, i.e., B^2 ($\lambda = 1$), equals B_m^2.

4-4 Fast Fission

The high-energy end of the neutron spectrum in a reactor is of particular importance since practically all heavy nuclei are fissionable at the high energies of interest. Fission in the megaelectron volt energy domain is called "fast fission." The fission rate obviously depends on the spectrum.

The most important nucleus for fast fission is ^{238}U, since all reactors that use natural or slightly enriched uranium contain high concentrations of ^{238}U [for example, light water reactors (LWRs), D_2O, and most graphite-moderated reactors]. Figure 4-8 shows the fission cross sections of ^{238}U, ^{235}U, and ^{232}Th between 100 keV and 10 MeV. Fission in ^{238}U is a threshold reaction with a threshold at ~700 keV. But the cross section rises to significant values only above 1 MeV. However, even above its threshold, $\sigma_f(^{238}U)$ is still less than one-half of $\sigma_f(^{235}U)$. But, if $\sigma_f(^{238}U)$ is multiplied by the concentration ratio of $^{238}U/^{235}U$ (~60 in LWRs[d]), one sees that the total fission cross section per ^{235}U atom (the dotted line in Fig. 4-8) has a large hump above 1 MeV. Figure 4-8 also shows that there will be a certain number of "intermediate" fissions in ^{235}U (between 1 MeV and thermal).

---

[d]This number is for the Dresden I boiling water reactor with 1.5 wt% ^{235}U. Larger LWRs may have lower ratios, as the ^{235}U enrichment is in the 3 to 4% range.

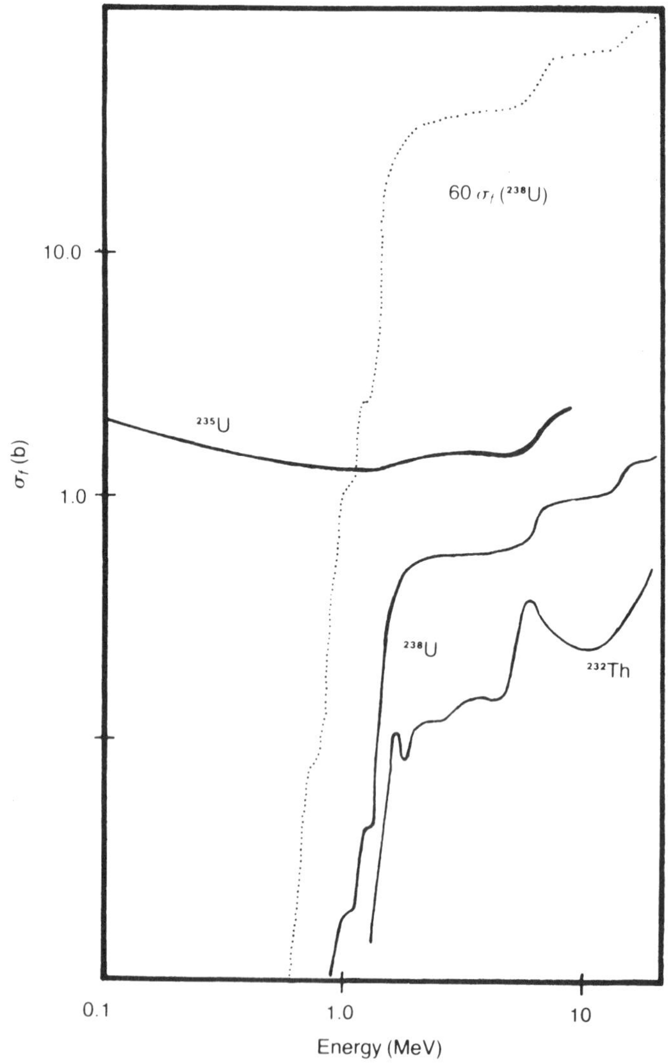

Fig. 4-8. Fission cross sections for ^{238}U, ^{232}Th, and ^{235}U at high energies.

In some reactor types, for example, high-temperature gas-cooled reactors (HTGRs), ^{232}Th is used instead of ^{238}U. The fission threshold in ^{232}Th occurs at ~1.2 MeV and the cross section becomes significant above ~1.6 MeV, but still remains substantially below the cross section of ^{238}U. An increase of the fission threshold beyond 1 MeV quickly reduces the fast fission since the fission neutron source and, even more, the resulting spectrum rapidly decreases above 1.5 MeV. Consequently, fast fission in ^{232}Th is of lesser importance than in ^{238}U, due to its higher threshold and its lower cross section.

Fast breeder reactors (FBRs) benefit from the fast fission of ^{238}U and of the two plutonium isotopes, ^{240}Pu and ^{242}Pu, which have an even lower threshold than ^{238}U. Therefore, fast fission contributes significantly to the fission neutrons produced in a FBR.

In approximate criticality estimates, based on the four-factor formula for k_∞, a special factor describes the fast fission, ε (see Sec. 2-5). In the reaction-rate-based factorization of k_∞, Eq. (2.127), ε is given by the ratio of the total fission neutron production rate to the production rate by thermal neutron fission. The fast fission factor in this reaction rate definition is denoted by ε':

$$\varepsilon' = \frac{R_p}{R_p(\text{th})} \quad . \tag{4.79}$$

The total fissions are split into those occurring at thermal and those taking place at fast energies. The intermediate energy fissions are either neglected or are considered as a modification of the resonance escape probability, p. This gives:

$$\varepsilon' = 1 + \delta\varepsilon' \quad , \tag{4.80a}$$

with

$$\delta\varepsilon' = \frac{R_p(\text{fast})}{R_p(\text{th})} = \frac{R_p(\text{fast})}{R_p - R_p(\text{fast})} \quad . \tag{4.80b}$$

The total neutron production rate, R_p, equals the source magnitude in the slowing down equation, Eq. (4.63). The resulting spectrum, and thus R_p(fast), are proportional to R_p. Therefore, R_p will cancel after inserting R_p(fast) into Eq. (4.80b).

In a medium that appears to be approximately homogeneous for fast neutrons (e.g., an LWR core), the high-energy end of the spectrum can be found by solving the first few equations of equation system (4.73). In the simplest treatment, the fast energy range is combined into just one group. If it is also assumed that all fission neutrons are generated in this group (i.e., $\chi_1 = 1$ and all other χ_γ's are zero) and leakage is neglected, one obtains simply:

$$\varphi_1 = \frac{S_1}{\Sigma_{r1}} = \frac{\chi_1 R_p}{\Sigma_{r1}} = \frac{R_p}{\Sigma_{r1}} \quad . \tag{4.81}$$

The fast fission rate of ^{238}U is obtained by multiplying Eq. (4.81) with $\nu\Sigma_{f1}$;

$$R_p(\text{fast}) = (\nu\Sigma_{f1})^{238}\frac{R_p}{\Sigma_{r1}} \quad , \tag{4.82}$$

and for $\delta\varepsilon'$:

$$\delta\varepsilon' = \frac{(\nu\Sigma_{f1})^{238}/\Sigma_{r1}}{1-(\nu\Sigma_{f1})^{238}/\Sigma_{r1}} = \frac{(\nu\Sigma_{f1})^{238}}{\Sigma_{r1}-(\nu\Sigma_{f1})^{238}} \quad . \tag{4.83}$$

The simple derivation leading to Eq. (4.83) should be understood as an illustration. Formulas that were practically applied for ε calculations employed two or even three neutron groups.[8]

In reactors with widely separated fuel rods, such as graphite or heavy-water-moderated reactors, simple homogeneous approximations are inapplicable. For the calculation of ε in such heterogeneous configurations, multicollision methods (compare Sec. 6-3) are frequently employed.

4-5 Slowing Down in Hydrogen

4-5A The Balance Equations

The slowing down of neutrons by scattering on hydrogen is the only case that can be treated analytically without approximation. The analytical solution is investigated in this section. The reason for the simplicity of the slowing down problem in hydrogen is that a neutron may lose *all* of its initial energy E in a single head-on collision with a proton; i.e., the differential scattering cross section is given by

$$\Sigma_s^{\text{H}}(E'\rightarrow E) = \frac{\Sigma_s^{\text{H}}(E')}{E'} \quad \text{for } 0\leq E\leq E' \quad . \tag{4.84}$$

This follows from the elastic scattering "transfer kernel" defined by Eq. (4.28) with $\alpha=0$ for hydrogen. The integral with respect to all final energies E is obtained just by multiplication with E', i.e., with the length of the scattering interval.

On the other hand, the rate of neutrons arriving at E from higher energies E' is given by an integral,

$$\int_E^\infty \Sigma_s^{\text{H}}(E'\rightarrow E)\varphi(E')\,dE' \quad ,$$

which formally extends to infinity.

Following an inductive approach, the slowing down problem is treated four times, with increasing complexity:

1. *Constant (energy-independent) scattering cross section and zero absorption cross section without leakage:* The scattering cross section of hydrogen, depicted in Fig. 4-6, is fairly constant throughout most of the slowing down range. In this approximation, it is assumed to be constant for all E. The corresponding balance equation follows from Eq. (2.138); under the above assumption, it simplifies to:

$$\Sigma_s^{\mathrm{H}}\varphi(E) - \Sigma_s^{\mathrm{H}}\int_E^{\infty}\frac{\varphi(E')}{E'}\,dE' = \chi(E)s_0 \quad . \qquad (4.85a)$$

The term s_0 denotes the fission source magnitude in the purely energy-dependent equation, as opposed to S_f in the complete balance equations. The fission source, s_0, may include a λ eigenvalue as in Eq. (4.76a).

2. *Energy-dependent scattering $[\sigma_s(E)]$ without absorption or leakage:* The balance equation is given by:

$$\Sigma_s^{\mathrm{H}}(E)\varphi(E) - \int_E^{\infty}\frac{\Sigma_s^{\mathrm{H}}(E')\varphi(E')}{E'}\,dE' = \chi(E)s_0 \quad . \qquad (4.85b)$$

3. *Energy-dependent scattering with absorption by hydrogen only:*

$$\Sigma_t^{\mathrm{H}}(E)\varphi(E) - \int_E^{\infty}\frac{\Sigma_s^{\mathrm{H}}(E')\varphi(E')}{E'}\,dE' = \chi(E)s_0 \quad , \qquad (4.85c)$$

where $\Sigma_t^{\mathrm{H}}(E) = \Sigma_s^{\mathrm{H}}(E) + \Sigma_a^{\mathrm{H}}(E)$.

4. *Energy-dependent scattering on hydrogen with absorption by all isotopes:*

$$\Sigma_t(E) = \Sigma_s^{\mathrm{H}}(E) + \Sigma_a^{\mathrm{H}}(E) + \Sigma_a^{\mathrm{others}}(E) \quad . \qquad (4.86a)$$

Only the absorption of the higher isotopes is included in that cross section. Thus, Eq. (4.86a) is actually not a complete total cross section. However, no special notation is introduced for Eq. (4.86a) since the resulting formulas will also be applied later on (see, for example, Sec. 5-4) for complete total cross sections.

In this approximation, one allows for absorption of neutrons in nuclei other than hydrogen, but one neglects the scattering action of these nuclei as compared to that of hydrogen. In Sec. 5-4, this approximation will be identified as the "narrow-resonance infinite-mass" approximation for hydrogen as moderator. The balance equation is formally the same in both cases 3 and 4; Σ_t containing the additional absorption. Therefore, the solution of cases 3 and 4 will be formally identical.

5. *Energy-dependent scattering, absorption, and leakage:* The balance equation is also of the form of Eq. (4.85c), but with Σ_t replaced by Σ_{tl}:

$$\Sigma_{tl}(E) = \Sigma_t(E) + D(E)B^2 \quad , \tag{4.86b}$$

where the leakage loss, described by the now familiar $D(E)B^2$ term, is included.

Each of these four slowing down models in hydrogen is discussed in detail in the next several sections. Solutions for the flux are obtained and physically interpreted in relation to the neutron slowing down process.

The advantage of the investigation of four different stages of this problem is not only the fact that it provides a more in-depth understanding of the effects of the individual phenomena (see Sec. 4-1)—it also provides an indication of the inaccuracies encountered if certain phenomena are neglected, and prepares the discussion of common approximations for the treatment of resonance absorption.

4-5B Slowing Down in Hydrogen with a Constant σ_s

The balance equation is an integral equation, but the integrand is independent of E and the variable E appears only in one integration limit. Therefore, the integral equation (4.85a) can be converted into a differential equation by differentiation:

$$\Sigma_s^{\mathrm{H}} \frac{d\varphi(E)}{dE} + \Sigma_s^{\mathrm{H}} \frac{\varphi(E)}{E} = s_0 \frac{d\chi(E)}{dE} \quad . \tag{4.87a}$$

Since Eq. (4.87a) is a first-order differential equation, it has to be completed with a boundary condition. The original balance equation was an integral equation that did not require a boundary condition. The fact that in one version a boundary condition is needed and in the other version it is not suggests that the boundary condition for the differential equation must directly follow from the corresponding integral equation. The consideration of the integral equation (4.85a) at both ends of the interval shows that the desired boundary condition is obtained from the upper end. Obviously, $\varphi(0)$ cannot be deduced to form the scattering integral from 0 to ∞ without knowing the complete solution. But, for $E \rightarrow \infty$, the scattering integral disappears. Furthermore, $\chi(\infty)$ is zero; this requires:

$$\varphi(E) \rightarrow 0 \text{ for } E \rightarrow \infty \quad , \tag{4.87b}$$

since Σ_s^H is finite. Equation (4.87b) is, therefore, the required boundary condition.

The first-order differential equation, Eq. (4.87a), can be readily solved. As described in Appendix D, the differential equation,

$$\frac{df(E)}{dE} = a(E)f(E) + g(E) \quad , \tag{4.88}$$

has the solution:

$$f(E) = \int_{\infty}^{E} g(E') \exp\left[\int_{E'}^{E} a(E'')\, dE''\right] dE' \quad , \tag{4.89}$$

if the boundary condition is $f(E=\infty) = 0$.

For applying the general solution (4.89), Eq. (4.87a) is rewritten:

$$\frac{d\varphi(E)}{dE} = -\frac{1}{E}\varphi(E) + \frac{s_0}{\Sigma_s^H}\frac{d\chi(E)}{dE} \quad . \tag{4.90a}$$

From the solution, Eq. (4.89) follows at first:

$$\varphi(E) = \frac{s_0}{\Sigma_s^H}\int_{\infty}^{E}\frac{d\chi(E')}{dE'}\exp\left(\int_{E}^{E'}\frac{dE''}{E''}\right) dE' \quad . \tag{4.90b}$$

The exponential function can be simplified:

$$\exp\left(\int_{E}^{E'}\frac{dE''}{E''}\right) = \exp\left(\ln\frac{E'}{E}\right) = \frac{E'}{E} \quad . \tag{4.90c}$$

This gives:

$$\varphi(E) = \frac{s_0}{\Sigma_s^H}\cdot\frac{1}{E}\cdot\int_{\infty}^{E}\frac{d\chi}{dE'}E'\, dE' \quad . \tag{4.90d}$$

The appearance of a derivative in the integrand suggests integration by parts:

$$\int_{E}^{\infty}\frac{d\chi}{dE'}E'\, dE' = [\chi(E')E']_{E}^{\infty} - \int_{E}^{\infty}\chi(E')\, dE'$$

$$= -E\chi(E) - \int_{E}^{\infty}\chi(E')\, dE' \quad . \tag{4.91}$$

This gives as the solution for the slowing down in hydrogen with constant σ_s and an extended source (the minus signs cancel):

$$\varphi(E) = \frac{s_0}{\Sigma_s^H}\chi(E) + \frac{s_0}{E\Sigma_s^H}\int_{E}^{\infty}\chi(E')\, dE' \quad . \tag{4.92}$$

The first term represents the virgin flux (or uncollided flux), i.e., the flux that comes directly from the source. This term follows directly from Eq. (4.85a) by omitting the scattering integral. The second term describes all scattered neutrons.

In treating slowing down in thermal reactors, one often approximates $\chi(E)$ by a δ function. This is more accurate than it might appear at first glance; a plot of $E\chi(E)$, on the logarithmic energy scale, is shown in Fig. 4-9. Since the slowing down spectrum is needed particularly for the description of the resonance absorption (the largest part of it occurs below 50 keV), the finite width of $E\chi(E)$ is not very important. From Eq. (4.92), the solution is directly obtained by shrinking $\chi(E)$ down to a δ function at $\overline{E}_f$, the average fission energy:

$$\chi(E) \rightarrow \delta(E - \overline{E}_f) \quad . \tag{4.93}$$

This gives as a solution of the slowing down problem:

$$\varphi(E) = \frac{s_0}{\Sigma_s^{\mathrm{H}}}\delta(E - \overline{E}_f) + \frac{s_0}{E\Sigma_s^{\mathrm{H}}}, \text{ for } E \leq \overline{E}_f \quad . \tag{4.94}$$

In many textbooks on thermal reactors, the position of the source δ function is called "E_0." This is not done here in order to distinguish $\overline{E}_f$ from E_0 in the definition of the lethargy ($u = \ln E_0/E$). If an extended source is treated explicitly, one chooses for E_0 (in the lethargy) an energy "above" the extended source. For fission reactors, $E_0 = 10$ MeV is nor-

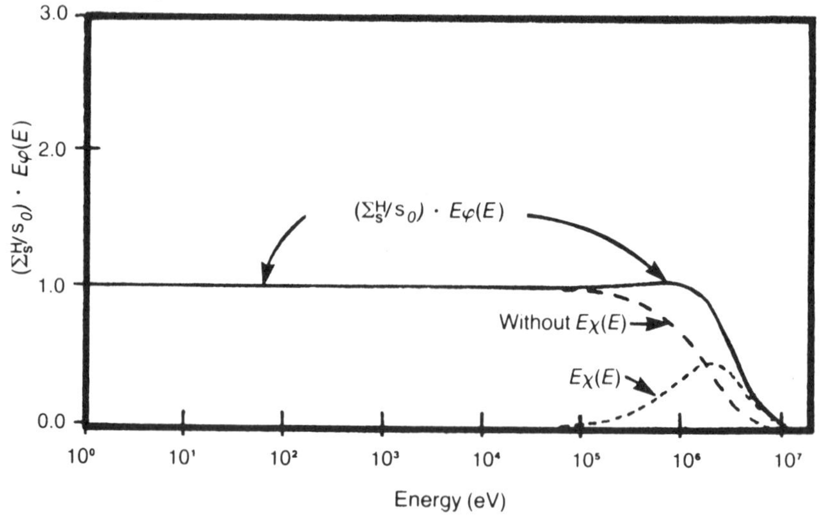

Fig. 4-9. Slowing-down spectrum in hydrogen for constant σ_s.

mally used. In fusion reactors where 14-MeV neutrons are the primary source of energy, one would use $E_0 \geq 14$ MeV.

Equation (4.94) describes the $1/E$ spectrum of the neutrons during slowing down. In the source area, there is in addition the virgin flux as illustrated in Fig. 4-9.

At energies well below the source energy, i.e., below ~50 keV, the contribution of the virgin flux vanishes and the integral over the source above that energy approaches unity. Then one obtains the basic spectral shape in the slowing down energy range, the $1/E$ spectrum:

$$\varphi(E) = \frac{s_0}{\Sigma_s^{\mathrm{H}}} \cdot \frac{1}{E} \text{ below the source} \quad . \tag{4.95}$$

Thus, $E\varphi(E)$, i.e., the flux per unit lethargy, becomes constant (see Fig. 4-9).

4-5C Slowing Down with an Energy-Dependent Hydrogen Scattering Cross Section

If Σ_s^{H} depends on E, it cannot be taken out of the scattering integral in the balance equation, Eq. (4.85b). The fact that $\Sigma_s^{\mathrm{H}}(E)$ in Eq. (4.85b) always appears multiplied by $\varphi(E)$ suggests the use of this product, the "scattering density,"[e] $F_s(E)$, as the unknown function:

$$F_s(E) = \Sigma_s^{\mathrm{H}}(E)\varphi(E) \quad . \tag{4.96}$$

The resulting balance equation for the scattering density is:

$$F_s(E) - \int_E^{\infty} F_s(E') \frac{dE'}{E'} = s_0 \chi(E) \quad . \tag{4.97}$$

The solution can then be copied directly from the previous problem [Eq. (4.92)]:

$$F_s(E) = s_0 \chi(E) + \frac{s_0}{E} \int_E^{\infty} \chi(E')\, dE' \quad . \tag{4.98}$$

At energies well below the fission source, one again obtains a $1/E$ dependence but only for the scattering density and not for the spectrum. The spectrum in this approximation is obtained by inserting F_s from Eq. (4.96) and solving for $\varphi(E)$:

---

[e]If the space dependence of the flux is eliminated (by separation or by considering an infinite medium) $F_s(E)$ is not actually a density.

$$\varphi(E) = \frac{s_0}{\Sigma_s^{H}(E)}\chi(E) + \frac{s_0}{E\Sigma_s^{H}(E)} \cdot \int_E^{\infty} \chi(E')\, dE' \quad . \tag{4.99}$$

Well below the fission source, the spectrum has the form:

$$\varphi(E) = \frac{s_0}{E\Sigma_s^{H}(E)} \quad . \tag{4.100}$$

Above several kiloelectron volts, the scattering cross section decreases for increasing energies, and at 1 MeV, it is only one-fifth of its plateau value (see Fig. 4-6). Then $E\varphi(E)$ increases with increasing energy. Around the fission source, the spectrum starts to decrease due to the decrease of $\chi(E)$ and of the χ integral in Eq. (4.99).

The spectrum calculated with constant σ_s has a slight hump in the source range due to the virgin flux [first term in Eq. (4.94)]. This hump is substantially increased and is carried down to lower energies if the decrease of $\sigma_s^{H}(E)$ (with increasing E) is taken into account (see Fig. 4-10). However, this hump is exaggerated due to the neglect of inelastic scattering in the analytical treatment. Including inelastic scattering, primarily on ^{238}U, reduces the hump at high energies by a substantial factor: The hump of the spectrum in LWRs is a factor of ~3 lower than indicated in Fig. 4-10, primarily due to the action of inelastic scattering.

4-5D Slowing Down with Energy-Dependent Scattering and Absorption

The capture cross section for neutrons by protons varies from 330 mb at thermal energies to 33 mb at 2.5 eV and decreases further with increasing E. This small capture cross section has no significant effect on the spectrum, and need not be considered in the slowing down problem.

Neutron absorption predominantly occurs in heavy nuclei for which the energy loss due to elastic scattering is quite small compared to the energy loss in collisions with hydrogen. A quantitative comparison of the loss in energy due to elastic scattering for protons and ^{238}U nuclei is shown below. It is particularly the "slowing down power," $\xi\sigma_s$, that determines the relative scattering action per nucleus:

$$\sigma_s(\mathrm{H}) \simeq 20 \text{ b} \qquad \xi(\mathrm{H}) = 1.0 \qquad \xi\sigma_s(\mathrm{H}) \simeq 20 \text{ b}$$

and

$$\sigma_s(^{238}\mathrm{U}) \simeq 10 \text{ b} \qquad \xi(^{238}\mathrm{U}) = 0.0084 \qquad \xi\sigma_s(^{238}\mathrm{U}) \simeq 0.09 \text{ b} \quad .$$

Thus, protons are ~200 times more effective than ^{238}U in slowing down neutrons elastically. In the range of inelastic scattering, i.e., above 45 keV

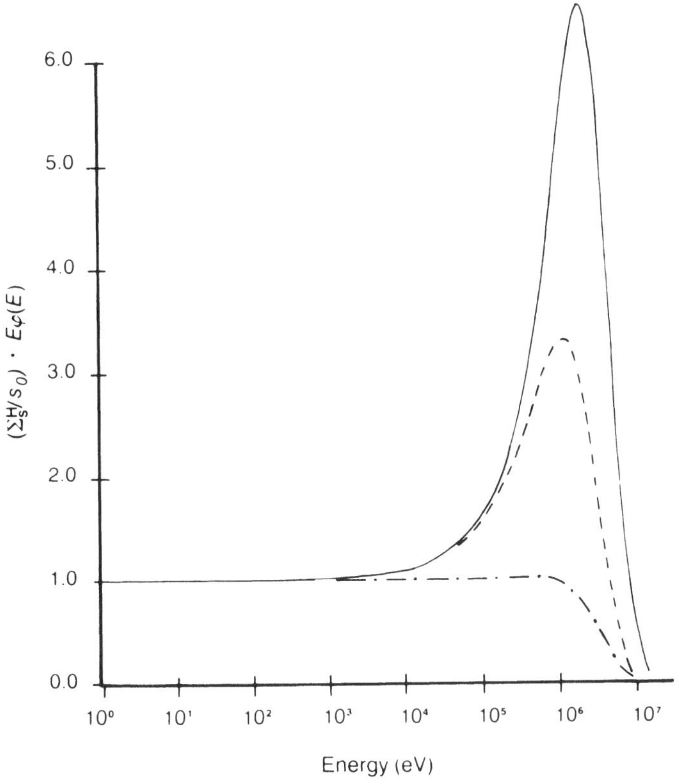

Fig. 4-10. Slowing down spectrum in hydrogen for constant (—·—·—) and energy-dependent (— — — and ———) scattering cross sections. The solid line includes the uncollided spectrum.

and particularly above ~1 MeV, the situation is different. Above 1 MeV, the situation is nearly reversed, namely ^{238}U slows neutrons down much more effectively than hydrogen. But for resonance absorption, one essentially has to consider the range below ~50 keV, and the slowing down model should be accurate in this range. For simplicity, however, the solution derived for the range below 50 keV is formally extended to higher energies, although the solution presented is not accurate in that range.

The balance equation for this important slowing down approximation is given by Eq. (4.85c) (see also the narrow-resonance infinite-mass approximation in Sec. 5-4):

$$\Sigma_t(E)\varphi(E) - \int_E^\infty \Sigma_s^H(E')\varphi(E')\,\frac{dE'}{E'} = s_0\chi(E) \quad . \tag{4.101}$$

The artifice that was applied previously, namely, to introduce the scattering density as an unknown function, is not directly applicable since one φ is multiplied by Σ_s^{H}, the other by Σ_t. But one can arrive at the same type of equation if the "collision density,"[f]

$$\Sigma_t(E)\varphi(E) = F(E) \quad , \tag{4.102}$$

is introduced as the unknown function. This leads to

$$F(E) - \int_E^{\infty} \frac{\Sigma_s^{\mathrm{H}}(E')}{\Sigma_t(E')} F(E') \frac{dE'}{E'} = s_0 \chi(E) \quad . \tag{4.103}$$

Differentiating Eq. (4.103) with respect to E yields an equation of the same form as the first and very simple approximation [compare Eq. (4.90a)]:

$$\frac{dF(E)}{dE} + \frac{1}{E} \frac{\Sigma_s^{\mathrm{H}}(E)}{\Sigma_t(E)} F(E) = s_0 \frac{d\chi(E)}{dE} \quad . \tag{4.104}$$

The only differences are that F appears instead of φ and $1/E$ is replaced by the more complicated expression:

$$\frac{1}{E} \frac{\Sigma_s^{\mathrm{H}}(E)}{\Sigma_t(E)} = \frac{1}{E}\left[1 - \frac{\Sigma_a(E)}{\Sigma_t(E)}\right] = -a(E) \tag{4.105}$$

in the general formula of Eq. (4.88).

In the solution equation, Eq. (4.89), the expression, Eq. (4.105), will appear under the integral in the exponent:

$$\int_E^{E'} \left(1 - \frac{\Sigma_a}{\Sigma_t}\right) \frac{dE''}{E''} = \ln\frac{E'}{E} - \int_E^{E'} \frac{\Sigma_a(E'')}{\Sigma_t(E'')} \frac{dE''}{E''} \quad . \tag{4.106}$$

The exponential function of the logarithm is carried through in the same way as before. The second term on the right side of Eq. (4.106) is related to what is called the "resonance integral." To avoid interrupting the solution procedure of Eq. (4.104), the extensive discussion of the resonance integral is postponed until Chapter 5. The following abbreviation is introduced:

$$h(E) = \int_E^{E_u} \frac{\Sigma_a(E')}{\Sigma_s(E') + \Sigma_a(E')} \frac{dE'}{E'} \propto \text{resonance integral} \quad . \tag{4.107}$$

---

[f]Again, with the pure spectrum $\varphi(E)$, the quantity $F(E)$ is not actually a density; one has first to multiply by $\phi(\boldsymbol{r})$ to obtain the collisions per cubic centimeter and second. The index "t" on F is omitted as is usual. (See footnote e.)

The resonance integral is a measure of the neutron losses in the capture and fission resonances above the energy E up to a reasonable[g] upper limit, E_u. Inserting Eqs. (4.106) and (4.107) into the formula for the general solution yields:

$$F(E)=\frac{s_0}{E}\int_{\infty}^{E}\frac{d\chi(E')}{dE'}\cdot E'\exp[h(E')-h(E)]\,dE'$$

$$=\left\{\int_{\infty}^{E}\frac{d\chi(E')}{dE'}\cdot E'\exp[h(E')]\,dE'\right\}\cdot\frac{s_0}{E}\exp[-h(E)] \quad . \qquad (4.108)$$

The resonance integral in the source range is very small,[g] because E' is close to E_u and Σ_a is small compared to Σ_s. Therefore, $h(E')$ can be neglected in the integral and the integration can be carried out by parts the same way as before in Eq. (4.91):

$$\int_{\infty}^{E}\chi'(E')E'\exp[h(E')]\,dE'\simeq\int_{\infty}^{E}\chi'(E')E'\,dE'$$

$$=E\chi(E)+\int_{E}^{\infty}\chi(E')\,dE' \quad . \qquad (4.109)$$

This gives for $F(E)$ [note that $h(E)$ is consistently neglected in the source range]:

$$F(E)=s_0\chi(E)+\frac{s_0}{E}\exp[-h(E)]\int_{E}^{\infty}\chi(E')\,dE' \quad . \qquad (4.110)$$

Well below the source range, $\chi(E)$ can again be neglected, and the integral over χ is practically one. This gives the very important formula for the neutron spectrum as it results from slowing down in hydrogen, with absorption of neutrons by heavy nuclei, for energies well below the source energies:

$$\varphi(E)=\frac{s_0}{E\Sigma_t(E)}\exp[-h(E)] \quad ; \qquad (4.111)$$

or explicitly:

$$\varphi(E)=\frac{s_0}{E\Sigma_t(E)}\exp\left[-\int_{E}^{E_u}\frac{\Sigma_a(E')}{\Sigma_t(E')}\frac{dE'}{E'}\right] \quad . \qquad (4.112)$$

---

[g]The actual choice of the upper energy limit E_u is not important; it may be a value above the resonance range, below the fast fission range.

The in-depth understanding of the structure and the content of this formula is of great importance for the understanding of a substantial part of reactor physics. The solution equation, Eq. (4.112), is therefore discussed extensively in Sec. 5-3, where the subject of resonance absorption and its impact on the neutron spectrum are presented. The exponential function in Eq. (4.112) describes the probability for a neutron *not* being absorbed during slowing down between E_u and E. Since all absorption is concentrated in "resonances," the exponential function in Eq. (4.112) is called the "resonance escape probability," $p(E)$, given here in the narrow-resonance infinite-mass approximation for hydrogen as the moderator.

4-5E Slowing Down with $\sigma_s(E)$, Absorption, and Leakage

As it was shown in the derivation of the balance equation for a separated neutron spectrum, the losses through leakage of neutrons out of a bare system are described by:

$$D(E)B^2\varphi(E) \quad , \tag{4.113}$$

with $B^2 = B_m^2$ if $\varphi(E)$ is to be found for a critical system, or $B^2 = B_{\text{geo}}^2$ if one calculates $\varphi(E)$ for a given geometry (a proper average B^2 may have to be found if the flux is not separable).

The term DB^2 can be formally included to yield the "total loss" macroscopic cross section:

$$\Sigma_t(E) \rightarrow \Sigma_{tl}(E) = \Sigma_t(E) + D(E)B^2 \quad , \tag{4.114}$$

where Σ_t may or may not include the scattering on heavier atoms (see Sec. 5-4).

The neutron spectrum for these conditions can be copied from the previous solution by replacing Σ_t by Σ_{tl} and Σ_a by $\Sigma_a + DB^2$; thus, well below the source range, one obtains:

$$\varphi(E) = \frac{s_0}{E\Sigma_{tl}(E)} \exp\left[-\int_E^{E_u} \frac{\Sigma_a(E')\,dE'}{\Sigma_{tl}(E')E'} - \int_E^{E_u} \frac{D(E')B^2}{\Sigma_{tl}(E')} \frac{dE'}{E'}\right] \quad . \tag{4.115}$$

Comparing the solutions of Eqs. (4.115) and (4.112) shows that there is an additional term in the exponent of the exponential function. Furthermore, all components of the infinite medium solution, Eq. (4.112), are altered by the inclusion of the leakage in the denominators. For large reactors, however, in which the leakage and thus B_{geo}^2 and B_m^2 are small, the leakage contribution to the total loss cross section in Eq. (4.114) is small.

A neutron in a thermal reactor may have to be scattered 25 to 50 times in order to reach the thermal range; but a neutron can leak out only once. This means that in thermal reactors, Σ_s is much larger than DB^2 on the average along the energy scale. Therefore, one can approximate Σ_{tl} by Σ_t in the denominators of Eq. (4.115):

$$\Sigma_{tl}(E) \simeq \Sigma_t(E) \quad . \tag{4.116}$$

But, Eq. (4.116) is a poor approximation for small high-leakage reactors. In these cases, the DB^2 term has to be retained in the denominators of Eq. (4.115).

Application of Eq. (4.116) gives then the approximate slowing down spectrum for large reactors (well below the source range):

$$\varphi(E) \simeq \frac{s_0}{E\Sigma_t(E)} \exp[-h(E)] \exp[-h_l(E)] \quad , \tag{4.117}$$

with $h(E)$ describing the resonance absorption according to Eq. (4.107) (see also Chapter 5) and $h_l(E)$ describing the leakage-loss during the slowing down process:

$$h_l(E) = B^2 \int_E^{E_u} \frac{D(E')}{\Sigma_t(E')} \frac{dE'}{E'} \simeq B^2 \tau \quad , \tag{4.118}$$

with τ being the approximate "Fermi age" (see, for example, Sec. 5-5A).

The second exponential function in Eq. (4.117) describes the nonleakage probability for fast neutrons, i.e., between the E and E_u:

$$P_f(E) = \exp[-h_l(E)] \quad , \tag{4.119}$$

with P_f denoting the nonleakage probability for "fast" neutrons. Of particular interest is the probability of neutrons not leaking out of the system during the entire slowing down process, i.e., $P_f(E_{th})$, the nonleakage probability at the brink of the thermal energy range:

$$P_f(E_{th}) = \exp[-h_l(E_{th})] \quad . \tag{4.120}$$

In large thermal power reactors, the probability for fast leakage,

$$1 - P_f(E_{th}) \quad , \tag{4.121}$$

does not amount to more than several percent ($\leq 10\%$). This also emphasizes that the neglect of DB^2 compared to Σ_t in the denominators of solution equation (4.115) is well justified [i.e., Eq. (4.116) is a good approximation].

Thus, the results of the spectrum calculation presented in this section show that:

> In large thermal reactors, the leakage during slowing down and the resonance absorption appear additive in an exponent.

Therefore, both features can be described by individual factors, the resonance escape probability, and the fast nonleakage probability. This also implies that the resonance absorption can be calculated without considering leakage, i.e., for an infinite system.

It is of interest to compare Eq. (4.119) with other formulas for the fast neutron energy group nonleakage probability in two-group theory (for the derivation, see Chapter 7):

$$P_f = \frac{1}{1 + L_f^2 B^2} \quad \text{(two-group theory)} \tag{4.122a}$$

and

$$P_f = \exp(-\tau B^2) \quad \text{(Fermi age theory)} \quad . \tag{4.122b}$$

The neutron "age" (Fermi age), τ, has the dimensions of length squared; it is approximately given by the integral in Eq. (4.118). The "Fermi age" theory is a model for the treatment of the combined space-energy dependence of the flux. The space dependence is treated in the diffusion approximation; the slowing down is approximated as a quasicontinuous process, which is not, however, accurate enough in the presence of resonance absorbers (see Sec. 5-5A).

4-6 Slowing Down in Nonhydrogeneous Materials Without Absorption

Nonhydrogeneous moderators such as D_20 or graphite are considered in this section. All moderators have negligible neutron capture cross sections in the slowing down region and the scattering cross section consists of potential cross sections only. Furthermore, the resonances (see Sec. 5-2) in these materials occur only at energies >>50 keV. The problem is treated at first for a single moderator isotope; the case of a mixture of moderator isotopes is treated later.

The balance equation, under the assumptions of constant Σ_s and neglect of inelastic scattering, is given by:

$$\Sigma_s \varphi(E) - \int_E^{E/\alpha} \frac{\Sigma_s}{1-\alpha} \varphi(E') \frac{dE'}{E'} = s_0 \chi(E) \quad , \tag{4.123}$$

using the general scattering kernel of Eq. (4.28). Differentiating[h] this integral equation leads to a difference-differential equation:

[h]Compare Appendix D for the general formula for the derivative of an integral.

$$\Sigma_s \frac{d\varphi(E)}{dE} - \frac{\Sigma_s}{1-\alpha}\left[\varphi\left(\frac{E}{\alpha}\right)\frac{1}{E} - \varphi(E)\frac{1}{E}\right] = s_0 \frac{d\chi(E)}{dE} \quad . \tag{4.124}$$

The fact that the flux argument appears at two different values $\left(\text{i.e., } E \text{ and } \frac{E}{\alpha}\right)$ makes Eq. (4.124) a "difference" equation. Difference equations are much more difficult to solve mathematically than differential equations. Even this simple difference-differential equation (4.124) does not have a known analytical solution. Thus, for further analytical treatment, the balance equation for a nonhydrogeneous moderator has to be approximated.

Since the integration interval (the maximum change in energy due to one elastic scattering) is much smaller than for hydrogen, one may try to expand the integrand at the lower limit into a Taylor expansion. Generally, a Taylor expansion of an integrand yields:

$$\int_x^{x_1} f(x')\, dx' = \int_x^{x_1} \left[f(x) + (x'-x)\frac{df(x)}{dx} + \ldots\right] dx' \tag{4.125}$$

and

$$\int_x^{x_1} f(x')\, dx' = f(x)(x_1 - x) + \frac{df(x)}{dx}\frac{(x_1-x)^2}{2} + \ldots \quad . \tag{4.126}$$

By application of such an expansion, the difference equation, Eq. (4.124), is avoided and a differential equation of infinite order is formally obtained. However, with the small number of terms of Eq. (4.126), the integral can often be represented with sufficient accuracy.

The accuracy of such an expansion can generally be improved either by explicitly introducing the major dependency or by a proper transformation of variables as shown in the following examples.

1. If $f(x)$ is roughly proportional to x, then one can set

   $$f(x) = xg(x)$$

   and $g(x)$ will vary slower than $f(x)$.
2. If $f[x(y)] = g(y)$ varies slower in y than f in x, then better convergence of the expansion is obtained using the new function $g(y)$ rather than $f(x)$.

As to which of these methods can be best applied to the specific integral in Eq. (4.123), observe that $\varphi(E)$ is expected to vary below the source approximately like $1/E$. In other words, the function $E\varphi(E)$ and thus the spectrum as a function of lethargy $[\varphi(u)]$,

$$E\varphi(E) = \varphi(u) \quad , \tag{4.127}$$

will be nearly constant over a wide range, or at least slowly varying. Therefore, the balance equation is transformed into lethargies and $\varphi(u)$ is expanded. Equation (4.123) is multiplied by E and the energy-lethargy relationship, Eq. (4.127), is introduced for the transformation of the integral (see Sec. 4-2B):

$$\Sigma_s\varphi(u) - \int_{u-a}^{u} \Sigma_s(u' \rightarrow u)\varphi(u')\, du' = s_0\chi(u) \quad , \tag{4.128}$$

with

$$\chi(u) = E\chi(E) \quad . \tag{4.129}$$

The integration limits in Eq. (4.128) have been interchanged to account for a minus sign in the differential. Now, $\varphi(u')$ is expanded in a Taylor series at the upper integration limit:

$$\varphi(u') = \varphi(u) + (u' - u)\frac{d\varphi(u)}{du} + \ldots \quad . \tag{4.130}$$

Inserting this expression into the scattering integral gives:

$$\int_{u-a}^{u} \Sigma_s(u' \rightarrow u)\varphi(u')\, du' \simeq \varphi(u)\int_{u-a}^{u} \Sigma_s(u' \rightarrow u)\, du'$$

$$+ \frac{d\varphi(u)}{du}\int_{u-a}^{u} (u' - u)\Sigma_s(u' \rightarrow u)\, du' \quad . \tag{4.131}$$

The two integrals in Eq. (4.131) are carried out over the starting lethargy, not over the final lethargy. But the lethargy distribution of the scattered neutrons depends only on the difference of lethargies—provided scattering is isotropic in the CMS as has been assumed throughout this discussion:

$$\Sigma_s(u' \rightarrow u) = \Sigma_s(u') \cdot K(u - u') \quad , \tag{4.132}$$

with $K(u - u')$ being the scattering kernel. Therefore, the integrals on the right side of Eq. (4.131) can be transformed into a simpler form, if one assumes in addition that $\Sigma_s(u')$ is constant:

$$\Sigma_s(u') = \Sigma_s(u) = \text{constant} \quad . \tag{4.133}$$

This is fulfilled for potential scattering in the energy range of interest.

Employing Eqs. (4.132) and (4.133), the two integrals on the right side of Eq. (4.131) become:

$$\int_{u-a}^{u} \Sigma_s(u' \rightarrow u)\, du' = \Sigma_s\int_{u-a}^{u} K(u - u')\, du' \tag{4.134a}$$

and

$$\int_{u-a}^{u} (u' - u)\Sigma_s(u' \rightarrow u)\, du' = \Sigma_s \int_{u-a}^{u} (u' - u)K(u - u')\, du' \quad . \tag{4.134b}$$

Substituting into Eqs. (4.134),

$$w = u - u' \quad , \tag{4.135}$$

gives for the integrals:

$$\Sigma_s \int_{u-a}^{u} K(u - u')\, du' = \Sigma_s \int_0^a K(w)\, dw \tag{4.136a}$$

and

$$\Sigma_s \int_{u-a}^{u} (u' - u)K(u - u')\, du' = -\Sigma_s \int_0^a wK(w)\, dw \quad . \tag{4.136b}$$

The integrals in terms of w (where $w = u - u'$) in Eqs. (4.136) are carried out over the starting lethargy u' with the final lethargy u being a constant. The same integrals but with a simpler physical meaning are obtained from the integrals over the final lethargy.

The direct integration over all final lethargies yields the total scattering cross section. Therefore, the integral over the kernel is equal to one [u' is the final lethargy in Eqs. (4.137)]:

$$\int_{u}^{u+a} K(u' - u)\, du' = 1 \quad . \tag{4.137a}$$

Using as w the same value as Eq. (4.135), but with u' denoted by u and u by u',

$$w = u' - u \quad , \tag{4.137b}$$

gives:

$$\int_0^a K(w)\, dw = 1 \quad , \tag{4.138a}$$

since the distribution over the final lethargies is normalized to one. In the same way, one obtains the analog for the second integral on the right side of Eq. (4.131):

$$\int_{u}^{u+a} \Sigma_s(u \rightarrow u')(u' - u)\, du' = \Sigma_s \int_0^a wK(w)\, dw = \xi\Sigma_s \quad , \tag{4.138b}$$

with ξ, the average lethargy change defined by Eq. (4.51).

Inserting the integrals of Eqs. (4.138) in Eqs. (4.136) yields:

$$\Sigma_s \int_{u-a}^{u} K(u-u')\, du' = \Sigma_s \tag{4.139a}$$

and

$$\Sigma_s \int_{u-a}^{u} (u'-u)K(u-u')\, du' = -\xi\Sigma_s \quad . \tag{4.139b}$$

With the integral from Eqs. (4.139), the expansion, Eq. (4.131), becomes:

$$\int_{u-a}^{u} \Sigma_s(u' \rightarrow u)\varphi(u')\, du' = \Sigma_s\varphi(u) - \xi\Sigma_s \frac{d\varphi(u)}{du} \quad . \tag{4.140}$$

Insertion of Eq. (4.140) into Eq. (4.128) yields, after cancellation of the two Σ_s terms, the simple balance equation (for a constant Σ_s):

$$\xi\Sigma_s \frac{d\varphi(u)}{du} = s_0\chi(u) \quad . \tag{4.141}$$

Equation (4.141) is to be solved with the boundary condition $\varphi(u \rightarrow -\infty) = 0$. This gives

$$\varphi(u) = \frac{s_0}{\xi\Sigma_s} \int_{-\infty}^{u} \chi(u')\, du' \quad . \tag{4.142}$$

Transforming Eq. (4.142) back into energy yields the slowing down spectrum due to scattering with nonhydrogenous isotopes:

$$\varphi(E) = \frac{s_0}{E \cdot \xi\Sigma_s} \int_{E}^{\infty} \chi(E')\, dE' \quad . \tag{4.143}$$

Several aspects of Eq. (4.143) are discussed in the following:

1. In the derivation of Eq. (4.143) from a Taylor expansion, it was assumed that $E\varphi(E)$ is slowly varying. Equation (4.143) shows that $E\varphi(E)$ is constant below the source if Σ_s is constant. This gives some validity to the two-term Taylor series expansion [Eq. (4.130)] and it is certainly fulfilled below the source.

The integral of χ is equal to unity below the source; one then obtains the same type of an asymptotic spectrum as for hydrogen:

$$\varphi(E) = \frac{s_0}{E\xi\Sigma_s} \tag{4.144}$$

below the source. For $\xi = 1$, Eq. (4.144) automatically turns into the result for scattering by hydrogen [compare Eq. (4.95)].

2. Since the assumption of slowly varying $\varphi(u)$ is exactly true below the source, one may expect to have the correct solution there. The

simplest proof consists of two parts. At first, it is shown that a $1/E$ spectrum satisfies the balance equation below the source [i.e., substitute $\varphi(E) = 1/E$ into Eq. (4.123)]. This gives:

$$\Sigma_s \frac{1}{E} - \frac{\Sigma_s}{1-\alpha} \int_E^{E/\alpha} \frac{dE'}{E'^2} = 0 \quad . \tag{4.145}$$

Since the balance equation below the source is homogeneous, it provides no information regarding the validity of the factor $s_0/\xi\Sigma_s$. The factor can be proved by checking that no neutrons have been lost. It has been assumed that no neutrons are lost by absorption or leakage; thus, the number of neutrons slowed down below any energy E has to be equal to s_0 for all E below the source. This quantity, generally referred to as the "slowing down density," $q_{sd}(E)$, should not be confused with $S_s(E)$ the scattering source at energy E. It should also be noted that $q_{sd}(E)$ does not have the dimension of a "density" if—as is the case here—$\phi(\mathbf{r})$ has been eliminated by separation.

The fraction of the total number of scattering events at E', which describes neutrons scattered below E, is given by the ratio of the two intervals:

$$\frac{E - \alpha E'}{E' - \alpha E'} = \frac{1}{1-\alpha} \frac{E}{E'} - \frac{\alpha}{1-\alpha} \quad . \tag{4.146}$$

Multiplication of Eq. (4.146) with the scattering rate at E' and integration over all possible E', which contribute to the slowing down density at E, should yield the source strength s_0. By inserting Eq. (4.144) for $\varphi(E)$, one obtains:

$$\begin{aligned} q_{sd}(E) &= \int_E^{E/\alpha} \Sigma_s \varphi(E') \frac{E - \alpha E'}{E' - \alpha E'} \, dE' \\ &= \frac{s_0}{\xi} \int_E^{E/\alpha} \frac{E - \alpha E'}{1-\alpha} \frac{dE'}{E'^2} \end{aligned} \tag{4.147a}$$

and

$$q_{sd}(E) = \frac{s_0}{\xi} \left(1 - \frac{\alpha}{1-\alpha} \ln \frac{1}{\alpha} \right) = s_0, \text{ for } E \text{ below source} \quad . \tag{4.147b}$$

The term in parentheses is equal to ξ, according to Eq. (4.36). Thus, Eq. (4.145) gives the correct solution of Eq. (4.123) below the source.

3. The assumption of slowly varying $\varphi(u)$, made for this derivation, is not correct in the source range. Comparison with the solution for hydrogen shows that the virgin spectrum has been lost by extending the assumption of slowly varying $E\varphi(E)$ into the source range.

4. For more than one slowing down isotope, one obtains in the same way, instead of Eq. (4.141):

$$\overline{\xi\Sigma}_s \frac{d}{dE}[E\varphi(E)] = -s_0\chi(E) \quad , \tag{4.148}$$

with

$$\overline{\xi\Sigma}_s = \sum_i \xi_i\Sigma_{si} = \bar{\xi}\Sigma_s \quad . \tag{4.149}$$

Equations (4.143) and (4.144) describe the spectrum in the normally occurring multi-isotope case; $\xi\Sigma_s$ is to be replaced by $\overline{\xi\Sigma}_s$ or $\bar{\xi}\Sigma_s$, Eq. (4.149).

The absorption and leakage during the slowing down process are treated in Chapter 5.

4-7 The Thermal Neutron Spectrum

4-7A General Considerations

In the slowing down range, the energy of the neutrons is so large that they always *lose* energy in a scattering event. This implies that the scattering kernel is zero for final neutron energies larger than the initial one. However, slowing down cannot continue indefinitely. The scattering nuclei are in thermal motion corresponding to the temperature of the scattering medium; when the neutron momentum eventually becomes comparable to that of the scattering nuclei, the neutron can gain as well as lose energy in a collision. In addition, the neutron energy in the thermal range becomes comparable with binding energies in chemical compounds, crystals, etc. This can lead to complicated transfer cross sections for elastic scattering.

The neutrons produced by fission are emitted at relatively high energies. Therefore, the neutron fission source term, $s_0\chi(E)$, is essentially zero in the thermal energy range. Neutrons can reach the thermal range only through scattering out of an intermediate energy range, which provides neutrons through a scattering source.

The calculation of the neutron spectrum in the thermal range is a difficult task if all aspects are considered that result from the complicated scattering cross sections and the heterogeneous structure of the core. The basic features, however, can be derived by considering the idealized situation that corresponds to neutrons interacting with a gas at a given temperature. If neutron losses (absorption and leakage) are neglected, the so-called "principle of detailed balance" holds. This principle requires that the spectrum be a Maxwell distribution, independent of the complexities of the scattering cross section.

4-7B Separation of "Fast" and "Thermal" Neutrons

The observations presented in the previous section suggest that because of the very different behavior of slower neutrons, the neutron energy range can be divided into two broad parts; a "fast" range, in which the neutron only loses energy by collisions with scattering nuclei, and a "thermal" range, in which neutrons both gain and lose energy in collisions. The neutron population in this thermal range *will then be in some kind of equilibrium with the thermal motion of the scattering nuclei.* A sharp distinction between "fast" neutrons and "thermal" neutrons is, of course, artificial and cannot be physically realized. Nevertheless, it is conceptually quite useful to define an energy, E_{th}, that separates "fast" and "thermal" neutrons. Both the fast and thermal neutrons are further subdivided into several neutron groups (see Chapter 7). If a distinction is needed between neutrons within the thermal range and those above the thermal range, the latter is often called the "epithermal" range.

The maximum "thermal" energy, E_{th}, is chosen in various ways, depending on the type of application and the degree of sophistication of a computational treatment.

To physically separate thermal and epithermal neutrons, one usually uses a sheet of cadmium, which has a neutron capture cross section of several thousand barns in the thermal range. Cadmium absorbs the thermal neutrons below ~0.4 eV. Thus, a difference measurement, with and without a cadmium shield, provides information on neutrons below 0.4 eV. This suggests a definition for E_{th} near 0.4 eV (the so-called "cadmium cutoff energy"), at least for comparison with experimental results.

In simple analytical treatments, relatively narrow thermal groups are defined to allow the application of simplifying assumptions (such as a $1/v$ cross section. Note that ^{235}U and ^{239}Pu have their first resonance near 0.3 eV (i.e., below the cadmium cutoff energy). This suggests E_{th} to be ~0.2 eV.

In sophisticated analytical treatments, the opposite consideration is applied. One chooses the thermal energy range so large that only an absolutely negligible amount of neutrons are scattered beyond E_{th}. The average thermal energy is kT, with $kT \simeq 0.025$ eV at room temperature, i.e., at ~300 K. The temperature of the scattering materials in a reactor core is much higher than room temperature: The temperature of the coolant is higher by a factor of ~2 to 3 and the fuel temperature by a factor of ~3 to 6 (in degrees Kelvin). But there is a wide variation of energies around the average value, which corresponds to the temperature. For E_{th} to fully contain all upscattered neutrons, values larger than the cadmium cutoff energy, e.g., 0.6 eV, are chosen. This selection of

E_{th}, based on neutron upscattering considerations, facilitates the numerical solution of the multigroup model (see Chapter 7).

4-7C Balance Equation for Thermal Neutrons in an Infinite Absorbing System

Based on the observations discussed in the previous sections, the balance equation for the source-free thermal region for $E \leq E_{th}$ is:

$$\Sigma_t(E)\varphi(E) = \int_0^\infty \Sigma_s(E' \to E)\varphi(E')\, dE' \quad . \tag{4.150}$$

The integral in Eq. (4.150) describes both the scattering into, as well as the scattering within, the thermal group. It can be separated as follows:

$$\Sigma_t(E)\varphi(E) = \int_0^{E_{th}} \Sigma_s(E' \to E)\varphi(E')\, dE' + \int_{E_{th}}^\infty \Sigma_s(E' \to E)\varphi(E')\, dE' \quad . \tag{4.151}$$

If E is within the thermal range, then the latter integral represents the source of neutrons that arrives in the thermal range due to scattering in the epithermal range. Denoting the source of thermal neutrons by $s_{th}(E)$,

$$s_{th}(E) = \int_{E_{th}}^\infty \Sigma_s(E' \to E)\varphi(E')\, dE'; \quad E \leq E_{th} \quad , \tag{4.152}$$

and rearranging terms gives the balance equation in the form:

$$\Sigma_t(E)\varphi(E) - \int_0^{E_{th}} \Sigma_s(E' \to E)\varphi(E')\, dE' = s_{th}(E) \quad . \tag{4.153}$$

From the differential balance equation, Eq. (4.153), one can readily derive the corresponding integral neutron balance. Integration of Eq. (4.153) over the thermal range gives:

$$\int_0^{E_{th}} \Sigma_t(E)\varphi(E)\, dE - \int_0^{E_{th}}\int_0^{E_{th}} \Sigma_s(E' \to E)\, dE\varphi(E')\, dE' = \int_0^{E_{th}} s_{th}(E)\, dE \quad . \tag{4.154}$$

If E_{th} is chosen to be so large that no upscattering occurs beyond E_{th}, then the integration with respect to E comprises *all* scattered neutrons in the thermal range. Consequently, the E integration yields the scattering cross section:

$$\int_0^{E_{th}} \Sigma_s(E' \rightarrow E)\, dE = \Sigma_s(E') \quad . \tag{4.155}$$

The remaining integral,

$$\int_0^{E_{th}} \Sigma_s(E')\varphi(E')\, dE' \quad , \tag{4.156}$$

is the scattering rate in the thermal group, which cancels the corresponding term in the total reaction rate. This gives the obvious integral balance of thermal neutrons in an infinite medium:

$$\int_0^{E_{th}} \Sigma_a(E')\varphi(E')\, dE' = \int_0^{E_{th}} s_{th}(E)\, dE = s_{th} \quad , \tag{4.157}$$

i.e., all neutrons scattered into the thermal group are absorbed in this group (since none are scattered back).

In the theoretical treatment of neutron slowing down, one often applies the concept of "slowing down density," $q_{sd}(E)$, which was introduced in the previous section. Application of Eq. (4.147b) for E_{th} shows that the source in Eq. (4.157) is the slowing down density at E_{th}:

$$s_{th} = q_{sd}(E_{th}) \quad . \tag{4.158}$$

Applying Eq. (4.144) near E_{th} [i.e., considering s_{th}, the residual neutron source at E_{th}, to be the equivalent of s_0 in Eq. (4.144)] gives:

$$\varphi(E_{th}) = \frac{s_{th}}{E_{th}\xi\Sigma_s} = \frac{q_{sd}(E_{th})}{E_{th}\xi\Sigma_s} \quad . \tag{4.159}$$

If one further expresses the left side of Eq. (4.157) by a thermal flux and an average thermal cross section,

$$\varphi_{th} = \int_0^{E_{th}} \varphi(E)\, dE \tag{4.160a}$$

and

$$\Sigma_{a,th} = \frac{1}{\varphi_{th}} \int_0^{E_{th}} \Sigma_a(E)\varphi(E)\, dE \quad , \tag{4.160b}$$

one can write the balance equation, Eq. (4.144), in the form:

$$\Sigma_{a,th}\varphi_{th} = q_{sd}(E_{th}) = \overline{\xi\Sigma_s}(E\varphi)_{E_{th}} \quad . \tag{4.161a}$$

By inserting Eq. (4.159), one obtains instead:

$$\frac{\text{spectrum integral}}{\text{"boundary" value of } E\varphi(E)} = \frac{\varphi_{th}}{E_{th}\varphi(E_{th})} = \frac{\overline{\xi\Sigma_s}}{\Sigma_{a,th}} = \frac{1}{\Delta} \quad . \tag{4.161b}$$

Note that φ_{th} is a spectrum integral and therefore has the dimension of $E\varphi(E)$.

According to Eq. (4.161b), the ratio of the thermal flux, φ_{th}, and epithermal flux level, $\varphi(E_{th})$, is determined by the ratio of the epithermal slowing down power and the thermal absorption cross section, in a sense, by the ratio of source-sink cross sections. The same ratio also determines other important thermal neutron characteristics (see, for example, Secs. 4-7E and 4-7F). The abbreviating denotation Δ is widely used for this quantity.

4-7D Thermal Neutron Balance in a Source-Free Nonabsorbing Infinite System

As a limiting case, it is important to consider the balance relationship for thermal neutrons in a source-free nonabsorbing infinite medium.

For this situation, the general balance expression of Eq. (4.151) simplifies to:

$$\Sigma_s(E)\varphi(E) = \int_0^\infty \Sigma_s(E' \rightarrow E)\varphi(E')\, dE' \quad . \tag{4.162}$$

The scattering cross section, $\Sigma_s(E)$, is related to the differential cross section, $\Sigma_s(E \rightarrow E')$, by integration over all possible final energies:

$$\Sigma_s(E) = \int_0^\infty \Sigma_s(E \rightarrow E')\, dE' \quad . \tag{4.163}$$

Inserting this into Eq. (4.162) brings the balance equation into the form:

$$\int_0^\infty \Sigma_s(E \rightarrow E')\varphi(E)\, dE' = \int_0^\infty \Sigma_s(E' \rightarrow E)\varphi(E')\, dE' \quad . \tag{4.164}$$

The physical interpretation of Eq. (4.164) is the same as that of Eq. (4.162), namely, all neutrons scattered to energy E on the right side are scattered away from energy E on the left side; otherwise the spectrum φ would change with time and would not be in an equilibrium. Actually, the condition for a thermal equilibrium state is much more severe and detailed than it appears from the integrals in Eq. (4.164). The balance of scattering to and from an energy must hold for all E and E':

$$\Sigma_s(E \rightarrow E')\varphi(E) = \Sigma_s(E' \rightarrow E)\varphi(E') \quad . \tag{4.165}$$

This is the famous "principle of detailed balance"; it holds for all statistical phenomena in physics, provided the system is free of sources and sinks. It requires that $\varphi(E)$ be a Maxwellian distribution $[\varphi_M(E)]$. If its energy integral is normalized to unity, one has

$$\varphi_M(E) = \frac{E}{(kT)^2} \exp\left(-\frac{E}{kT}\right) \quad . \tag{4.166}$$

Combining Eqs. (4.165) and (4.166) brings the principle of detailed balance to the form:

$$\Sigma_s(E \rightarrow E')\varphi_M(E) = \Sigma_s(E' \rightarrow E)\varphi_M(E') \quad , \tag{4.167}$$

i.e., in thermal equilibrium, the same number of neutrons scatter from E to E' as scatter from E' to E. This is a more detailed statement than the one expressed by Eq. (4.164), since the latter requires only that the number of neutrons scattered from *all* energies (E') to E is the same as the number scattered from E to *any other energy*. Whereas that latter statement is obvious for a stationary state, the principle of detailed balance is a physical law, which was discovered in the last century. Several proofs of that principle can be found in textbooks on thermodynamics and in Ref. 3.

4-7E The Shape of the Thermal Neutron Spectrum

Establishing a thermal equilibrium requires that the medium be source and loss free. In this case, the neutron spectrum in the thermal range is always a Maxwellian distribution, independent of the form and the complexity of the scattering transfer cross section.

A large pool of heavy water and a large block of pure graphite are media in which absorption and leakage losses are very small. By placing an external source near such a system, one can practically establish a stationary spectrum that is very close to a Maxwellian distribution.

A reactor core contains fuel and therefore has a much larger absorption rate than a D_2O pool. With increasing absorption and leakage loss, compensated by an increasing neutron source, the spectrum deviates more and more from the Maxwellian spectrum of the thermal equilibrium. Consequently, its shape becomes more and more dependent on the specifics of the absorption and scattering cross sections.

Normally, thermal range absorption cross sections have a $1/v$ shape, which results from the low-energy tail of the nearest resonance. If the resonance nearest to $E = 0$ happens to be in the thermal range, the absorption cross sections may be substantially deformed.[5] Important resonances in the thermal range occur in isotopes such as ^{235}U, ^{239}Pu, ^{113}Cd, and ^{135}Xe, but not in ^{238}U and ^{232}Th. An unusually wide $1/v$

absorption cross-section range occurs in ^{10}B, which is used as a reactor control material. It is important for a nuclear engineer to have some familiarity with the shapes of nuclear cross sections in the thermal range. A good presentation and display of cross sections in the thermal range can be found in Ref. 9. More accurate cross sections can be obtained from Refs. 5 and 10.

In the *absence* of absorption and leakage, thermal neutrons are distributed as a Maxwellian spectrum, independent of the characteristics of the scattering nuclei. Alternately, in the *presence* of absorption, leakage, and sources, the thermal neutron spectrum depends on the velocity (or momentum) distribution of the scattering nuclei. Therefore, one must first describe their thermal motion. The velocity distribution of the atoms, and thus of the nuclei, depends on the physical state of the matter in which the nuclei are found.

Molecules in a gas, in collisions, can move freely from one translatory state to another; in a confined volume, there is no loss or source of gas molecules. The principle of detailed balance then applies. Thus, the energy distribution of the gas molecules is Maxwellian:

$$n_M(E) = \frac{2}{\sqrt{\pi}} \frac{\sqrt{E}}{(kT)^{3/2}} \exp\left(-\frac{E}{kT}\right) \quad . \tag{4.168}$$

Note that Eq. (4.168), which is a Maxwellian for a number of particles, is proportional to $\sqrt{E}$ (E = gas molecule energy), but Eq. (4.166), which is Maxwellian for a flux, i.e., velocity times the number of neutrons, is proportional to E.

In the thermalization physics of neutrons, room temperature is commonly defined as 293.6 K (≃20.4°C) in order to obtain

$$v(kT) = v(E = 0.0253 \text{ eV}) = 2200 \text{ m/s} \quad , \tag{4.169a}$$

with

$$kT = 0.0253 \text{ eV and } T = 293.6 \text{ K} \quad . \tag{4.169b}$$

The energy value, kT, is the most probable neutron energy in the Maxwellian distribution, Eq. (4.166). The neutron speed of 2200 m/s with an energy of 0.0253 eV (corresponding to $T = 293.6$ K) is used as a standard value for the "thermal" neutron velocity.

The scattering between neutrons and nuclei is influenced by the chemical and crystal binding effects, and thus by the possible modes of motion of the nuclei. In a gas, the dominating form of motion is translational. In a solid, it is vibrational. More complicated, however, is a liquid in which all three modes of motion occur superimposed: vibration, hindered translation, and rotation.

The next logical step after the consideration of the kinematics of the scattering nuclei is the formulation of a scattering law, in terms of the transfer cross section:

$$\Sigma_s(E' \rightarrow E) \quad . \tag{4.170}$$

The internal kinematics of the scatterer must be worked into the scattering kernel so that it describes the average scattering of neutrons with nuclei in thermal motion. Note that the resulting scattering kernel will be a superposition of scattering kernels between nuclei and neutrons of a *given* velocity; it describes scattering by a system of particles where only the system is at rest in the laboratory system, not the particles themselves. The slowing down kernel is a special example for scattering with individual nuclei of a *given* velocity, namely, $v = 0$ in the laboratory system. This does not mean that the velocity is actually zero; but it is very small compared to the neutron velocity.

The complexities of the scattering kernels for thermal neutrons are beyond the scope of this text. A scattering kernel for H_2O was derived by Nelkin.[11] The Nelkin kernel describes scattering by H_2O molecules as a superposition of vibration, translation, and rotation; it agrees very well with experimental data. The details of development and form of this kernel can be found in more advanced textbooks.[3,12]

The simplest example is the scattering on a hydrogen gas, although even in this case a rather complicated scattering kernel results.

After the scattering kernel is known, one can, in principle, calculate the neutron energy spectrum in the thermal range. The calculation of the thermal spectra is more complicated than the calculation of the slowing down spectra because of the neutron upscattering. If the balance equation is written in finite difference form, it is not triangular (see Sec. 4-3), and the scattering matrix is full. Only a qualitative discussion of the basic results is presented below.

The simplest case is the one with no absorption or leakage (which in practice means negligibly small absorption and leakage, e.g., a large volume of graphite or D_2O). Neutrons injected into such a system will become thermalized; they live long enough in the thermal region to establish a practically unperturbed equilibrium with scattering nuclei. The resultant neutron spectrum is then a Maxwellian distribution.

The average neutron energy is to be calculated from $n_M(E)$, the number of neutrons at E, Eq. (4.168). The result is

$$\bar{E} = \frac{3}{2}kT \quad , \tag{4.171}$$

i.e., $kT/2$ per degree of freedom of the three-dimensional translatory motion.

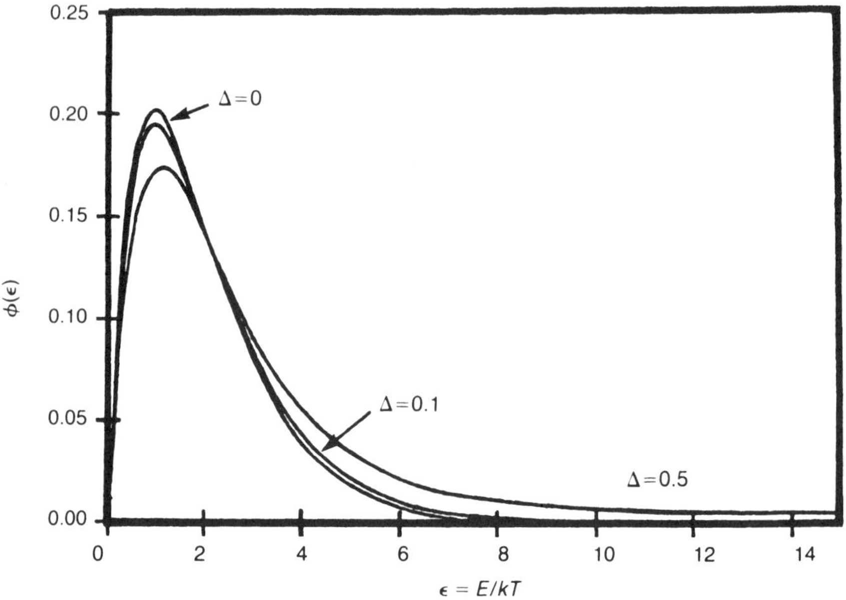

Fig. 4-11. Thermal neutron spectra calculated with the heavy gas model for three values of Δ.

As mentioned above, the Maxwellian distribution follows from fundamental laws of thermodynamics. Its fundamental nature is emphasized by the fact that it is independent of the details of the scattering kernels, and requires only that the scattering medium be homogeneous, infinitely large,[i] nonabsorbing, and source free.

In more realistic neutronics problems, one has capture and fission cross sections in the thermal range, which are approximately $1/v$, i.e.:

$$\sigma_a(E) = \sigma_a(kT_r)\left(\frac{kT_r}{E}\right)^{1/2} \left(\frac{1}{v} - \text{cross section}\right) , \qquad (4.172)$$

with $T_r = 293.6$ K = room temperature. The scattering collisions try to establish a Maxwellian; neutron absorption, however, tends to deform the spectrum.

Figure 4-11 shows a Maxwellian flux spectrum, $\varphi_M(E)$, and two spectra deformed by $1/v$ absorption cross sections. The deformed spectra are calculated with the "heavy gas" model.[3,13] Figure 4-11 indicates that

[i]Or finite, with ideally reflecting walls.

relatively more low- than high-energy neutrons are being absorbed, which leads to a deformation of the Maxwellian distribution. One effect of this deformation is to raise the average energy of the remaining neutrons. This effect is called "absorption hardening." The hardened spectrum appears to be the spectrum at some temperature, T_h, higher than the temperature of the original Maxwellian distribution. One can approximate the hardened spectrum with a Maxwellian at the higher temperature:

$$T_h = T\left(1 + \alpha_h \frac{\Sigma_{a,th}}{\xi\Sigma_s}\right) = T(1 + \alpha_h \Delta) \quad . \tag{4.173}$$

Various values for the constant α_h have been calculated.[3,13] A typical value is $\alpha_h = 1.46$ (Ref. 13). However, since the deformed shape is not actually Maxwellian, it is better to use more exact methods to treat the thermal region and to consider the "temperature shift," given by Eq. (4.173), only as a qualitative indication of the impact of absorption on the thermal spectrum. Examples of computer programs for the calculation of thermal neutron spectra are given in Refs. 14 and 15.

4-7F The Magnitude of the Thermal Spectrum

In this section, a simple estimate of the magnitude of the thermal spectrum relative to that of the slowing down spectrum is discussed. The estimate is simplified by neglecting neutron leakage and capture.

From thermal neutron balance considerations, presented in Sec. 4-7C, a relation between the thermal neutron flux, φ_{th}, and the magnitude of the spectrum above E_{th} has been obtained. The former determines the absorption within the thermal range, the latter is proportional to the scattering source entering that range from above [see Eqs. (4.159), (4.160), and (4.161)]:

$$\frac{\varphi_{th}}{\varphi(E_{th})} = \frac{E_{th}\overline{\xi\Sigma_s}}{\Sigma_{a,th}} = \frac{E_{th}}{\Delta} \quad . \tag{4.174}$$

If the thermal spectrum is approximated by a Maxwellian,

$$\varphi(E) = \varphi_{th}\frac{E}{(kT)^2}\exp(-E/kT) \quad , \tag{4.175}$$

its peak, at $E = kT$, has the value

$$\varphi_{th}^{\max} = \varphi(kT) = \frac{\varphi_{th}}{ekT} \quad . \tag{4.176a}$$

Replacing φ_{th} in Eq. (4.174) by $\varphi_{th}^{\max}$ gives for the rise of the spectrum from its value at E_{th} to its maximum:

$$\frac{\varphi_{th}^{\max}}{\varphi(E_{th})} = \frac{E_{th}\overline{\xi\Sigma_s}}{ekT\Sigma_{a,th}} = \frac{E_{th}}{ekT\Delta} \quad . \tag{4.176b}$$

Note that this ratio becomes infinite if $\Sigma_{a,th}$ is zero. In this case, then, the thermal spectrum becomes a pure Maxwellian.

The actual calculation of the spectrum at the transition region around E_{th} is very complicated and beyond the scope of this text. The reader is referred to more advanced texts, or for practical applications to respective computer programs, such as in Refs. 14 and 15.

4-7G Average Thermal Absorption Cross Section $\Sigma_{a,th}$ and the η Coefficient

For absorption cross sections that vary more or less inversely with the velocity, ($1/v$ cross sections), one can readily calculate the reaction rates with a Maxwellian distribution as an approximation to the thermal spectrum. The absorption cross section is used in the form of Eq. (4.172):

$$\sigma_a(E) = \sigma_a(kT)\left(\frac{kT}{E}\right)^{1/2} = \sigma_a\sqrt{\varepsilon}^{\,-1} \quad , \tag{4.177}$$

where $\varepsilon = E/kT$. Then,

$$\overline{\sigma}_a = \sigma_a(kT)\int_0^{\varepsilon_{th}} \frac{\varepsilon}{\sqrt{\varepsilon}} e^{-\varepsilon}\, d\varepsilon \tag{4.178}$$

and

$$\simeq \sigma_a(kT)\int_0^{\infty} \sqrt{\varepsilon} e^{-\varepsilon}\, d\varepsilon = \sigma_a(kT)\cdot\frac{1}{2}! = \frac{\sqrt{\pi}}{2}\sigma_a(kT) \quad . \tag{4.179}$$

is obtained as the average over a Maxwellian ($\varepsilon e^{-\varepsilon}$). Thus, the average cross section, $\overline{\sigma}_a$, for a $1/v$ absorber in a Maxwellian spectrum is $\frac{\sqrt{\pi}}{2}$ times its kT value, the value corresponding to the energy at the maximum of the Maxwellian distribution.

The average value of the η factor is defined in a similar manner, always reaction-rate-weighted:

$$\overline{\eta} = \frac{\int_0^{E_{th}} \nu\Sigma_f(E)\varphi(E)\, dE}{\int_0^{E_{th}} \Sigma_a(E)\varphi(E)\, dE} \quad . \tag{4.180}$$

If $\varphi(E)$ is Maxwellian and $\nu\Sigma_f(E)$ and $\Sigma_a(E)$ are $1/v$, then $\bar{\eta}$ may be evaluated as:

$$\bar{\eta} = \frac{\nu\Sigma_f(kT)}{\Sigma_a(kT)} , \tag{4.181}$$

i.e., $\bar{\eta}$ is simply given by the ratio of the kT values of $\nu\Sigma_f$ and Σ_a, since the factor $\sqrt{\pi}/2$ cancels. Deviations from (4.181) result from deviations from the $1/v$ cross section and from deviations in the spectrum from the Maxwell distribution.

In the thermal region for the calculation of flux spectra, the basic computer model has been the SOFOCATE code [14] or the more detailed integral transport code THERMOS (Ref. 15). The previously mentioned slowing down model MUFT has been combined with SOFOCATE in a single computer program called LEOPARD (Ref. 16), which has found widespread usage in the design of commercial pressurized water reactors.

4-8 Summarizing Remarks

The treatment of the energy dependency of the neutron flux in this chapter is of great importance in the description of the physics of nuclear reactors. The fact that neutrons are born at high energies requires that their energy fate be traced as they undergo scattering collisions. How well neutrons survive this journey in energy largely determines the ability of the reactor to sustain criticality. This slowing down process, while important in all reactor types, is of particular importance in thermal reactors where the majority of fission events occur in the thermal energy range. Moderation and thermalization of neutrons play a key role in thermal reactors because of the large energy range through which neutrons must travel, typically from 2 MeV down to 0.1 eV. The energy separability assumed in a first evaluation of the neutron spectra has another calculational significance as the separated energy spectra are used in generating group cross-section sets needed in multigroup neutron calculations to be treated in Chapter 7.

In this chapter, the study of the neutron flux dependencies was continued with attention focused on the basics of the energy dependencies. The presentation in this chapter concentrated on the simplest approximations with the more realistic slowing down models (e.g., including resonance absorption) deferred until Chapter 5. In addition to slowing down theory, the very important neutron thermalization subject was also briefly presented in this chapter.

Fundamental to the neutron slowing down theory are the scattering processes that occur during neutron/nucleus interactions. Two types of collisions occur; elastic and inelastic. Elastic collisions merely involve the exchange of energy and momentum, whereas in the case of inelastic scattering, a neutron is absorbed and subsequently another neutron reemitted from the nucleus. Thermal reactors rely on the elastic scattering process to slow the fission neutrons down to thermal energies. Inelastic scattering events are possible only above rather high energy thresholds and occur primarily in heavy materials. Inelastic scattering reactions are a major down mechanism in fast reactors.

The description of the slowing down or moderation of high-energy neutrons requires the determination and understanding of the elastic scattering cross sections and scattering kernels defined in Chapter 1. Use was made of classical collision dynamics theory to describe the elastic interaction between a neutron and a free atomic nucleus. It was advantageous to use both the laboratory frame of reference as well as the CMS to arrive at the scattering relationships. The angular neutron scattering behavior was obtained; in light nuclei, scattering in the forward direction is strongly preferred, whereas with heavy nuclei, the scattering is nearly isotropic in the laboratory system. The average energy loss from an elastic collision was shown to decrease with increasing atomic weight and also to increase linearly with increasing initial energy. The fractional energy change, however, is constant with energy, which suggests the introduction of a logarithmic energy scale for the slowing down process (the so-called "lethargy scale," u). An important parameter is the average lethargy change, ξ, which is an independent constant.

The energy-dependent behavior of neutrons is treated by first considering the calculation of the general slowing down problem. A method of solution for the source-sink problem and eigenvalue problem is described in terms of numerical procedures. Next, after briefly discussing the fast fission effect, neutron slowing down in the simplest and most effective of moderators, namely, hydrogen, was studied. Four cases were treated analytically, each with increasing complexity and the resulting neutron slowing down spectrum was compared. The slowing down in nonhydrogeneous materials without absorption resulted in a spectrum analogous to the asymptotic spectrum obtained in hydrogen. In these derivations, the solution for the flux spectrum is facilitated by the introduction of scattering density, collision density, and slowing down density as the dependent variable replacement for the neutron flux.

In the concluding sections of this chapter, the very important thermal neutron region was studied. The neutron balance was obtained assuming thermal equilibrium with the scattering nuclei and with the distinction made between "fast" and "thermal" neutrons for both an infinite absorbing system as well as in a source-free nonabsorbing infinite system. In the latter case, the concept of the principle of detailed balance was evoked with the introduction of the famous Maxwellian distribution. The shape and magnitude of the thermal spectrum was presented in general terms along with a brief overview of complicating effects (i.e., absorption hardening). Specifics of the complexities (for example, the generation of scattering kernels) in the thermal region are beyond the level of this presentation and hence were not discussed.

Homework Problems

1a. Derive the angular distribution in the LS for elastic scattering that is isotropic in the CMS.

b. Derive the corresponding average scattering angle, $\bar{\mu}_s$.

2a. Derive the angular distribution in the LS for elastic scattering with a linear anisotropy in the CMS:

$$p^c(\mu^c) = \frac{1}{2}(1 + a_1\mu^c) \quad ,$$

with $a_1 < 1$.

b. Derive the corresponding $\bar{\mu}_s$.

3. Prove that the average fractional loss of energy per elastic collision for CMS isotropic scattering between a neutron and a nucleus is given by $\frac{1}{2}(1 - \alpha)$.

4. How many elastic scattering events are required to slow a neutron down from 2.0 MeV to 0.025 eV at room temperature in graphite and in hydrogen if s-wave scattering with constant σ_s is assumed in the entire energy range.

5. In presenting neutron flux spectra from reactor design calculations, usually $\varphi(u)$ or $E\varphi(E)$ versus log E is plotted rather than $\varphi(E)$ versus log E. Explain why this former method of presenting spectra is preferred.

6. Transform $\chi(E)$ as given in Problem 6 of Chapter 1 from E to u.

7. Derive a formula for the fast fission contribution, $\delta\varepsilon'$ (note $\varepsilon = 1 + \delta\varepsilon'$), for two energy groups in the fast fission range. Neglect fission in the resonance range.

8a. Replace $\chi(E)$ in the balance equation for neutron slowing down in hydrogen (with constant σ_s and no absorption) by the δ function, $\delta(E - E_f)$. Solve the balance equation in the same way as shown in the text for a continuous $\chi(E)$. Note that the derivative of the δ function is needed.

b. Use the solution obtained in problem 8a (i.e., the solution of an inhomogeneous differential equation with a δ function in the inhomogeneous term) to obtain the solution for an extended source, $\chi(E)$, by superposition.

9a. Present the main steps of the derivation for the slowing down spectrum for hydrogen with $\sigma_s = \sigma_s(E)$ and no absorption.

b. Derive the analytical solution for the simple χ approximation:

$$\chi(E) = Ee^{-E} \quad ,$$

with E in megaelectron volts.

c. Plot the resulting spectrum for $\sigma_s = 20$ b $=$ constant.

d. Plot several points of this spectrum for $\sigma_s = 20$ b below $E = 20$ keV; and for $\sigma_s(E) =$ 17, 13, 8, 4.4, 2.2, and 1 b for $E =$ 30, 100, 300 keV, 1, 3, and 10 MeV, respectively.

e. Find the average energy of the emitted neutrons, E_f, for the approximation given in problem 9b.

10. Find the neutron speed at the maximum of the neutron flux in a Maxwellian distribution.

11. A cylindrical BF_3 counter is placed in a neutron flux with a Maxwellian distribution at a temperature of 25°C. The BF_3 in the counter is at a pressure of 25 cm of mercury (at 20°C) and physically is 0.6 cm in diameter and 25 cm long.

The counter has an efficiency of 1% (i.e., detects 1 out of every 100 incident neutrons). Placed in an isotropic neutron flux it registers 10 000 count/min.

a. What is the magnitude of the incident total neutron flux?

b. If the Maxwellian neutron distribution is at 200°C, what would the neutron flux level with the same 10 000 count/min counting rate be?

c. Assuming the same total neutron flux, what would the relative count rate between neutrons at 25 and 200°C be?

Review Questions

1. Why is the energy dependence of the flux so important for reactor calculations?

2. Which conservation laws determine the kinematics of neutron/nucleus scattering?
3. What is *elastic* scattering?
4a. Sketch the velocity vector diagram of the transformation from CMS into LS quantities.
  b. Give v_{CMS} in terms of the initial neutron velocity, v_n', and A.
5a. Derive the relations of μ^c and μ_s and the neutron and CMS velocities.
  b. Convert the velocities in both relations to energies and express μ^c and μ_s in terms of energies.
6a. Give the angular distribution of CMS isotropic scattering.
  b. Give the transformation of $p^c(\mu^c)$ that yields the angular distribution in the LS.
  c. Present a formal derivation of the scattering kernel $K_s(E'\rightarrow E)$.
7a. Give the angular distribution $p(\mu_s)$; give explicitly only the term that determines $\bar{\mu}_s$; for the rest, you need to give only the typical dependence.
  b. Derive $\bar{\mu}_s$.
8a. Sketch the angular distribution for CMS isotropic scattering for hydrogen, carbon, and uranium (*a*) as a polar diagram and (*b*) as $p(\mu_s)$ as function μ_s.
  b. What is *s*-wave scattering?
  c. Why and how does the angular distribution deviate from 8a for high energies?
9a. Give and sketch the scattering kernel $K_s\ (E'\rightarrow E)$ for *s*-wave scattering.
  b. Sketch the typical deviation resulting from anisotropic scattering.
10a. Give the definition of the lethargy, sketch the u scale, and comment on the choice of E_0.
  b. Why is the lethargy used instead of E in reactor calculations and spectrum representations?
  c. Give the energy-dependent spectrum that corresponds to $\varphi(u)$.
11. Transform from E to u: (a) $\sigma(E)$, (b) $\varphi(E)$, and (c) $K(E'\rightarrow E)$.
12. Give the energy-dependent neutron balance equation in lethargy.
13a. Give the definition of ξ.
  b. Give the values of ξ for hydrogen and the formula of its asymptotic behavior.
14. In which form do you plot an energy-dependent spectrum over $\log E$? Why?
15a. Describe the three kinds of scattering.
  b. Discuss the typical energy variations of the cross sections.
  c. Discuss the typical energy variations of the respective kernels, $K(E'\rightarrow E)$.

16. Give typical energy ranges in which $\sigma_p(E)$ is practically constant.
17. Give typical values for the first two threshold energies for $\sigma_{in}(E)$ for carbon, oxygen, ^{56}Fe, ^{235}U, and ^{238}U.
18. Sketch a typical $K_{in}(E' \rightarrow E)$ for $E' \simeq 2$ MeV in a heavy isotope.
19. Give the balance equation for the spectrum in lethargies and convert it to a discrete u value form.
20a. Give the very-many-group balance equation for φ_γ in the slowing range.
b. Describe the numerical solution procedure for a source-sink problem; for a λ eigenvalue problem; and for a B^2 eigenvalue problem.
21. Sketch the fission cross sections in the megaelectron volt range (0.1 to 10 MeV) for ^{235}U, ^{238}U, and ^{232}Th.
22. Derive the fast fission factor, ε', for a one-group model in the fast fission range.
23. Solve the slowing down problem in hydrogen* and sketch the solution:
(a) for $\Sigma_S^{\mathrm{H}}(E)$ = constant, $\Sigma_{tl} = \Sigma_S^{\mathrm{H}}$
(b) for $\Sigma_S^{\mathrm{H}}(E)$ = energy dependent, $\Sigma_{tl} = \Sigma_S^{\mathrm{H}}(E)$. Discuss the impact of inelastic scattering on the spectrum.
(c) for $\Sigma_t^{\mathrm{H}}(E)$ = energy dependent, $\Sigma_{tl} = \Sigma_t(E)$
(d) for $\Sigma_t^{H}(E)$ = energy dependent, $\Sigma_{tl} = \Sigma_t(E) + D(E)B^2$.

*Note: The solution of

$$F'(E) = a(E)F(E) + g(E), \text{ with } F(\infty) = 0$$

is

$$F(E) = \int_\infty^E g(E') \exp\left\{\int_{E'}^E a(E'')\, dE'' \right\} dE' \quad .$$

24. Give P_f in two-group, continuous slowing down, and Fermi age theories.
25. Find the uncollided component of the spectrum for all four cases of question 23.
26a. Give the balance equation for the spectrum for slowing down on a nonhydrogeneous scatterer in E and u ($\Sigma_a = 0$).
b. Convert it into a differential equation.
c. Describe the approximate approach for its solution.
d. Give the resulting approximate balance equation for $\varphi(u)$.
27a. Give the solution for $\varphi(E)$ for the slowing down on a nonhydrogeneous material without absorption.
b. Compare it with the corresponding solution for hydrogen and discuss the comparison.

28. What is the slowing down density $q_{sd}(E)$ in words and in a defining formula?
29. Give three typical values for the upper boundary of the thermal group and the reasons for their choice.
30a. Consider a source- and loss-free medium.
b. Give the principle of detailed balance.
c. What follows from that principle for the neutron spectrum in the thermal range?
d. Are there special conditions for the scattering kernel that must be satisfied for the principle of detailed balance to yield that particular spectrum?
31a. Give the Maxwell (normalized) distribution for the spectrum $\varphi_M(E)$, and for the number of neutrons $n_M(E)$.
b. What is the average neutron energy?
c. What is the average E in $\varphi_M(E)$?
32a. In which manner is $1/v$ absorption influencing the spectrum?
b. How can the impact of the $1/v$ absorption on the spectrum be described approximately? Also give the formula for the modification of the key parameter that reflects that spectral modification.
33a. Give the Maxwellian average of a $1/v$ cross section.
b. Give the Maxwell-spectrum-averaged η.

REFERENCES

1. D. Halliday and R. Resnick, *Physics for Students of Science and Engineering,* Chapter 10, John Wiley & Sons, New York (1965).
2. I. Kaplan, *Nuclear Physics,* 2nd ed., Chapter 18, Addison-Wesley Publishing Co., Reading, Massachusetts (1964).
3. M. M. R. Williams, *The Slowing Down and Thermalization of Neutrons,* North-Holland Publishing Co., Netherlands (1966).
4. R. D. Evans, *The Atomic Nucleus,* Chapter 11, McGraw Hill Book Co., New York (1955).

5. D. Garber, Ed., "ENDF/B Summary Documentation," 2nd ed., BNL 17541 (ENDF-201), Brookhaven National Laboratory (Oct. 1975).
6. H. Bohl, E. M. Gelbard, and G. H. Ryan, "MUFT 4—Fast Neutron Spectrum Code for the IBM-704," WAPD-TM-72, Westinghouse Electric Corp. (July 1957).
7. H. Henryson II, B. J. Toppel, and C. G. Steinberg, "MC^2-2: A Code To Calculate Fast Neutron Spectra and Multigroup Cross Sections," ANL-8144, Argonne National Laboratory (June 1976).
8. H. Soodak, Ed., *Reactor Handbook,* 2nd ed., Vol. II, Part A: "Physics," Wiley Interscience Publishers, New York (1962).
9. Donald J. Hughes and John A. Harvey, *Neutron Cross Sections,* BNL-325, Brookhaven National Laboratory (July 1955).
10. D. I. Garber and R. R. Kinsey, "Curves," Vol. II of *Neutron Cross Sections,* 3rd ed., BNL-325, Brookhaven National Laboratory (Jan. 1976).
11. M. S. Nelkin, "Scattering of Slow Neutrons by Water," *Phys. Rev.,* **119,** 741 (1960); also see D. E. Parks, M. S. Nelkin, N. F. Wikner, and J. R. Beyster, *Slow Neutron Scattering and Thermalization with Reactor Applications,* Benjamin/ Cummings Publishing Co., Menlo Park, California (1970).
12. G. I. Bell and S. Glasstone, *Nuclear Reactor Theory,* Chapter 7, Van Nostrand Reinhold Co., New York (1970).
13. K. H. Beckurts and K. Wirtz, *Neutron Physics,* Springer Verlag, Berlin and New York (1964).
14. H. Amster and R. Suarez, "The Calculation of Thermal Constants Averaged Over a Wigner-Wilkins Flux Spectrum/Description of the SOFOCATE Code," WARD-TM-39, Westinghouse Electric Corp. (Jan. 1957).
15. H. C. Honeck, "THERMOS, A Thermalization Transport Theory Code for Reactor Lattice Calculations," BNL-5826, Brookhaven National Laboratory (Sep. 1961).
16. R. F. Barry, "LEOPARD: A Spectrum-Dependent Nonspatial Depletion Code for the IBM-7094," WCAP-3269-26, Westinghouse Electric Corp. (Sep. 1963) (revised Aug. 30, 1968); also see L. E. Strawbridge and R. F. Barry, "Criticality Calculations for Uniform Water-Moderated Lattices," *Nucl. Sci. Eng.,* **23,** 58 (1965).

Five

NEUTRON SLOWING DOWN THEORY WITH RESONANCE ABSORPTION

5-1 Introduction

In the previous chapter, the basic elements of slowing down theory were presented and the energy dependence of the neutron flux was examined in considerable detail. The starting point of this development was the energy-dependent separated neutron balance equation obtained from diffusion theory. To bring out the essential fundamental principles, this study of the neutron energy flux relationships concentrated on the general features of the slowing down process. Flux expressions were derived starting with the simplest case of neutron slowing down in hydrogen and then expressions were obtained for the slowing down neutron spectrum under various approximations of increasing sophistication. The treatment and incorporation of the important aspects of neutron resonance absorption into slowing down theory were, however, not considered in Chapter 4; therefore, the expressions developed were only first approximations. In this chapter, this important absorption phenomenon is discussed and a more unified description of the neutron slowing down process presented.

The neutron absorption process plays a crucial role in determining the criticality of a nuclear reactor. Recall that in a thermal reactor, the competition of radiative capture and fission play the dominating role for criticality. In the heavy elements, common to all reactor fuels, radiative capture is quite large in the so-called "resonance energy region." Although generally the capture cross sections follow an average $1/v$ trend from the electron volt to the kiloelectron volt range, the resonances determine the actual energy dependence. As the fast fission neutrons slow down into the thermal region in a thermal reactor in order to cause more fissions, many neutrons are captured in this resonance region. This problem is also of great importance in fast reactors as well, where resonance absorption causes such high neutron losses that practically no neutrons reach the thermal domain.

A brief section on the important area of resonance escape calculations in the presence of heterogeneous media is also presented. This is an area that has an important impact on the criticality prediction in thermal reactors. The neutron "shadowing" influences of the cell structure on the resonance escape probability are incorporated in the cross-section library sets used in the multigroup calculation approach that is presented in Chapter 7.

Finally, in order to develop more realistic neutron slowing down models, which incorporate neutron resonance absorption, the effect of temperature on the resonance absorption process (the Doppler effect) is discussed. This phenomenon is very important in thermal reactors, but even more so in fast reactors where it has been identified as a major inherent shutdown mechanism for reactor transients.

5-2 Resonance Cross Sections

5-2A Basic Considerations

Neutrons can react with nuclei basically in two different ways. A neutron can react with the nucleus as a whole, i.e., with the joint forces of all nuclei. This leads to potential scattering, which was discussed in Sec. 4-2C. Potential scattering is not sensitive to the neutron energy; consequently, the cross sections are a smooth function of energy. Table 4-III lists energy ranges in which the potential cross section of the moderators is virtually constant. Figure 4-6 shows the scattering cross section of hydrogen, which also illustrates the energy dependency typical of all potential cross sections, even though the hydrogen nucleus consists of only one nucleon, the proton.

Potential scattering cross sections are relatively constant if the neutron wavelength is greater than the size of the scattering nucleus. This condition is satisfied for most of the slowing down range with the cross section decreasing at higher energies, similar to $\sigma_s(E)$ for hydrogen. These values vary between ~3 and 20 b. Hydrogen, with 20 b, has the highest potential cross section of all nuclei. The magnitude of the potential cross sections is not closely related to the size of the nuclei (compare Table 4-III).

All other neutron reactions of interest in a reactor consist of two steps. At first, the neutron and target nucleus, A, form a so-called "compound" nucleus, $(A + 1)^*$, in an excited state. Since nuclei have a discrete level scheme of excited states, these reactions occur only at certain energy values, as illustrated in Fig. 5-1. Cross sections of this type are called "resonance cross sections." At low energies, these resonances are well separated. At higher energies, they are generally broader and more

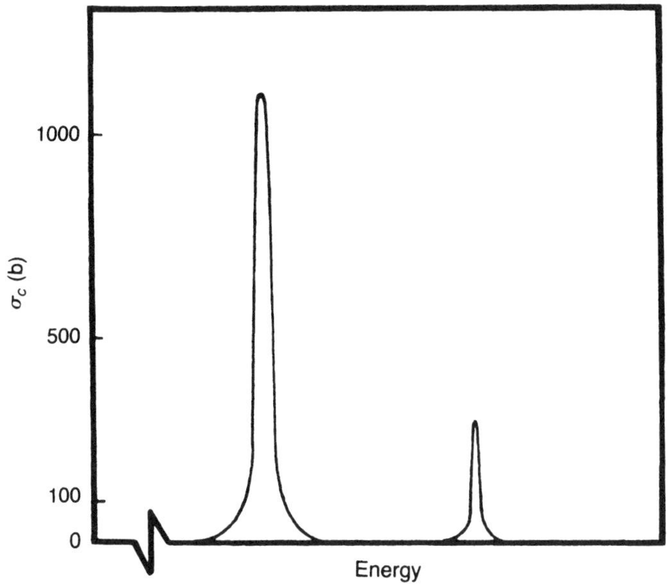

Fig. 5-1. Illustration of a capture (or fission) cross section with resonances.

numerous. The cross sections, which result from a superposition of these resonances, may then have the appearance of smoothly varying cross sections.

Formation of a compound nucleus occurs if the sum of the colliding neutron energy and the binding energy of the neutron in the nucleus correspond to discrete quantum states or resonances of the compound nucleus. Then, sharp peaks at these discrete energies appear in the cross sections. Their width at half of their height is called the "half-width"; it is denoted by Γ.

The compound nucleus can decay in several different ways called "channels" (e.g., fission, neutron, or gamma emission). Each decay process has a certain probability of occurrence, given by the ratio of partial and total resonance half-widths (the so-called "branching ratio"):

$$\Gamma_j/\Gamma = \lambda_j/\sum_{j'}\lambda_{j'} = \lambda_j/\lambda \quad , \tag{5.1}$$

where

$\Gamma_j = \hbar\lambda_j$ = partial width for j'th channel
$\Gamma = \sum_j \Gamma_j$ = total level width for all channels
λ_j = decay constant of nucleus through j'th channel
$\lambda = \sum_{j'} \lambda_{j'}$ = total decay constant

$\hbar$ = (Planck's constant)/$2\pi \simeq 6.58 \times 10^{-16}$ eV·s.

The various partial widths corresponding to different modes of decay nuclear reactions, for example, (n,α), (n,γ), (n,p), fission, and neutron scattering [(n, n) reactions], are expressed as Γ_α, Γ_γ, Γ_p, Γ_f, and Γ_n, respectively, with the sum of all of the partial widths equal to the total width, Γ. Without resonance deformation through interference (see below), E_R is the energy corresponding to the peak of the resonance curve. The resonance width, Γ, which is the width of the resonance as measured at the half resonance height, is the parameter that determines the shape of the resonance. This is further discussed in Sec. 5-2B.

Physically, the resonance width can be interpreted in terms of the mean lifetime, τ, of the compound nucleus by the relationship:

$$\tau = \hbar/\Gamma = \frac{1}{\lambda} \quad . \tag{5.2}$$

Hence, wide resonances imply short compound nucleus decay times.

Equation (5.2) is a consequence of Heisenberg's uncertainty principle:

$$\Delta E \cdot \Delta t \gtrsim \hbar \quad ,$$

which expresses the fact that the energy resolution, ΔE, obtainable from a measurement of duration Δt, is at best equal to $\hbar/\Delta t$. The maximum time available for the determination of a resonance energy E_R is the lifetime τ of the corresponding excited state. If one uses all that time, the best energy resolution obtained is $\Delta E = \Gamma$. Thus, the equal sign of the uncertainty principle applies in the relation of τ and Γ of Eq. (5.2).

The major characteristics of a resonance absorption cross section are depicted in Fig. 5-2 for ^{238}U. The dips appearing on the lower energy side of many resonances result from interference between potential and resonance scattering (see Sec. 5-2B). Interference generally occurs if elementary particles or gamma quanta have two ways to achieve the same final state.

The energy distance between resonances is called "spacing." The average spacing between resonances is fairly constant below ~50 keV. As far as resonance absorption is concerned, the material of greatest reactor interest, ^{238}U, has ~3000 major resonances between 0 and 54 keV (s-wave resonances). This gives an average spacing of ~18 eV. The spacing decreases toward higher energies; further, *p*- and *d*-wave resonances appear in addition and dominate *s*-wave resonances at higher energies.[a]

---

[a]See Appendix B for an explanation of the terminology "*s*-," "*p*-," and "*d*-wave," etc.

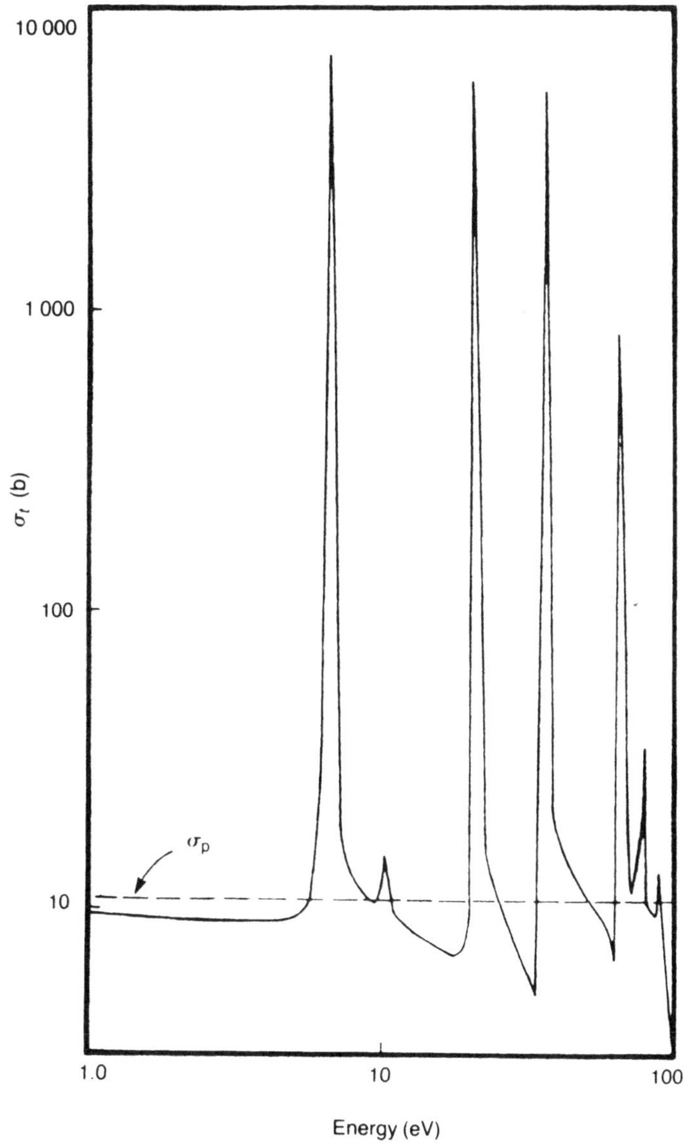

Fig. 5-2. The total cross section of ^{238}U between 1 and 100 eV.

In high-enrichment reactors, the resonance absorption in ^{235}U is important. The average spacing of these resonances is ~2 eV, with, for example, ~2500 resonances below 5 keV. The ^{235}U resonances are generally much lower than the ^{238}U resonances.

Resonances in structural materials are generally much broader than in heavy elements. They are of little importance for resonance absorption of neutrons in thermal reactors; primarily they are scattering resonances. Excited states in light nuclei generally occur at much higher energies than in heavy nuclei.

To evaluate the resonance effects, use is made—in a qualitative manner—of a parameter called the "practical width" (see Sec. 5-4B). This quantity is defined as that energy interval for which the macroscopic resonance cross section is larger than the nonresonance macroscopic cross section. With this definition, then, the practical width depends on the composition and type of elements in the system.

In the calculation of the resonance absorption during neutron slowing down, the width of the resonance relative to the neutron energy loss is of importance. For most reactor systems and the resonances present, the practical width is much smaller than the average neutron energy loss in a scattering event with a moderator atom, i.e., the resonances are "narrow" compared to a scattering width. A single average elastic collision is capable of removing neutrons from the resonance. This situation is exploited by devising the so-called "narrow-resonance" (NR) approximation. For many resonances, especially in thermal reactors, a different approach is advised, the "narrow-resonance infinite-mass" (NRIM) approximation. Both approximations are described in more detail in Sec. 5-4.

5-2B Resolved Resonance Region

In the energy region where the resonances are well separated, the cross section for the reaction is given by the famous Breit-Wigner formula. This formula describes the cross section in terms of the resonance parameters, the de Broglie neutron wavelength, and the neutron energy. The same basic shape holds for most resonance phenomena in nature.

A more general formula accounts for the contributions from many energy levels of the compound nucleus. This complete description is complex and difficult to apply in actual reactor analysis practice. Fortunately, most of the resonance absorption comes from the *resolved* resonance region, in which the resonances are widely separated, allowing the simplified description by a "single-level" expression, the *single-level Breit-Wigner formula* for a reaction j:

$$\sigma_j(E) = 4\pi \lambda\!\!\!^{-2} g_R \frac{\Gamma_n \Gamma_j}{\Gamma^2} \frac{(\Gamma/2)^2}{(E - E_R)^2 + (\Gamma/2)^2} , \qquad (5.3)$$

where

$ƛ$ = the reduced de Broglie wavelength of the neutron ($ƛ = h/2\pi mv$) (note that the symbol λ is commonly used for wavelengths and decay constants)
E_R = resonance energy
E = neutron energy
g_R = the spin parameter for resonance R
Γ_n = neutron width
Γ_j = width for process j
Γ = total width for all processes.

The spin parameter g_R, a quantum-mechanics-derived parameter, is sometimes referred to as the "statistical weight"; it is given by:

$$g_R = \frac{(2J_c + 1)}{2(2J_0 + 1)(2l + 1)} \quad , \tag{5.4}$$

where J_0, J_c, and l are the integer or half-integer quantum numbers of the angular momentum:

J_0 = total angular momentum of the target nucleus (orbital plus spin momentum)
J_c = total angular momentum of the compound nucleus (orbital plus spin momentum)
l = orbital angular momentum of the neutrons with respect to the nucleus.

The "statistical factor" g can be interpreted as the probability of a neutron and target nucleus adding their angular momenta such that the total angular momentum of the compound nucleus becomes J_c for that particular energy level. For neutrons from $E = 0$ to a few tens of kilo-electron volts, l is mostly zero. For ^{238}U, $J_0 = 0$, with $l = 0$, $J_c = 1/2$; thus, $g = 1$ for the most important resonances in ^{238}U (i.e., for s-wave resonances). For intermediate and fast neutrons, the g factor in the Breit-Wigner formula may be modified as l may take on larger values (1, 2, etc.).

The Breit-Wigner formula, Eq. (5.3), describes the cross sections around a resonance. As evident from Eq. (5.3), the cross section $\sigma_j(E)$ has a maximum when $E = E_R$ and, as expected from the definition of Γ, reaches half of this maximum energy for $(E - E_R) = \Gamma/2$. It should be noted that for a given resonance, all components such as scattering, capture, and fission have the same half-width Γ. The components differ only in their magnitudes.

The neutron width—for s-wave neutrons—is proportional to neutron velocity; it is often expressed in terms of its value at 1 eV:

$$\Gamma_n = \Gamma_n^0 \sqrt{E \text{ in electron volts}} \quad . \tag{5.5}$$

The quantity Γ_n^0 is called the "reduced neutron width."

By substituting this expression for Γ_n into the Breit-Wigner formula, Eq. (5.3), and applying proper approximations at $E \ll E_R$ of the first resonance, one obtains the expected $1/v$ cross-section behavior for the capture and fission cross sections and the energy-independent trend for scattering cross sections (see the first homework problem at the end of this chapter).

Capture or fission cross sections are given directly by the Breit-Wigner formula, Eq. (5.3). Scattering cross sections, however, have two additional terms:

$$\sigma_s = 4\pi\lambda\!\!\!^{-2} g_R \frac{(\Gamma_n/2)^2}{(E-E_R)^2 + (\Gamma/2)^2} + \sigma_p + 4\pi\lambda\!\!\!^{-} g_R \frac{\Gamma_n(E-E_R)R_N}{(E-E_R)^2 + (\Gamma/2)^2} \quad . \tag{5.6}$$

The second term in Eq. (5.6), σ_p, describes the "potential scattering" component; it dominates at energies between resonance. Often, σ_p is expressed in terms of an effective radius, R_N, by

$$\sigma_p = 4\pi R_N^2 \quad . \tag{5.7}$$

In addition to the first two terms in Eq. (5.6), there is a term that accounts for the *interference* between the two basic scattering modes (potential and resonance scattering). It is typified by a pronounced asymmetry in σ_s on either side of the resonance. This asymmetry is very apparent in Fig. 5-2.

If neighboring resonances are overlapping to a significant extent, there will also be resonance-to-resonance interference, described by the so-called "multilevel Breit-Wigner formulations" (see, for example, Refs. 1 and 2). The presentation in this text is restricted to the single-level description, which suffices for most reactor applications.

5-2C Unresolved Resonance Region

In the previous section on the resolved energy region, it was assumed that the resonances are well separated and remote from each other. This is the case for even-numbered isotopes, such as ^{238}U and ^{232}Th. For the odd-numbered fissile nuclei, such as ^{235}U and ^{239}Pu, the resonances are much closer together. In this energy region, several or many levels of the compound nucleus may be within the resolution width, ΔE^{res}, of the experimental equipment since it increases nearly linearly with E; e.g.,

$\Delta E^{res} = 0.005E$. Furthermore, at higher energies, the resonances in these heavy isotopes, especially in fissile isotopes, become more numerous and broader, but also considerably lower. Then, individual resonance parameters cannot be determined experimentally; hence, the term "unresolved" resonance region.

For light nuclides, resonances are generally widely spaced with several kiloelectron volts between individual resonances. Thus, resonance overlapping is only a problem for heavy nuclides (see Table 4-IV).

Several procedures exist for determining the resonance effects in the unresolved region. Generally, they employ statistical distributions for the widths and energy spacing, as extrapolated from known lower energy parameters. (See, for example, Refs. 3, 4, and 5 for statistical resonance parameters and Ref. 6 for application of these data.)

5-3 Resonance Absorption During Slowing Down in Hydrogen in an Infinite Medium

In this section, the basic features of the slowing down expression obtained in Sec. 4-5D for scattering in hydrogen with absorption will be explored. The goal of this review is to generalize the main features of the analytic solution for hydrogen as sole moderator. This generalization will then provide the understanding and the basis to construct the respective solution for more realistic cases in which the scattering on other moderators and on intermediate and heavy nuclei is taken into account (see Sec. 5-4). This generalization is especially required for cases without hydrogen, for example, D_2O- or graphite-moderated reactors.

5-3A Basic Slowing Down Spectrum Features

Recall that the starting point for this development was the energy-dependent component of the neutron slowing down equation with hydrogen as the scatterer. The approximation treated above is expressed in the following balance equation:

$$\Sigma_t(E)\varphi(E) - \int_E^{\infty} \Sigma_s^H(E')\varphi(E')\,\frac{dE'}{E'} = 0 \text{ below the source} \quad , \qquad (5.8)$$

with $\Sigma_t(E)$ given by Eq. (4.86a). Without leakage, i.e., without including DB^2 per Eq. (4.86b), Eq. (5.8) is the general neutron balance equation in an infinite medium. There is then no difference between diffusion theory and the Boltzmann equation.

The solution to this equation for the neutron spectrum in an infinite medium for energies below the source is [$q_{sd}(E_u) = q_r$ = slowing down density at the "entry energy" of the resonance range]:

$$\varphi(E) = \frac{q_r}{E\Sigma_t(E)} \exp\left[-\int_E^{E_u} \frac{\Sigma_a(E')}{\Sigma_t(E')} \frac{dE'}{E'} \right] , \qquad (5.9a)$$

with E_u being the upper limit of the resonance absorption range below the source. Here, Σ_a describes practically only resonance absorption on heavy nuclei. Since resonance absorption decreases strongly with increasing E (because of their decreasing height), the contribution from higher energies to the integral in Eq. (5.9a) is minimal. Therefore, only an approximate value for E_u is needed. We use $E_u = 100$ keV. Denoting the exponent in Eq. (5.9a) by $h_a(E)$, Eq. (5.9a) can be expressed in the simpler form:

$$\varphi(E) = \frac{q_r}{E\Sigma_t(E)} \exp[-h_a(E)] , \qquad (5.9b)$$

which consists of factors, one being the exponential function.

The factor in front of the exponential function is denoted here by:

$$\varphi_0(E) = \frac{q_r}{E\Sigma_t(E)} . \qquad (5.10a)$$

Its meaning becomes clear when it is compared to the slowing down spectrum without absorption, as derived in Sec. 4-5C:

$$\varphi(E) = \frac{q_r}{E\Sigma_s^{\mathrm{H}}(E)} . \qquad (5.10b)$$

Notice that both spectra [Eqs. (5.10)] are of the same form. Actually, since $\Sigma_t(E) = \Sigma_s^{\mathrm{H}}(E)$ between well-separated resonances, the spectra, as given by Eqs. (5.10a) and (5.10b), are numerically the same between the resonances. At the resonance energies, however, the spectrum shows dips as described by the inverse of the total cross section in Eq. (5.10a). A section of such a spectrum is depicted in Fig. 5-3.

In particular, just above the thermal energy range, both Eqs. (5.10) yield the same spectrum between the resonances, independent of the neutron losses that may have occurred at higher energies. Thus, the factor in front of the exponential function in $\varphi(E)$ in Eqs. (5.9) describes the spectrum without neutron losses, called here the "loss-free spectrum."

In spite of the lack of a "lasting" effect from neutron losses, the loss-free spectrum is obviously deformed by the resonances. Wherever $\Sigma_t(E)$ assumes large values due to a resonance, the spectrum via $1/\Sigma_t(E)$ shows a corresponding dip or *spectral depression.* This results in a reduced

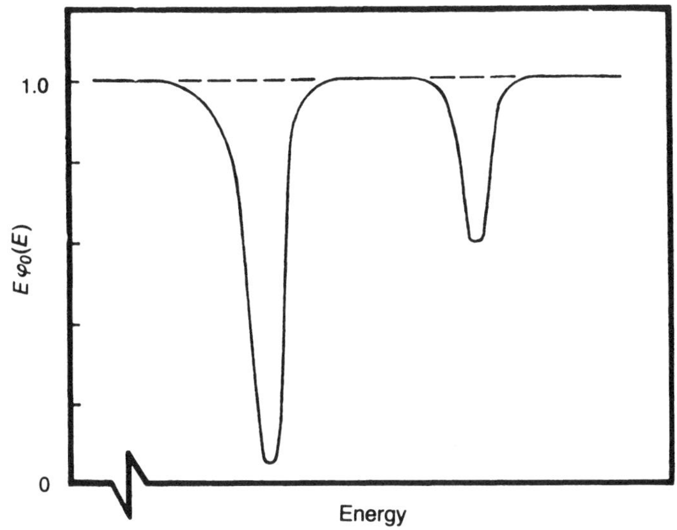

Fig. 5-3. Illustration of the loss-free spectrum with resonances [Eq. (5.10a)] and without resonances [Eq. (5.10b)].

absorption effect of the resonances, a phenomenon called "self-shielding."

The spectral depressions, in Eqs. (5.9a), (5.9b), and (5.10a) and thus the proportionality to $1/\Sigma_t(E)$, result from the solution of the neutron balance equation in the slowing down range. If the scattering on higher isotopes in $\Sigma_t(E)$ is not neglected [as it was in Eq. (4.86a) on which Eqs. (5.9) are based], one has different total cross sections and thus different resonance self-shielding. It will turn out (see Secs. 5-4 and 5-5) that improvements on the "self-shielding denominator" [here Σ_t of Eq. (4.86a)] are the main aspect of analytical treatments of the neutron spectra in the resonance region and of the determination of the absorption rates.

The magnitude and the importance of resonance self-shielding are illustrated in the following numerical example, depicted in Fig. 5-4. The resonances are idealized by rectangular shapes in order to allow the evaluation of the self-shielding effect by multiplication and division rather than by complicated integrals. The self-shielding denominator is approximated by using only the absorption part of the resonance ($\Sigma_s = \Sigma_p =$ potential cross section):

$$E\varphi(E) \propto \frac{\Sigma_p}{\Sigma_p + \Sigma_a^{\text{res}}(E)} \quad \text{within}$$

and

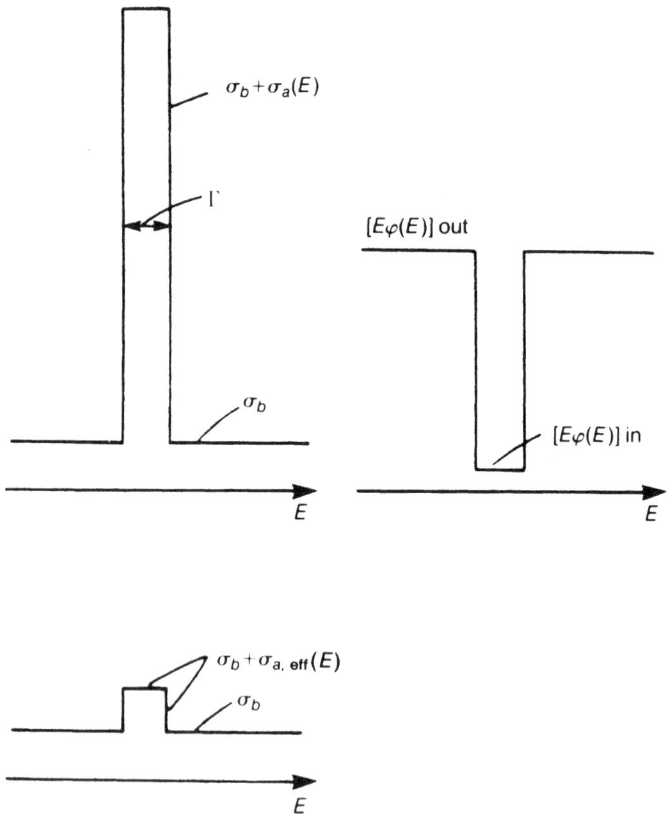

Fig. 5-4. Illustration of resonance self-shielding: $\sigma_a = 9\sigma_b$; $[E\varphi(E)]_{in} = 0.1 \cdot [E\varphi(E)]_{out}$ of resonance; $\sigma_{a,eff}(E) = 0.1\sigma_a(E)$ within the resonance.

$$\propto \frac{\Sigma_p}{\Sigma_p} \text{outside of the resonance} \quad .$$

In this example, the absorption rate is reduced by a factor of 10 through self-shielding, i.e., without considering self-shielding, the result would be wrong in any practical sense.

The concept of a loss-free spectrum lends itself readily to an application for nonhydrogeneous moderators. As shown in Sec. 4-6, the spectrum without absorption is given by

$$\varphi(E)_{\text{without } \Sigma a} = \frac{q_r}{E\overline{\xi\Sigma_s}} = \frac{q_r}{E\bar{\xi}\Sigma_s} \quad , \tag{5.11}$$

where $\overline{\xi\Sigma_s}$, calculated by Eq. (4.149), may include a hydrogen component. Equation (5.11) is the obvious generalization of Eq. (5.10b). It describes the loss-free spectrum between the resonances. The effect of absorption is considered in Secs. 5-4 and 5-5.

5-3B Resonance Absorption and Escape Probabilities

In the previous chapter, especially Sec. 4-5, the neutron spectrum was obtained as it results from slowing down on hydrogen under the influence of heavy absorbers. Having the spectrum makes possible the calculation of the probability, $\bar{p}(E)$, for neutron absorption in resonances, which is also the probability of "not escaping" the resonances. Together with the probability $p(E)$ of "escaping" the resonances, it sums to one:

$$\bar{p}(E) + p(E) = 1 \quad . \tag{5.12}$$

In Eq. (5.12), $\bar{p}(E)$ is just the absorption rate (between E and E_u) per neutron entering the resonance absorption range:

$$\bar{p}(E) = \frac{R_a(E' \geq E)}{q_r} = \frac{1}{q_r}\int_E^{E_u} \Sigma_a(E')\varphi(E')\, dE' \quad . \tag{5.13a}$$

Inserting the slowing down spectrum below the source range, Eq. (5.9b), into Eq. (5.13a) yields:

$$\bar{p}(E) = \int_E^{E_u} \frac{\Sigma_a(E')}{\Sigma_t(E')} \cdot \frac{1}{E'} \exp[-h(E')]\, dE' \quad . \tag{5.13b}$$

Using

$$\frac{\Sigma_a(E')}{\Sigma_t(E')}\frac{dE'}{E'} = -dh \quad , \tag{5.14}$$

which follows from the definition of $h(E)$ [as obtained from Eqs. (5.9)], transforms the integral in Eq. (5.13a) into:

$$\bar{p}(E) = -\int_{h(E)}^{0} \exp(-h)\, dh = 1 - \exp(-h) \quad . \tag{5.15}$$

Thus, the probability for absorption in resonances is given by:

$$\bar{p}(E) = 1 - \exp[-h(E)] \quad . \tag{5.16}$$

The complement of Eq. (5.16), the "resonance escape probability," is given by:

$$p(E) = \exp[-h(E)] = \exp\left[-\int_E^{E_u} \frac{\Sigma_a(E')}{\Sigma_t(E')}\frac{dE'}{E'}\right] \quad . \tag{5.17}$$

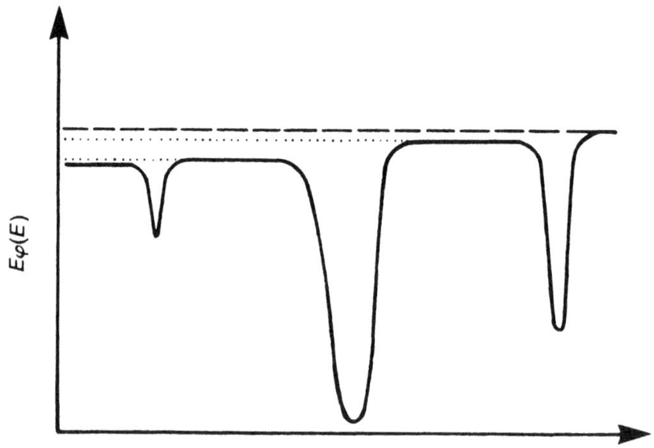

Fig. 5-5. Illustration of the basic energy dependence of the neutron spectrum about resonances during slowing down.

The slowing down spectrum, Eq. (5.9b), can thus be factored into $\varphi_0(E)$, the loss-free spectrum, and a probability, $p(E)$, which describes the accumulation of neutron losses due to absorption in resonances above energy E. Figure 5-5 illustrates the basic energy dependence of the neutron spectrum during slowing down. The spectral depressions are already described by $\varphi_0(E)$, but the increasing reduction of the spectrum between the resonances results from a corresponding decrease in $p(E)$. The decrease of $E\varphi(E)$ over a single resonance is due to absorption in this resonance. This is described by $p(E)$, which is smaller below than above a resonance as illustrated in Fig. 5-6. Thus, $p(E)$ accounts for the spectrum reductions in individual resonances as well as for the proper accrual of all the losses in the resonances.

To better understand this important aspect of the resonance absorption, the exponent in $p(E)$ is considered separately from the exponential function.

The exponent of $p(E)$ is just the absorption probability, $\bar{p}_0(E)$, in the loss-free spectrum $\varphi_0(E)$:

$$\bar{p}_0(E) = \frac{1}{q_r}\int_E^{E_u} \Sigma_a(E')\varphi_0(E')\,dE' = \frac{R_a^0(E' \geq E)}{q_r}$$

$$= \int_E^{E_u} \frac{\Sigma_a(E')}{\Sigma_t(E')}\frac{dE'}{E'} = \bar{p}[\varphi_0(E)] \quad . \tag{5.18}$$

Thus, the spectrum $\varphi(E)$ can be expressed in terms of the loss-free spectrum $\varphi_0(E)$ as

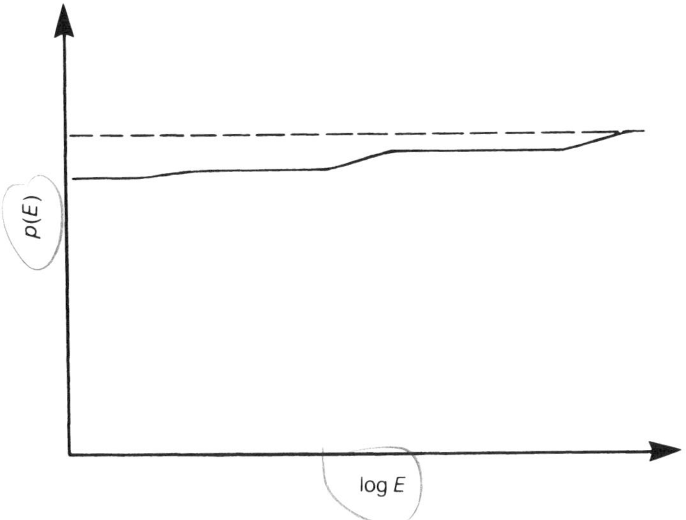

Fig. 5-6. Illustration of the typical energy dependence of the resonance escape probability, $p(E)$, around the resonances.

$$\varphi(E) = \varphi_0(E)p(E) = \varphi_0(E)\exp\{-\bar{p}[\varphi_0(E)]\} \quad .$$

It may be surprising that the absorption rate in the exponent of $p(E)$ is to be calculated with the loss-free spectrum. Since this particular absorption rate [Eq. (5.18)] is the basis for the definition of the very important "resonance integral" concept (see below), the reason for the use of the loss-free spectrum is illustrated by a different consideration, i.e., by exploring the mechanism of accumulating the contributions of individual resonances to absorption and to the spectral shape.

For the discussion of the accrual mechanism, the resonance absorption probability, Eq. (5.18), is first decomposed into components of individual resonances. Distinguishing the resonances with an index, k, Eq. (5.18) can be written as a sum over all resonances above E (indicated by $k>E$, which means all resonances for which $E_k>E$):

$$\bar{p}_0(E) = \sum_{k>E} \int_{\text{res}\,k} \frac{\Sigma_a(E')}{\Sigma_t(E')} \frac{dE'}{E'} = \sum_{k>E} \pi_k \quad , \tag{5.19}$$

where π_k equals the respective integrals in the middle of this equation. The π_k are equal to the absorption probability in the resonance k for neutrons arriving at that resonance with a $1/E$ spectrum, as they do if the resonances are isolated.

Inserting the decomposition equation, Eq. (5.19), into the exponential function, Eq. (5.17), allows the factoring of the "resonance escape probability," $p(E)$, into a product of factors resulting from the individual resonances above E:

$$p(E) = \exp\left(- \sum_{k>E} \pi_k \right) = \prod_{k>E} \exp(-\pi_k) \quad , \tag{5.20}$$

with $\prod_k$ denoting the product of the respective factors $[\exp(-\pi_k)]$.

The ratio of $E\varphi(E)$ above and below the k'th resonance can obviously be calculated without taking into account previous reductions. Since $\exp(-\pi_k)$ gives the ratio of two spectra, one has to accumulate the spectrum reduction by multiplying the individual ratios. Let $\varphi_b(u_k)$ be the value of $E\varphi(E)$ "between" the k'th and the $(k-1)$'th resonance. One obtains the spectrum below the k'th resonance by multiplying with $\exp(-\pi_k)$:

$$\varphi_b(u_{k+1}) = \exp(-\pi_k)\varphi_b(u_k) = \frac{\varphi_b(u_{k+1})}{\varphi_b(u_k)} \cdot \varphi_b(u_k) \quad , \tag{5.21}$$

since

$$\exp(-\pi_k) = \frac{\varphi_b(u_{k+1})}{\varphi_b(u_k)} \quad . \tag{5.22}$$

The subscript b indicates that $\varphi_b(u)$ describes the spectrum "between" two resonances.

This section can be summarized as follows:

1. The absorption probability per neutron entering the resonance absorption range during slowing down below energy E, i.e., $\bar{p}(E)$, is to be calculated with the actual spectrum, $\varphi(E)$:

$$\bar{p}(E) = \frac{1}{q_r} \int_E^{E_u} \Sigma_a(E')\varphi(E')\, dE' \quad . \tag{5.13a}$$

2. Employing the analytic expression for hydrogen as the main scatterer in Eq. (5.13a) yields for the complement to $\bar{p}(E)$, i.e., for the resonance escape probability $p(E)$, an exponential function with an absorption probability in the exponent that is calculated with the loss-free spectrum, $\varphi_0(E)$:

$$1 - \bar{p}(E) = p(E) = \exp\left[-\frac{1}{q_r} \int_E^{E_u} \Sigma_a(E')\varphi_0(E')\, dE' \right] \quad . \tag{5.23}$$

3. Assuming isolated and well-separated resonances allows for an interpretation of Eq. (5.23) in terms of absorptions in single resonances. It also illustrates the accrual of the respective contributions:

a. The individual relative spectrum reductions, $\exp(-\pi_k)$, are correctly calculated by using the loss-free spectrum in π_k, with π_k being the absorption in resonance k per neutron arriving at that resonance.

b. The proper way to accumulate the individual relative spectrum reductions is by multiplication of the individual reduction factors:

$$p(E) = \prod_{k>E} \exp(-\pi_k) \quad . \tag{5.20}$$

c. Forming the product of Eq. (5.20) automatically yields the sum of the quantities π_k, which are individually calculated with the loss-free spectrum, in the exponent. Therefore, it has to be the absorption rate of the loss-free spectrum which appears in the exponential function in the slowing down spectrum $\varphi(E)$.

4. The spectrum $\varphi(E)$ can be expressed as a product of the loss-free spectrum and the resonance escape probability. If all source neutrons enter the resonance range, then:

$$\varphi(E) = \varphi_0(E)\exp\left[\frac{-1}{s_0}\int_E^{E_u} \Sigma_a(E')\varphi_0(E')\,dE'\right] = \varphi_0(E)p(E) \quad , \tag{5.24}$$

where s_0 is the magnitude of the fission source (see Sec. 4-5A).

5. If the fast fission range is treated explicitly, Eq. (5.24) is to be replaced by

$$\varphi(E) = \varphi_0(E)\exp\left[\frac{-1}{q_r}\int_E^{E_u} \Sigma_a(E')\varphi_0(E')\,dE'\right] \quad , \tag{5.25}$$

where q_r is the slowing down density at E_u.

6. It should be noted that although the derivations presented above assume isolated and well-separated resonances, the main results, such as Eq. (5.25), are not invalidated by the practically occurring overlapping of resonances. It is the identification of the individual contributions, π_k, that becomes imprecise. But these contributions had been introduced primarily to enhance the semiquantitative understanding of the resonance integral and the accrual effects.

5-4 Basic Approximations for the Treatment of Resonance Absorption

It has been shown in Sec. 4-3 that the neutron spectrum during slowing down can be calculated numerically in just one "sweep." i.e., without iterations. Since no assumption had to be made on the energy dependence of the cross sections, this must also hold for resonance cross sections. For high accuracy, however, the number of energy intervals will have to be very large (10^5 to 10^6) because of the strong variation of $\sigma(E)$ in every small energy interval that contains a resonance (see Fig. 5-2). The use of a very large number of intervals implies a large computation time. There is, therefore, an incentive to devise approximate methods for the calculation of the spectrum within resonances, the resulting self-shielding, and the self-shielded reaction rates. The basic characteristics of spectra with resonances and the resulting absorption rates have been explored in detail in Sec. 5-3 for the idealized case of scattering on hydrogen only. This allowed an exact solution of the energy-dependent balance equation, from which the key characteristics could be deduced. This section deals with the case of scattering—at least in part—with nonhydrogeneous substances, which is the normal case. Since no exact solution is available for this case, approximate methods need to be devised. The most common ones are presented in this section. In practical applications, approximate methods of this type are applied, although mostly in a more sophisticated form.

5-4A Illustration of the Two Basic Approximations

The approximate methods for the calculation of the spectrum in resonances depend on the width of the resonance as compared to the energy loss of neutrons during elastic scattering (ΔE). The two basic methods assume the resonances to be "narrow" compared to ΔE. Making this assumption for scattering on *all* nuclei leads to the so-called "narrow-resonance" (NR) approximation, i.e., the resonances are assumed to be narrow compared to the energy loss in scattering on light *and* heavy nuclei. Alternately, the energy loss in the heavy nuclei scattering is neglected and resonances are assumed to be narrow only compared to the energy loss in scattering on the lighter nuclei; one then obtains the "narrow-resonance infinite-mass" approximation (NRIM).

Figure 5-7 illustrates the scattering into and out of a resonance in a case that justifies the application of the NR approximation. Practically all neutrons scattered into the resonance are coming from energies far above that particular resonance. Thus, the energy density of the scattering source within the resonance cannot be influenced by the resonance itself. It must be practically energy independent across the resonance.

Only one of the neutrons scattered into the resonance in Fig. 5-7 is not scattered out. It is being absorbed in the resonance. The other neutrons are scattered way out of the energy interval covered by this resonance. Then, these neutrons are scattered over a wide range and the

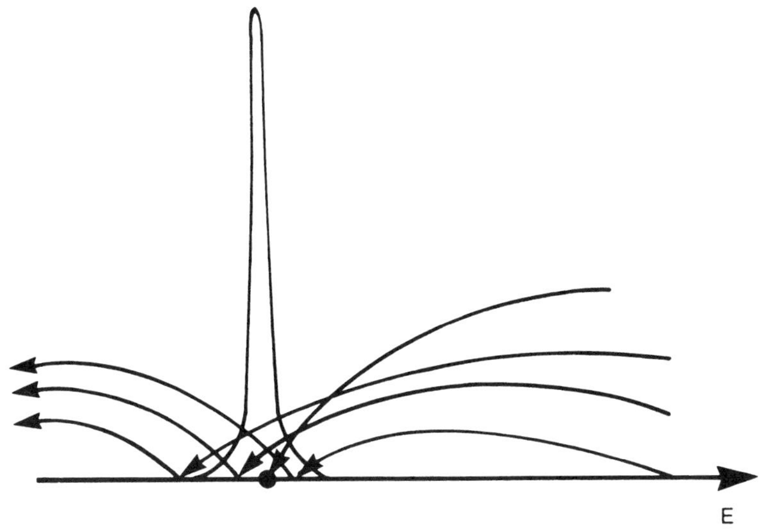

Fig. 5-7. Illustration of scattering into and out of a resonance in a case for which the NR approximation is a good approximation.

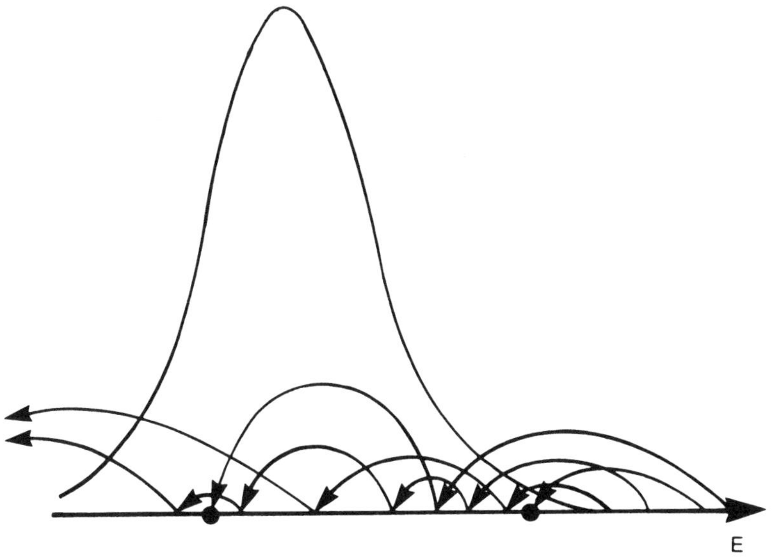

Fig. 5-8. Illustration of scattering in a wide resonance. Four neutrons enter the resonance and, after several collisions, two are absorbed and two are scattered.

subsequent effect of that resonance is strongly smoothed, especially when one considers that the neutron may be scattered on any of the various nuclei in the medium, which includes light, medium, and heavy nuclei.

Figure 5-8 illustrates the scattering in a resonance that is still narrow if compared with the scattering energy loss on light nuclei, but wide compared to the energy loss in scattering on heavy nuclei. Apparently, some deformation of the source spectrum within the resonance by the resonance itself is to be expected.

Both approximations are discussed in more detail below.

5-4B Loss-Free Spectrum in the NR Approximation

The balance equations for the neutron spectrum below the source where resonance absorption is of importance state that the scattering source, $S_s(E)$, equals the collision density, $F(E)$ [see, for example, Eqs. (4.85)]. Both $S_s(E)$ and $F(E)$ are practically proportional to $1/E$ if one disregards the small effect of the accruing resonance absorption described in $p(E)$. For hydrogen as a moderator, this follows, for example, from Eq. (4.112). For nonhydrogenous materials, the $1/E$ dependence of $F(E)$ is indicated by Eq. (4.144) for the case without absorption. The $1/E$ dependence holds also between well-separated resonances for the spectrum with absorption, as illustrated in Figs. 5-3 and 5-4. For the analytical evaluation of the NR approximation in a homogeneous medium, one considers an interval ΔE_R, which includes the resonance R at E_R below the fission source. If ΔE_R is small, the neutrons arrive in ΔE_R through scattering from a wide range above ΔE_R, as illustrated in Fig. 5-7. The source spectrum arriving in ΔE_R, the scattering source, $S_s(E)$, must have basically the energy distribution of the slowing down process without resonances, since the neutrons scattered into ΔE_R do not know that they are being scattered into a resonance. Then,

$$S_s(E) \propto \frac{1}{E} \quad , \text{ in } \Delta E_R \quad . \tag{5.26}$$

To preserve a stationary state, all the neutrons arriving in ΔE_R have to leave the interval again, either through absorption or scattering. From these arguments, the following balance is obtained:

$$S_s(E) \propto F(E) = \Sigma_t(E)\varphi(E) \propto \frac{1}{E} \simeq \text{constant, in } \Delta E_R \quad . \tag{5.27}$$

This gives the local spectrum of the NR approximation:

NR approximation spectrum:

$$\varphi_{NR}(E) = \frac{\text{constant}}{E\Sigma_t(E)} \simeq \frac{\text{constant}}{\Sigma_t(E)} \text{ in } \Delta E_R \quad . \tag{5.28}$$

It is important to note that $\Sigma_t(E)$ in Eqs. (5.27) and (5.28) contains the cross sections of all constituents in the homogeneous medium. This distinguishes the NR approximation from all other approximations, especially from the NRIM approximation discussed below.

The meaning of "narrow" can be characterized in terms of the practical resonance width, Γ_p [see Eq. (5.30)], and the energy range from which neutrons are scattered to $E = E_R$: A resonance with a practical width of Γ_p is narrow if for scattering on a material with an energy reduction factor α, the following inequality holds:

$$\Gamma_p \ll \frac{E_R}{\alpha} - E_R = E_R \frac{1-\alpha}{\alpha} = \Delta E \quad . \tag{5.29}$$

The practical width Γ_p, which is the width of that part of a resonance that is larger than Σ_p, may be much larger than the natural half-width Γ. The practical width can be estimated by equating $\Sigma_t^{\text{res}}(E)$, the resonance cross section for that isotope, with the potential scattering cross section of the entire medium:

$$\Sigma_t^{\text{res}}(E) = \Sigma_p \quad , \tag{5.30}$$

that is,

$$N\sigma_t^{\text{res}}\left(E_R \pm \frac{\Gamma_p}{2}\right) = \Sigma_p \quad . \tag{5.31}$$

Applying Eq. (5.31) to the Breit-Wigner formula, Eq. (5.3), for $\sigma_t^{\text{res}}(E)$ gives

$$N\sigma_t^{\text{res}}\left(E_R \pm \frac{\Gamma_p}{2}\right) = N\sigma_t^{\max}\frac{(\Gamma/2)^2}{(\Gamma_p/2)^2 + (\Gamma/2)^2} = \Sigma_p \quad . \tag{5.32}$$

For large resonances, dividing by N, introducing σ_b [Eq. (5.58)] for the resulting ratio, and neglecting Γ compared to Γ_p in the denominator gives:

$$\frac{\Gamma_p}{\Gamma} = \left(\frac{\sigma_t^{\max}}{\sigma_b}\right)^{1/2} = \left(\frac{\sigma_t^{\max}}{\Sigma_p}N\right)^{1/2} \quad . \tag{5.33}$$

Thus, if $\sigma_t^{\max}$ is, for example, ~25 times larger than the background cross section, σ_b, then Γ_p is ~5 times larger than the natural width Γ.

It should be noted, however, that the concept Γ_p is used only in qualitative considerations, involving inequalities with $\gg$ or $\ll$. Therefore, a more accurate evaluation of the resonance width, which would have to be calculated from Doppler-broadened resonance shapes (see below), is not required at that point.

The spectral segments for the vicinity of the individual resonances, as given by Eq. (5.28), can be readily connected. They then form the loss-free spectrum, $\varphi_0^{NR}(E)$, throughout the resonance range. Between the resonances, φ_0^{NR} must agree with the spectrum without Σ_a, given by Eq. (5.11); i.e., the constant in Eq. (5.28) must be equal to $q_r/\bar{\xi}$. Thus,

$$\varphi_0^{NR}(E) = \frac{q_r}{E\bar{\xi}\Sigma_t(E)} \quad , \tag{5.34}$$

with q_r being the slowing down density at the upper limit of the resonance range (E_u), i.e., q_r determines the number of neutrons entering the resonance range, which reaches from E_u down to E_{th}.

Apparently,

$$\varphi_0^{NR}(E)_{\substack{\text{between}\\\text{resonances}}} = \frac{q_r}{E\bar{\xi}\Sigma_s} = \varphi(E)_{\text{without } \Sigma_a} \quad . \tag{5.35}$$

It should be realized, however, that the NR approximation does not provide consistently accurate results for *all* resonances, especially at lower energies, where ΔE on heavy isotopes may become smaller than the Γ_p of many resonances. Then, one of the basic assumptions of the NR approximation is violated.

The spectra in NR approximations are characterized by the fact that Σ_t contains all cross sections; Σ_p contains the potential cross section of all isotopes, which have to be separately identified here: moderator (m), heavy nuclei (h), and others. The $\Sigma_t^{\text{res}}(E)$ contains the scattering contribution to the resonance cross section:

$$\Sigma_t(E) = \Sigma_p + \Sigma_t^{\text{res}}(E) \quad , \tag{5.36a}$$

$$\Sigma_p = \Sigma_p^m + \Sigma_p^h + \Sigma_p^{\text{others}} = \text{constant} \quad , \tag{5.36b}$$

and

$$\Sigma_t^{\text{res}}(E) = \Sigma_a^{\text{res}}(E) + \Sigma_s^{\text{res}}(E) \quad . \tag{5.37a}$$

Thus, in the treatment of resonance absorption, the scattering cross section $\Sigma_s(E)$ is split into the potential scattering cross section, denoted by Σ_p, and $\Sigma_s^{\text{res}}(E)$, the resonance scattering cross section. The latter consists of the respective contribution of the Breit-Wigner formula (Σ_s^{BW}) and the interference term (Σ^{interf}) [compare Eq. (5.6)]:

$$\Sigma_s^{\text{res}}(E) = \Sigma_s^{BW}(E) + \Sigma_s^{\text{interf}}(E) \quad . \tag{5.37b}$$

The effect of the interference term on resonance absorption is relatively small, so it is frequently neglected.

5-4C Loss-Free Spectrum in the NRIM Approximation

The derivation of the spectrum in the NRIM approximation is not as simple as for the NR approximation since the practical neutron width is not small compared to the energy loss on *all* constituents; Γ_p is rather just in between the ΔE_{max} for scattering on the two major constituents, the lighter moderator nuclei (index m) and the heavy nuclei (index h) in the fuel:

$$\Gamma_p \ll E\frac{(1-\alpha_m)}{\alpha_m} = \Delta E_m \tag{5.38a}$$

for the moderator nuclei, but

$$\Gamma_p \gg E\frac{1-\alpha_h}{\alpha_h} = \Delta E_h \tag{5.38b}$$

for the heavy absorber. The energy reduction factors for moderator and heavy nuclei are denoted with α_m and α_h, respectively.

The fact that Γ_p was much smaller than ΔE in *all* nuclei in the case of the NR approximation allowed the evaluation of the spectrum across a resonance from the approximate balance equation, Eq. (5.27). If Γ_p is between the various ΔE, the NR simplification does not apply. Then, the spectrum within the resonance can only be determined from the original balance equation (below the fission source):

$$\Sigma_t(E)\varphi(E) - \Sigma_p^m \int_E^{E/\alpha_m} \frac{\varphi(E')}{E'(1-\alpha_m)}\, dE' - \int_E^{E/\alpha_h} \frac{\Sigma_s^h(E')\varphi(E')}{E'(1-\alpha_h)}\, dE' = 0 \quad . \tag{5.39}$$

In Eq. (5.39), only one integral is used to represent the lighter and heavy constituents, respectively. In a complete derivation, one has to consider the sum of scattering integrals of all isotopes. The medium-weight materials, zirconium in light water reactors (LWRs), steel and sodium in fast breeder reactors (FBRs), are still light enough to be represented in Eq. (5.39) by the moderator term.

The total cross section, $\Sigma_t(E)$, in Eq. (5.39) contains the respective contributions of all isotopes:

$$\Sigma_t(E) = \Sigma_p^m + \Sigma_a^m(E) + \Sigma_s^h(E) + \Sigma_a^h(E) \quad , \tag{5.40a}$$

with

$$\Sigma_s^h(E) = \Sigma_p^h + \Sigma_s^{\text{res}}(E) \quad . \tag{5.40b}$$

In dealing with resonance absorption, the absorption in lighter nuclei

(Σ_a^m) is absolutely insignificant. Thus, Eqs. (5.40) can be replaced by

$$\Sigma_t(E) = \Sigma_p^m + \Sigma_s^h(E) + \Sigma_a^h(E) \quad . \tag{5.41}$$

Comparing Eq. (5.41) with Eq. (4.86a) shows that Σ_s^h still appears in Eq. (5.41). It had been omitted in Eq. (4.86a) with the argument that the scattering on the heavy nuclei is negligible compared to hydrogen. The derivation of the NR approximation, however, yielded a spectrum that was deformed analytically with $1/\Sigma_t(E)$, with $\Sigma_t(E)$ containing all cross sections, also $\Sigma_s^h(E)$. Apparently, the subject of including or deleting $\Sigma_s^h(E)$ in $\Sigma_t(E)$ needs special attention. It needs to be directly related to the treatment of the scattering integral:

$$\int_E^{E/\alpha_h} \frac{\Sigma_s^h(E')\varphi(E')}{(1-\alpha_h)E'}\, dE' = S_s^h(E) \quad . \tag{5.42}$$

If the scattering interval E to E/α_h is large compared to the resonance width, the interval ΔE may contain several resonances. Through integration over ΔE, their effect on $S_s^h(E)$ is smeared out; then $S_s^h(E)$ is a smooth function, proportional to $1/E$, or about constant in a small interval (say from 1000 to 1010 eV).

If, however, the scattering interval ΔE_h is much smaller than Γ_p, ΔE_h covers only a small part of a resonance. Then many ΔE_h-size intervals fit under a single resonance. The spectrum $\varphi(E')$ in the integral Eq. (5.42) follows then the deformed spectrum under the resonance. If the scattering interval is sufficiently small, the spectrum—or better—the quantity $\Sigma_s^h(E')E'\varphi(E')$ can be assumed not to vary much in the integrand; it is then approximately equal to its value at the lower integration limit:

$$\Sigma_s^h(E')E'\varphi(E') = \Sigma_s^h(E)E\varphi(E) = \text{constant in } E \le E' \le \frac{E}{\alpha} \quad . \tag{5.43}$$

This gives

$$S_s^h(E) = \Sigma_s^h(E)E\varphi(E) \cdot \int_E^{E/\alpha} \frac{dE'}{(1-\alpha)E'^2} \quad , \tag{5.44a}$$

which upon evaluation of the integral yields:

$$S_s^h(E) = \Sigma_s^h(E)E\varphi(E) \cdot \frac{1}{1-\alpha}\left(\frac{1}{E} - \frac{\alpha}{E}\right) \quad . \tag{5.44b}$$

One then obtains:

$$S_s^h(E) \simeq \Sigma_s^h(E)\varphi(E) \quad . \tag{5.45}$$

Thus, in the NRIM approximation, the scattering source approximately cancels the Σ_s^h contribution to the total cross section. Inserting Eq. (5.45) in Eqs. (5.41) and (5.39) then yields:

NRIM approximation balance equation:

$$[\Sigma_p^m + \Sigma_a^h(E)]\varphi(E) - \Sigma_p^m \int_E^{E/\alpha_m} \frac{\varphi(E')}{(1-\alpha_m)E'}\, dE'$$

$$= 0 \text{ below the source} \quad . \tag{5.46}$$

The discussion and treatment of the scattering integral on heavy nuclei presented in this subsection shows that the resonance absorption treated together with scattering on hydrogen in Sec. 4-5E actually made use of the NRIM approximation, since $\Sigma_t(E)$ was approximated by Eq. (4.86a). Also the discussions of resonance escape probabilities and of resonance integrals in Secs. 5-3B and 5-3C, respectively, which exploit the results of Sec. 4-5E, are then based on the same approximation. The subsequent presentation is concerned with a more refined treatment of resonance absorption, calculated with more accurate spectra.

The integral in Eq. (5.46) represents the neutron source, $S_s^m(E)$ due to scattering on the lighter nuclei. In the same way as in the NR approximation, $S_s^m(E)$ is proportional to $1/E$, which gives for each segment around a resonance the simplified balance equation in the NRIM approximation:

$$[\Sigma_p^m + \Sigma_a^{\text{res}}(E)]\varphi(E) \simeq \frac{\text{constant}}{E} \simeq \text{constant in } \Delta E_R \quad . \tag{5.47}$$

Here, Σ_p^m is assumed to be constant and Σ_a^h is denoted by $\Sigma_a^{\text{res}}(E)$.

For the spectrum segment around a resonance at E_R, this yields

$$\varphi(E) \propto \frac{\text{constant}}{E[\Sigma_p^m + \Sigma_a^{\text{res}}(E)]} \simeq \frac{\text{constant}}{\Sigma_p^m + \Sigma_a^{\text{res}}(E)} \text{ in } \Delta E_R \quad . \tag{5.48}$$

Both heavy element scattering contributions, Σ_p^h and $\Sigma_s^{\text{res}}(E)$ are omitted in the denominator of Eq. (5.48) as compared to Eq. (5.28). The deletion of these scattering terms represents the difference of the NR and the NRIM approximations, which in turn results in a modification of the self-shielding.

In the same way as for the NR approximation, the spectral segments of Eq. (5.48) can be connected to give the loss-free spectrum in the NRIM approximation:

$$\varphi_0^{\text{NRIM}}(E) = \frac{q_r \Sigma_p^m}{E\bar{\xi}\Sigma_p[\Sigma_p^m + \Sigma_a^{\text{res}}(E)]} \quad . \tag{5.49}$$

Again, the same reservations hold for Eq. (5.49) as they were expressed above for the corresponding NR approximation results. The spectrum for resonances for which the NRIM balance equation, Eq. (5.48), is not a good approximation may be quite inaccurate in the above formula. Thus, improved treatments to self-shielding and resonance absorption are required. Several methods are discussed in Sec. 5-5.

5-4D Applicability of the NR and NRIM Approximations

The spectrum segments, calculated in Secs. 5-4B and 5-4C, have been employed to find the respective loss-free spectra that will be used below to calculate the resonance absorption rate, particularly to the resonance integrals. First, however, it needs to be determined when to apply the NR approximation and when to apply the NRIM approximation.

The applicability of the NR or the NRIM approximations varies along the energy scale. It depends on the comparative values of ΔE and Γ_p.

The former quantity, ΔE, increases linearly with energy:

$$\Delta E = E\frac{1-\alpha}{\alpha} \quad .$$

The average ΔE in a core also depends on the composition; this average ΔE is larger in LWRs than in FBRs. The second quantity, Γ_p, depends on $\Gamma \cdot (\sigma_t^{\max})^{1/2}$, which slightly increases with E due to the increase of Γ_n [compare Eq. (5.5)]; Γ_n appears in Γ and $\sigma_t^{\max}$ [compare Eqs. (5.1) and (5.3)]. The increase of $\Gamma \cdot (\sigma_t^{\max})^{1/2}$ due to the increase in Γ_n is largely compensated by the decrease of $\sigma_t^{\max}$ resulting from the $1/E$ dependence of λ^2 in Eq. (5.5).

In addition to the continuous slight increase of Γ_p, there is a strong statistical variation of the Γ_n and thus of Γ and $\sigma_t^{\max}$. Therefore, the respective applicability of the NR and the NRIM approximations has both a general trend with energy and a statistical variation from resonance to resonance.

Furthermore, Γ_p is proportional to the concentration, N, of the resonance absorber [compare Eq. (5.32)]. However, the major resonance absorbers, for example, ^{238}U in LWRs or FBRs, appear in most systems in comparable concentrations; at least they are sufficiently close as far as the semiquantitative considerations on the applicability of the two approximations (NR and NRIM) are concerned. A composition dependence comes primarily from the average ΔE.

The general trend of the applicability of the NR versus NRIM approximations is illustrated by an example. Consider ΔE and Γ_p at two different energies, say 20 and 2000 eV. At 20 eV, ΔE for ^{238}U is ~0.34 eV (1.7%) and low-energy resonances may have an average Γ_p of

~0.5 eV. Thus, for statistically broad resonances, the NRIM approximation is applicable.

At 2000 eV, however, ΔE has increased to 34 eV with probably only a minor increase in Γ_p. Thus, the NR approximation should be applicable for nearly all resonances in that energy range. This holds even more so for higher energies.

Although $N(^{238}U)$ may not be drastically different in LWRs and FBRs, there is a very substantial difference in the overall neutron spectra in both reactors, which reflects on the applicability of the NR versus the NRIM approximations.

The spectrum in the isolated resonance range in LWRs is largely a $1/E$ spectrum, superimposed on the relatively small decrease due to resonance absorption and leakage: $[E\varphi(E)]_{\text{LWR}}^{\text{res range}}$ is slightly decreasing.

In FBRs, however, due to the much smaller slowing down power, $\overline{\xi\Sigma_s}$, neutrons diffuse in the resonance range much longer than in LWRs. Therefore, neutrons have a much larger chance to be absorbed or to leak out of the core. Consequently, the spectrum $[E\varphi(E)]$ strongly decreases with decreasing E.

In an LWR then, resonance absorption occurs predominately in the low-energy resonances, but in FBRs, it occurs predominately at much higher energies, where the NR approximation yields good results. Thus, in LWRs, both approximations are of importance, whereas in FBRs, the NR approximation suffices for most applications. Deviations from the NR approximation amount to only a relatively small correction, which is applied in sophisticated applications.

The above discussion does not address the question of the treatment of the resonances, which in width are such that neither the NR nor the NRIM approximations are reasonably applicable. Since there are, especially in LWRs, numerous resonances of these characteristics exposed to a large neutron flux, there is obviously a need to develop methods for the treatment of these "intermediate" resonances (compare Sec. 5-5).

5-4E Resonance Integrals in Homogeneous Media

The resonance absorption during slowing down, calculated with the loss-free spectrum, is a very fundamental quantity in the description of thermal reactors. As discussed in detail in Sec. 5-3B, the resonance absorption integral, calculated with φ_0, describes the sum of the absorptions in the individual resonances *without* considering prior losses. The neutron losses above a particular resonance are automatically accounted for by placing the φ_0-based integral in the exponential function. Such fundamental quantities often get special names, in this case, the "resonance integral," denoted here by L. For the definition of the resonance integral for isotope i, (I_i), the exponent in Eq. (5.23) is split into isotopic contributions:

$$p(E) = \exp\left[-\frac{1}{q_r}\int_E^{E_u} \Sigma_a(E')\varphi_0(E')\,dE'\right]$$

$$= \exp\left[-\sum_i \frac{1}{q_r}\int_E^{E_u} \Sigma_{ai}(E')\varphi_0(E')\,dE'\right] \quad . \tag{5.50}$$

For the loss-free spectrum, $\varphi_0(E)$, we use here the NR approximation, Eq. (5.34), although a more accurate approach is required in accurate numerical evaluations. The source magnitude, represented in Eq. (5.50) by q_r, cancels when φ_0 is inserted, as the exponent describes the φ_0-based resonance absorption per neutron entering the resonance range. Thus, upon inserting Eq. (5.34), one obtains

$$p(E) = \exp\left[-\sum_i \int_E^{E_u} \frac{\Sigma_{ai}(E')}{\bar{\xi}\Sigma_t(E')}\frac{dE'}{E'}\right] \quad . \tag{5.51}$$

Taking $\bar{\xi}$ and N_i out of the integral in Eq. (5.51) and expanding with Σ_p yields the form in which the common resonance integral is introduced:

$$p(E) = \exp\left[-\sum_i \frac{N_i}{\bar{\xi}\Sigma_p}\int_E^{E_u} \frac{\Sigma_p\sigma_{ai}(E')}{\Sigma_t(E')}\frac{dE'}{E'}\right] \quad . \tag{5.52}$$

For the entire resonance absorption ($E \rightarrow E_{th}$), one obtains from Eq. (5.52):

$$p = p(E_{th}) = \exp\left[-\sum_i \frac{N_i}{\bar{\xi}\Sigma_p}\int_{E_{th}}^{E_u} \frac{\Sigma_p\sigma_{ai}(E')}{\Sigma_t(E')}\frac{dE'}{E'}\right]$$

$$= \exp\left(-\frac{1}{\bar{\xi}\Sigma_p}\sum_i N_i I_i\right) \quad , \tag{5.53}$$

with

$$I_i = \int_{E_{th}}^{E_u} \frac{\Sigma_p\sigma_{ai}(E')}{\Sigma_t(E')}\frac{dE'}{E'} \tag{5.54}$$

being the resonance integral for isotope i (in the NR approximation).

For the calculation of the resonance integral, general use is made of two approximate assumptions:

1. The potential scattering cross section, Σ_p, is assumed to be independent of E. This is a fairly well-justified assumption, since the potential scattering has a significant energy dependence only at high ener-

gies, where the contribution to resonance absorption is very small.

2. For the definition of I_i, the resonance integral of isotope i, one considers in $\Sigma_t(E)$ only the resonance of that particular isotope, i.e., overlapping of resonance of different isotopes is neglected.

This gives the effective microscopic absorption cross section, called the "resonance integral" of isotope i:

$$I_i = \int_{E_{th}}^{E_u} \frac{\Sigma_p}{\Sigma_p + \Sigma_{ti}^{\text{res}}(E)} \sigma_{ai}(E) \frac{dE}{E} \quad . \tag{5.55}$$

In most nuclear reactors there is one dominating contributor to resonance absorption: ^{238}U or ^{232}Th. The fissile isotopes and the higher uranium chain and plutonium chain isotopes are much less concentrated than ^{238}U or ^{232}Th.

If only ^{238}U is important as a resonance absorber, the general expression for the resonance escape probability, i.e.,

$$p = p(E_{th}) = \exp\left(-\frac{1}{\bar{\xi}\Sigma_p}\sum_i N_i I_i\right) \quad ,$$

becomes simply:

$$p = \exp\left(-\frac{N_{238}}{\bar{\xi}\Sigma_p} I_{238}\right) \quad . \tag{5.56}$$

Dividing numerator and denominator in Eq. (5.55) by N_i gives an alternate expression for the resonance integral:

$$I_i = \int_{E_{th}}^{E_u} \frac{\sigma_b}{\sigma_b + \sigma_{ti}^{\text{res}}(E)} \sigma_{ai}(E) \frac{dE}{E} \quad . \tag{5.57}$$

Equation (5.57) contains the important parameter, σ_b, which is the "background" cross section:

$$\sigma_b = \frac{\Sigma_p}{N_i} = \frac{\text{potential scattering cross section}}{\text{resonance absorber atom density}} \quad . \tag{5.58}$$

Obviously, the resonance integral depends on the parameter σ_b. In addition, it depends on temperature through the respective dependence of the resonance shape (see Sec. 5-6).

The background cross section, σ_b of Eq. (5.58), is different for different isotopes. This is not commonly indicated in the notation.

The integrand in Eq. (5.57) is often called the "effective" cross section:

$$I_i(\sigma_b) = \int_{E_{th}}^{E_u} \sigma_{ai}^{eff}(E) \frac{dE}{E} \quad , \tag{5.59}$$

with the "effective" cross section given by σ^{eff} (also denoted by $\tilde{\sigma}$):

$$\sigma_{ai}^{eff}(E) = \frac{\sigma_b}{\sigma_b + \sigma_{ti}^{res}(E)} \cdot \sigma_{ai}(E) \quad . \tag{5.60}$$

The spectral depression factor, denoted by $f(E)$, is called the "self-shielding factor":

$$f(E) = \frac{\sigma_b}{\sigma_b + \sigma_t^{res}(E)} \quad . \tag{5.61}$$

The multigroup representations of the self-shielding factors are important concepts for practical calculations (see, for example, Ref. 6).

The background cross section, σ_b, tends to infinity with decreasing density of the resonance absorber. Then the self-shielding effect disappears:

$$f(E) = \frac{\sigma_b}{\sigma_b + \sigma_t^{res}} \to 1 \text{ for all } E \text{ if } \sigma_b \to \infty \quad . \tag{5.62}$$

The effective resonance integral, Eq. (5.53), then becomes the "infinite dilute resonance integral,"

$$I(\sigma_b \to \infty) = I^{\infty} = \int_{E_{th}}^{E_u} \sigma_a(E) \frac{dE}{E} \quad . \tag{5.63}$$

Thus, I^{∞} measures the entire resonance absorption in a $1/E$ spectrum without self-shielding.

Important values for I^{∞} are the ones for ^{235}U, ^{238}U, and ^{239}Pu:

$$I_c^{\infty}(235) = 144b,\ I_f^{\infty}(235) = 275b,\ I_c^{\infty}(238) = 277b,\ I_c^{\infty}(239) = 200b,$$

$$\text{and } I_f^{\infty}(239) = 301b \quad .$$

This subsection is concluded by emphasizing that for the definition of the resonance integral, one has to use the loss-free spectrum, $\varphi_0(E)$, since it is $\varphi_0(E)$ that appears in the exponent of the p factor:

$$I_i = \frac{\bar{\xi}\Sigma_p}{q_T} \int_{E_{th}}^{E_u} \sigma_{ai}(E)\varphi_0(E)\, dE \quad . \tag{5.64}$$

The loss-free spectrum itself may be calculated with various degrees of sophistication (compare Secs. 5-4 and 5-5).

5-5 Improved Treatment of Resonance Absorption

5-5A Slowing Down with Weak-Absorption Fermi Age Theory in an Infinite Medium

The so-called "age" theory for the treatment of neutron slowing down was developed by Fermi in the early days of nuclear reactor development. The term "age" indicates that—in this approximation—the neutrons are *continuously* slowed down (i.e., not in leaps); then, the energy loss of the neutrons gradually increases as does their "age."

The actual age theory treats both energy and space dependencies. Here, only the space-independent model is presented, primarily as an interim step to improved treatments of resonance absorption. The age theory itself is not an acceptable approximation to the resonance absorption phenomenon.

The age theory in an infinite medium is derived by application of the same expansion of the scattering integral, which was found to be successful for the case without absorption. The complete balance equation for several isotopes is given by:

$$[\Sigma_a(E) + \Sigma_s(E)]\varphi(E) - \sum_i \int_E^{E/\alpha_i} \frac{\Sigma_{si}(E')\varphi(E')}{(1-\alpha_i)E'}\, dE' = s_0\chi(E) \quad . \tag{5.65}$$

One rewrites this equation in lethargy units and inserts the linear expansion:

$$\Sigma_{si}(u')\varphi(u') = \Sigma_{si}(u)\varphi(u) + (u' - u)\frac{d}{du}[\Sigma_{si}(u)\varphi(u)] \quad . \tag{5.66}$$

The derivation closely follows the derivation without absorption that was presented in Sec. 4-6. Instead of Eq. (4.141), one obtains then:

$$\Sigma_a(u)\varphi(u) + \sum_i \xi_i \frac{d}{du}[\Sigma_{si}(u)\varphi(u)] = s_0\chi(u) \quad . \tag{5.67}$$

This is the Fermi age equation for an infinite medium. If Σ_s is assumed to be constant, Eq. (5.67) becomes simply:

$$\Sigma_a(u)\varphi(u) + \overline{\xi\Sigma_s}\frac{d}{du}\varphi(u) = s_0\chi(u) \quad . \tag{5.68}$$

The solution of Eq. (5.68), below the source, rewritten in E, is given by:

$$\varphi(E) = \frac{s_0}{\overline{\xi\Sigma_s}\cdot E}\exp\left[-\frac{1}{\overline{\xi\Sigma_s}}\int_E^{E_u}\Sigma_a(E')\frac{dE'}{E'}\right] \quad . \tag{5.69}$$

The spectrum, Eq. (5.69), does not show the self-shielding of resonances. It apparently was lost by the expansion, Eq. (5.66), because an approximation of $\varphi(u)$ by a straight line (in $\Sigma_s\varphi$) cannot follow or model a narrow spectral depression. Therefore, Eq. (5.69), the result of the Fermi age theory for an infinite medium, can only be used for weak absorption, where weak means, in particular, not causing substantial spectrum depressions in small energy intervals. Therefore, this approximation is applicable only in systems and energy ranges without high resonances in the macroscopic cross sections.

5-5B The Selengut-Goertzel and the Greuling-Goertzel Approximations

There are several important approximations that make better use of the expansion, Eq. (5.66), than does the age theory. These methods include the Selengut-Goertzel[7] and the Greuling-Goertzel[8] approximations.

The Selengut-Goertzel approximation[b] is in a sense a refinement of the NRIM approximation for hydrogen. The balance equation in lethargy, using only hydrogen and one heavier constituent for simplicity, is given by (neglecting absorption in hydrogen in the resonance region):

$$[\Sigma_s^{H} + \Sigma_a^h(u) + \Sigma_s^h(u)]\varphi(u) - \int_{-\infty}^{u} \Sigma_s^{H}(u'\rightarrow u)\varphi(u')\,du'$$

$$- \int_{u-a}^{u} \Sigma_s^h(u'\rightarrow u)\varphi(u')\,du' = s_0\chi(u) \quad . \tag{5.70}$$

The heavier constituents include here also nonhydrogenous moderators, for example, the oxygen in H_2O and in UO_2. Expanding the latter integral by applying Eq. (5.66) gives

$$\int_{u-a}^{u} \Sigma_s^h(u'\rightarrow u)\varphi(u')\,du' = \Sigma_s^h(u)\varphi(u) - \xi^h \frac{d}{du}[\Sigma_s^h(u)\varphi(u)] \quad . \tag{5.71}$$

Using only the first term, which cancels the corresponding term in the total cross section, yields the NRIM approximation. Including the second term also gives the balance equation of the Selengut-Goertzel approximation in which hydrogen is treated correctly but heavy atoms are treated by the age approximation:

---

[b]The Selengut-Goertzel approximation is presented here primarily for its historical significance and its position in the chain of increasingly better approximations.

$$[\Sigma_s^H + \Sigma_a^h(u)]\varphi(u) - \int_{-\infty}^{u} \Sigma_s^H(u' \to u)\varphi(u')\,du' + \xi^h \frac{d}{du}[\Sigma_s^h(u)\varphi(u)] = s_0\chi(u) \quad . \tag{5.72}$$

The solution to this equation will yield the spectral depressions at the resonances more accurately than the NRIM approximation. By using the additional term in Eq. (5.72), one obtains an improvement to the corresponding NRIM results. The NRIM approximation is obtained as the limit of $\xi(A) \to 0$ (for $A \to \infty$).

The second approximation is an improvement of the weak absorption model, which is achieved by a trick in the expansion. Greuling and Goertzel used the same expansion as in the age theory,

$$\Sigma_s^i(u')\varphi(u') = \Sigma_s^i(u)\varphi(u) + (u' - u)\frac{d}{du}[\Sigma_s^i(u)\varphi(u)] \quad , \tag{5.66}$$

but applied it to the slowing down density. Recall that the slowing down density, $q_{sd}(E)$, was defined by the integral [Eq. (4.147a)]:

$$q_{sd}(E) = \int_E^{E/\alpha} \Sigma_s(E')\varphi(E')\frac{E - \alpha E'}{E' - \alpha E'}\,dE' \quad . \tag{5.73}$$

This integral contains a weight factor that is zero at $E' = E/\alpha$, which is the far end of the expansion interval. Toward the far end, the error is increasing but its influence is strongly reduced by a vanishing weight factor. In the Greuling-Goertzel (G-G) approximation, Eq. (5.66) is substituted for the integrand of Eq. (5.73) (only the result is given):

$$\varphi(E) = \frac{s_0}{E \cdot [\xi\Sigma_s(E) + \gamma\Sigma_a(E)]} \cdot \exp\left[-\int_E^{E_u} \frac{\Sigma_a(E')}{\xi\Sigma_s(E') + \gamma\Sigma_a(E')}\frac{dE'}{E'}\right] \quad , \tag{5.74}$$

with

$$\gamma = \frac{1}{\xi} \cdot \frac{1}{1 - \alpha}\left(1 - \alpha - \alpha a - \alpha\frac{a^2}{2}\right) \tag{5.75a}$$

and

$$a = \ln\frac{1}{\alpha} \quad . \tag{5.75b}$$

The slowing down parameters ξ and γ are given in Table 5-I for several typical moderators. When γ is approximated by ξ, the NR approximation is obtained. The fact that γ is smaller than ξ means a somewhat smaller self-shielding than in the NR approximation.

TABLE 5-I

The Average Logarithmic Decrement, ξ, and the Constant, γ, for Various Moderator Materials.

	γ	ξ
Hydrogen	1.000	1.000
Deuterium	0.583	0.725
Beryllium	0.149	0.209
Carbon	0.116	0.158

5-5C Individual Resonance Treatments

The approaches to an improved treatment of resonance absorption presented in the previous sections result in the same modification of the self-shielding denominator for all resonances. However, the size of the resonances varies statistically from resonance to resonance due to the corresponding straggling of the Γ_n values. Therefore, the correct treatment of resonance absorption should reflect that aspect.

In 1962, Goldstein and Cohen[9] devised an approach between NR and NRIM approximations. These authors assumed, speaking in the terminology used here, a loss-free spectrum of the form:

$$\varphi_0(E) = \frac{s_0}{\xi\Sigma_p \cdot E} \cdot \frac{\Sigma_p^m + c\Sigma_p^h}{\Sigma_p^m + c\Sigma_p^h + c\Sigma_s^{\text{res}}(E) + \Sigma_a^{\text{res}}(E)} \quad . \tag{5.76}$$

Equation (5.76) gives the NR approximation for $c = 1$, and the NRIM approximation for $c = 0$. It is plausible that a very accurate description of the resonance absorption can be obtained with this formula. The form of Eq. (5.76) is inserted into the correct balance equation for each resonance and the best value of c is found numerically.

Comparing the G-G solution with the Goldstein-Cohen (G-C) approximation shows that in both approximations the components in the denominator have been modified. In the G-C approximation, the denominator is modified for each resonance individually, whereas in the G-G approximation,[c] the denominator remains the same over the entire

[c]Applied in MUFT-type computations (see Ref. 10).

slowing down range. Many modern computation methods that employ the G-G approximation can also be corrected by a side computation that finds c in an approximation such as the G-C method.[10] More sophisticated programs (for example, the HAMMER code[11]) include as an option an approximate numerical integration of the slowing down balance equation through resolved resonances that makes NR- or NRIM-type approximations unnecessary.

5-5D Summary of Resonance Absorption in Homogeneous Media

In the previous sections, the NR approximation, the NRIM approximation, and the intermediate resonance approximation (the G-C approximation) were presented. The emphasis was on deriving descriptions for the neutron spectrum for each of these approximations as well as on the conceptual understanding of the physics of the slowing down in homogeneous media. Formally, the regions were infinite in extent. The leakage during slowing down can be treated separately as shown in Sec. 4-5E. The concept of resonance escape probability, introduced as a factor in the criticality formulation in Chapter 2, was also provided with some firm physical and mathematical foundation.

The two major resonance absorption models developed were the NR and the NRIM approximations:

1. *The NR approximation:* It is assumed that the resonances are so narrow in width that a collision (even with a heavy absorber nucleus) scatters the neutron way out of the energy range of the resonance and that neutrons are scattered into the resonance out of an energy range much larger than the resonance width.
2. *The NRIM approximation:* For broader resonances, especially at low energies, collisions with heavy absorbers result in scattering within the range of a resonance. The respective energy loss is small compared to the resonance width.

While it is impossible to provide firm rules as to the applicability or validity of these two approximations, at high energies the energy loss through scattering (ΔE) tends to be much larger than the resonance widths, so that the NR approach is very useful. At low energies, ΔE_h is smaller than most resonance widths so that the NRIM approximation works well. This low-energy range is important in thermal reactors and hence plays a critical role in commercial reactor performance. For intermediate energies, neither approximation holds with much confidence. In this case, the resonance must be treated individually using sophisticated computer programs such as the HAMMER code.[11]

The various approximations to the treatment of the slowing down problems are listed in Table 5-II, which shows formulas comprising balance equations and approximations to the scattering terms. Not included are the methods that treat individual resonances in different ways, such as the G-C approximation.

5-6 Resonance Effects in Heterogeneous Systems

5-6A General Comments

In the previous section, the slowing down of neutrons in the presence of strong resonance absorbers was evaluated for homogeneous systems. However, all but a few special types of present-day reactors are composed physically of a great many fuel rods and associated moderator, coolant (which may also be the moderator), and structural material. These reactors then are strongly heterogeneous, and the effect of these repeated "cells" or subregions on the neutron slowing down process must be incorporated into the reactor calculations. Whereas in theory this can be done by retaining the space and energy dependency throughout the heterogeneous structure, the cost and time required of doing this even with present high-speed computers is prohibitive. Therefore, approximations or methods are developed that allow the heterogeneous system to be replaced by an equivalent homogeneous system. There exist several recipes and formulations that have been employed in reactor analyses for describing the resonance effects in reactor unit cells. Since these methods all rely heavily on neutron transport theory, they are not presented in detail. Only a brief sketch of some simple cookbook-type recipes employed in treating the effect of heterogeneities on the resonance escape probability are presented here. For a more detailed discussion of these effects, the reader is referred to the definitive book by Dresner,[2] the fine review article by Nordheim,[12] and textbooks on advanced nuclear reactor theory.[13–16]

Historically, the value of lumping fuel into a distinct heterogeneous form was recognized as a method of improving k_∞ over that possible with the same material in a uniform homogeneous system. Several factors come into play with this approach, the improvement in resonance escape probability being a prominent one. With discrete lumps of fuel at the resonance energies, penetration of neutrons into the central regions of the fuel is inhibited by "spatial self-shielding." This reduction in neutron absorption relative to what it would be in a homogeneous system results in an increase in the resonance escape probability. Although this effect also tends to reduce the thermal utilization factor, f, the net effect is an increase in the medium multiplication constant k_∞. In addition, lumping the fuel provides another beneficial effect by tending to increase the fast

TABLE 5-II
Survey of Slowing Down Treatment Methods

Balance Equations and Approximations	Methods
$[\Sigma_p^m + \Sigma_s^h(E) + \Sigma_a^h(E)]\varphi(E) - \int_E^{E/\alpha_m} \frac{\Sigma_p^m \varphi(E')}{(1-\alpha_m)E'}\, dE' - \int_E^{E/\alpha_h} \frac{\Sigma_s^h(E')\varphi(E')}{(1-\alpha_h)E'}\, dE' = \begin{matrix} s_0\chi(E) \\ 0 \text{ below } \chi(E) \end{matrix}$	Balance equation for numerical treatment
$[\Sigma_p^m + \Sigma_s^h(E) + \Sigma_a^h(E)]\varphi(E) \underbrace{- q_r \frac{\Sigma_p^m}{\overline{\xi\Sigma_p}} \cdot \frac{1}{E} - q_r \frac{\Sigma_p^h}{\overline{\xi\Sigma_p}} \cdot \frac{1}{E}}_{\sim \text{constant}/E} = 0 \quad .$	NR approximation
$[\Sigma_p^m + \underline{\Sigma_s^h(E)} + \Sigma_a^h]\varphi(E) - q_r \frac{\Sigma_p^m}{\overline{\xi\Sigma_p}} \cdot \frac{1}{E} \underline{- \Sigma_s^h(E)\varphi(E)} = 0 \quad .$ (underlined terms: → cancel ←)	NRIM approximation
$\overbrace{\underline{[\Sigma_p^m + \Sigma_s^h(E)} + \Sigma_a^h]\varphi(E) - \left\{ \underline{\Sigma_s(E)\varphi(E)} - \overline{\xi\Sigma_s} \frac{d}{dE}[E\varphi(E)] \right\}} = s_0\chi(E) \quad .$ (underlined terms: → cancel ←)	Fermi age equation in infinite medium
$[\Sigma_p^{\mathrm{H}}(E) + \underline{\Sigma_s^h(E)} + \Sigma_a^h(E)]\varphi(E) - \int_E^{\infty} \frac{\Sigma_p^{\mathrm{H}}(E')\varphi(E')}{E'}\, dE' - \left\{ \underline{\Sigma_s(E)\varphi(E)} - \overline{\xi\Sigma_s^h} \frac{d}{dE}[E\varphi(E)] \right\} = s_0\chi(E) \quad .$ (underlined terms: → cancel ←)	Selengut-Goertzel approximation

fission factor. This is due to the fact that neutrons born within the fuel have a higher chance for fast fission within the same fuel when the fuel is in a lumped form. This increases the fast fission effect.

5-6B Absorption Rates in Heterogeneous Media

In a bundle of fuel rods, the diffusion theory is inappropriate and transport theory must be used since the medium has many internal interfaces. The neutron flux depends strongly on space. Several approaches are possible, but the collision probability method using the integral transport equation is one technique that has proven to be most useful (for details, see Chapter 6).

The calculational problem is greatly simplified by assuming that the basic repeatable cell structure consists of only two distinct regions: a moderator and a fuel region. It is not necessary that the fuel region consist only of absorber isotopes; it may be, for example, a compound of resonance absorber plus moderator (the oxygen in UO_2). The basic intent is to include the spatial dependency as calculated by integral transport theory in quantities called "escape probabilities" or "collision probabilities." Rather complicated geometries can be treated this way. The derivation of this approach is not given in any detail but merely sketched (for particulars, see Refs. 13 and 14).

The basic approach is to describe the unit cell region by two distinct regions; a moderator region (denoted by superscript m) and a predominantly absorbing region (denoted by superscript f for fuel region). Escape probabilities $P_{f\to m}(u)$ and $P_{m\to f}(u)$ are then defined. They are geometry (region) dependent but not space dependent. In most calculations for a given design, once determined, the escape probabilities can be used over and over, hence greatly simplifying the overall analysis. These escape probabilities, $P_{f\to m}$ and $P_{m\to f}$, are the probabilities that a neutron having a scattering collision in the region considered will have its next collision in the other region.

The simplifying assumption usually employed in determining these escape probabilities is that the flux is flat in each region. This is clearly satisfied between the resonances where—due to the large mean-free-path of the neutrons—the flux is flat throughout the unit cell, i.e., it is the same in both media. See Sec. 5-6C for the escape probabilities.

The subsequent derivation is carried out for the NR approximation. A corresponding result can be obtained—in a similar fashion—for the NRIM approximation.

To relate the heterogeneous media results closely to the ones for a homogeneous medium, the latter result is rewritten such that it is fully analogous to the heterogeneous result. Let $\varphi^{out}(u)$ be the constant spectrum per unit lethargy between the resonances (the superscript "out" refers to the u range "outside" the isolated resonances):

$$\varphi^{out}(u) = \varphi_u^{out} = \text{constant} \quad . \tag{5.77}$$

The scattering source, S_s^{hom}, which furnishes the neutrons within the resonance, equals—in the NR approximation—the total collision density $\Sigma_t^{hom}(u)\varphi^{in}(u)$:

$$S_s^{hom} = \Sigma_p^{hom}\varphi_u^{out} = \Sigma_t^{hom}(u)\varphi^{in}(u) \quad , \tag{5.78}$$

where the superscript *in* refers to the small lethargy ranges that contain the resonances.

For the heterogeneous case, the derivation aims directly at the absorption rate. Therefore, Eq. (5.78), which describes the total reaction rate, is applied to compute the absorption rate in the fuel, $R_a^{f,hom}$:

$$R_a^{f,hom}(u) = \Sigma_a(u)\varphi^{in}(u) = \Sigma_a(u)\frac{\Sigma_p^{hom}}{\Sigma_t^{hom}(u)}\varphi_u^{out} \quad , \tag{5.79}$$

where $\varphi^{in}(u)$ is inserted as it results from Eq. (5.78). Note that Eq. (5.79) can also be obtained from S_s^{hom} times the probability $(\Sigma_a/\Sigma_t^{hom})$ that the next collision leads to an absorption.

Remember, φ_u^{out} is given by

$$\varphi^{out}(u) = \varphi_u^{out} = \frac{q_r}{\xi\Sigma_p} \quad . \tag{5.80}$$

In the two-media problem, the scattering source, Eq. (5.78), is replaced by two different sources, one in the fuel (S_s^f) and one in the moderator region (S_s^m). In applying collision probability methods, one generally works with integrated quantities:

$$\hat{S}_s^f = S_s^f V^f = \Sigma_p^f \varphi_u^{out} V^f \tag{5.81a}$$

and

$$\hat{S}_s^m = S_s^m V^m = \Sigma_p^m \varphi_u^{out} V^m \quad , \tag{5.81b}$$

where V^f and V^m are the respective volumes and φ_u^{out} is the same in both media.

The absorption rate in the fuel, R_a^f, is treated in the multiple collision method as consisting of two components, of neutrons that had their previous collision in the fuel ($R_a^{f\rightarrow f}$) or in the moderator ($R_a^{m\rightarrow f}$), respectively:

$$\hat{R}_a^f(u) = \hat{R}_a^{f\to f}(u) + \hat{R}_a^{m\to f}(u) \quad . \tag{5.82}$$

The two components are given by Eqs. (5.83) and (5.84):

$$\hat{R}_a^{f\to f}(u) = \hat{S}_s^f(1 - P_{f\to m})\frac{\Sigma_a^f(u)}{\Sigma_t^f(u)} \quad , \tag{5.83}$$

i.e., the total number of scattering source neutrons (at u) times the probability for not escaping from the fuel region $(1 - P_{f\to m})$ times the probability for the collision in the fuel region being an absorption (note they did not escape, i.e., they *must* collide).

In the same way,

$$\hat{R}_a^{m\to f}(u) = \hat{S}_s^m P_{m\to f}\frac{\Sigma_a^f(u)}{\Sigma_t^f(u)} \quad , \tag{5.84}$$

i.e., the total number of scattering source neutrons (at u) in the moderator times the probability for escaping the moderator and thus having the next collision in the fuel region $(P_{m\to f})$ times the probability for that collision being an absorption.

Note that if the moderator and fuel region merge, if the volume of the separated moderator region goes to zero, and if the properties and volume of the fuel region change accordingly, $P_{m\to f}$ and $P_{f\to m}$ approach one and zero, respectively. Then, after dividing by the volume, both Eqs. (5.83) and (5.84) become the same as Eq. (5.79), since

$$(\hat{S}_s^f/V^f) \longrightarrow \Sigma_p^{\text{hom}}\varphi_u^{\text{out}} \quad . \tag{5.85}$$

The two escape probabilities, $P_{f\to m}$ and $P_{m\to f}$, in Eqs. (5.83) and (5.84) could be calculated separately. But their values are related by the famous reciprocity relation for a two-region problem (see, for example, p. 112 of Ref. 14):

$$\Sigma_p^m V^m P_{m\to f} = \Sigma_t^f V^f P_{f\to m} \quad . \tag{5.86}$$

Note that $\Sigma_p^m \simeq \Sigma_t^m$. If, in Eq. (5.86), Σ_t^f increases as the energy moves closer and closer to the maximum of the resonance, the escape probability decreases accordingly so that the product remains about constant, especially for large Σ_t^f (see below).

Normally, $P_{m\to f}$ is expressed by $P_{f\to m}$ by invoking the reciprocity relation:

$$P_{m\to f} = \frac{\Sigma_t^f V^f}{\Sigma_p^m V^m} P_{f\to m} \quad . \tag{5.87}$$

Replacing $P_{m\to f}$ in Eq. (5.84) by Eq. (5.87) gives

$$\hat{R}_a^{m\to f}(u) = \hat{S}_s^m \frac{V^f}{V^m}\frac{\Sigma_a^f(u)}{\Sigma_p^m} P_{f\to m}(u) \quad . \tag{5.88}$$

Inserting the scattering sources of Eqs. (5.81) in Eqs. (5.83) and (5.88), and adding these two contributions yields

$$\hat{R}_a^f(u) = \left\{\Sigma_p^f[1 - P_{f\to m}(u)]\frac{\Sigma_a^f(u)}{\Sigma_t^f(u)} + \Sigma_a^f(u)P_{f\to m}(u)\right\}V^f\varphi_u^{\text{out}} \quad . \tag{5.89}$$

Rearranging terms and integrating from u_u to u_{th} gives:

$$\hat{R}_a^f = V^f\varphi_u^{\text{out}}\int_{u_u}^{u_{th}}\left\{\frac{\Sigma_p^f}{\Sigma_t^f(u)}\Sigma_a^f(u) + \frac{\Sigma_t^f(u) - \Sigma_p^f}{\Sigma_t^f(u)}\Sigma_a^f(u)P_{f\to m}(u)\right\} du \quad , \tag{5.90}$$

or in energy (using also $\Sigma_a = \Sigma_t - \Sigma_p$):

$$\hat{R}_a^f = V^f\varphi_u^{\text{out}}\int_{E_{th}}^{E_u}\left[\frac{\Sigma_p^f}{\Sigma_t^f(E)}\Sigma_a^f(E) + \frac{\Sigma_a^f(E)}{\Sigma_t^f(E)}\Sigma_a^f(E)P_{f\to m}(E)\right]\frac{dE}{E} \quad . \tag{5.91}$$

The first term in Eq. (5.90) is the same as in the respective homogeneous case [compare Eq. (5.79)]. It represents only energetic self-shielding. The third term describes the decrease of the reaction rate due to spatial self-shielding. The second term gives the added reactions due to neutrons entering from the moderator. The latter two terms disappear when the fuel medium is so large that neutrons cannot escape, as in the infinite homogeneous region. The reaction rate of Eq. (5.91) is smaller than the corresponding homogeneous one, although it is difficult to see, since all Σ's are different in the two cases:

$$\hat{R}_a^{f,\text{hom}} = (V^f + V^m)\varphi_u^{\text{out}}\int_{E_{th}}^{E_u}\frac{\Sigma_p^{\text{hom}}\Sigma_a^{\text{hom}}(E)}{\Sigma_p^{\text{hom}} + \Sigma_t^{\text{res,hom}}(E)}\frac{dE}{E} \quad . \tag{5.92}$$

The macroscopic cross sections in Eq. (5.92) refer to the uniform mixture of fuel and moderator in the joint volume $V^f + V^m$.

5-6C Approximate Escape Probabilities and Equivalence Relations

Several approximate methods or formulations exist for estimating the escape probability, $P_{f\to m}$, in fuel lumps of various shapes.

In the limiting case of a very large mean-free-path, $P_{f\to m}$ approaches unity. In the opposite limit, when the body is large compared with the mean-free-path, the following limit is obtained:

$$P_{f\to m} = \frac{1}{l\Sigma_t^f} \quad , \tag{5.93}$$

with $\bar{l}$ being the average "chord length" in the fuel. Evaluation of $\bar{l}$ for various convex fuel lump shapes gives the general formula

$$\bar{l} = \frac{4V^f}{S^f} \quad , \tag{5.94}$$

where the quantities S^f and V^f are the surface area and volume of the fuel lump.

For intermediate-size fuel regions, an approximation introduced by Wigner, the so-called "rational approximation," is frequently used:

$$P_{f\to m}(E) = \frac{1}{1 + \bar{l}\Sigma_t^f(E)} \quad . \tag{5.95}$$

Equation (5.95) can be expressed in an alternate form as:

$$P_{f\to m}(E) = \frac{\Sigma_l}{\Sigma_l + \Sigma_t^f(E)} \quad , \tag{5.96}$$

where Σ_l is defined as the "escape cross section":

$$\Sigma_l = 1/\bar{l} \quad . \tag{5.97}$$

Escape probabilities prescribed by Eq. (5.95) or (5.96) are quite accurate for large $\bar{l}\Sigma_t^f$, but are in general too low for intermediate values of $\bar{l}\Sigma_t^f$, with errors up to 20% when $\bar{l}\Sigma_t^f$ is near $\bar{l}\Sigma_t^f = 1$ (see Ref. 2).

The escape probability, Eq. (5.95), is dependent on the particular shape of the fuel lump (e.g., cylinder, flat plate, sphere) through the $\bar{l}\Sigma_t^f$ term. These relationships have been developed for various shapes of isolated lumps (for example, Ref. 17). If, however, several lumps are close together, a correction has to be employed, the so-called "Dancoff" or "Dancoff-Ginsburg" correction,[18] which is discussed below.

In the previous formula for $P_{f\to m}$, only single lumps of fuel were assumed to be present or acted as if they were completely isolated. However, in most real reactor unit cells, especially in LWRs, the fuel rods are so close together that adjacent fuel lumps interfere with the resonance absorption in the neighboring fuel rods. In a good approximation, it is possible to account for this interference merely by a redefinition of $P_{f\to m}$, to incorporate collisions in other fuel lumps. Therefore, one defines $P_{f\to m}^*$ as the probability that a neutron of a uniform and isotropic flux scattered within a fuel lump will not suffer its next collision in that fuel lump or in any other fuel lump. Substitution of $P_{f\to m}^*$ for $P_{f\to m}$ in the previous formulas then yields the reaction rate in a lattice of fuel.

In practice, however, it is very difficult to calculate $P_{f\to m}^*$ exactly. Monte Carlo calculations may be required; see Ref. 19 for an example.

Therefore, the simpler approach, proposed by Dancoff and Ginsburg,[18] is usually employed.

The basic approach taken by Dancoff and Ginsburg in accounting for the interaction of cells, i.e., the effect of a neighboring rod, was to reduce the number of neutrons incident from the direction of the neighboring rods. In other words, the spatial influence of one rod on another is treated as a reduction of the neutrons incident on that side of the fuel lump facing its neighbor, relative to the side of the fuel facing only the moderator. To incorporate this "surface efficiency" effect, an effective surface area S_{eff} is defined:

$$S_{\text{eff}} = S(1 - C_{\text{D}}) \quad , \tag{5.98}$$

where $(1 - C_D)$ represents the "Dancoff correction."

Employing the surface area reduction in a straightforward manner to calculate the respective modification of $\bar{l}$ and thus $P_{f \to m}$ yields:

$$\bar{l}^* = \frac{4V^f}{S^f_{\text{eff}}} \quad , \tag{5.99}$$

and Wigner's rational approximation

$$P^*_{f \to m}(E) = \frac{1}{1 + \bar{l}^* \Sigma^f_t(E)} \quad . \tag{5.100}$$

More accurate approximations to the resonance absorption problem in tightly packed resonance absorbers have been developed.[20, 21] Also, various approximate methods have been proposed for the calculation of C_{D}; see, for example, Ref. 14. The derivation and evaluation of these methods are beyond the scope of this book. More advanced texts, such as Ref. 14, need to be consulted.

The absorption rate, Eq. (5.91)—in a sense—is calculated with the loss-free spectrum. It also can be expressed in terms of heterogeneous resonance integrals (I_i^{het}) in the same way as in the homogeneous case:

$$I_i^{\text{het}} = \int_{E_{th}}^{E_u} \frac{\Sigma^f_p}{\Sigma^f_t(E)} \sigma_{ai}(E) \frac{dE}{E} + \int_{E_{th}}^{E_u} \frac{\Sigma^f_a(E)}{\Sigma^f_t} \sigma_{ai}(E) P_{f \to m}(E) \frac{dE}{E} \quad . \tag{5.101}$$

Inserting the rational approximation for $P_{f \to m}$, Eq. (5.100) allows a substantial simplification of the integrand of Eq. (5.101) and yields an important equivalence relation between heterogeneous and homogeneous systems. The factor of σ_{ai}/E is apparently the heterogeneous self-shielding factor $f^{\text{het}}(E)$: Self-Shielding Factor

$$f^{\text{het}}(E) = \frac{\Sigma^f_p}{\Sigma^f_t} + \frac{\Sigma^f_a}{\Sigma^f_t} \frac{1}{1 + \bar{l}^* \Sigma^f_t} = \frac{1 + \bar{l}^* \Sigma^f_p}{1 + \bar{l}^* \Sigma^f_t}$$

and

$$f^{\text{het}}(E) = \frac{\Sigma_l^* + \Sigma_p^f}{\Sigma_l^* + \Sigma_p^f + \Sigma_t^{f,\text{res}}(E)} = \frac{\Sigma_p^*}{\Sigma_p^* + \Sigma_t^{f,\text{res}}(E)} \quad . \tag{5.102}$$

Equation (5.102) shows that the heterogeneous self-shielding factor is of the same form as the homogeneous one, with

$$\Sigma_p^* = \Sigma_p^f + \Sigma_l^* \tag{5.103}$$

substituted for the homogeneous potential cross section, Σ_p^{hom}.

With the advent of the computer age, the resonance integral and escape methods have been primarily relegated to a supporting role in nuclear calculations. Their usefulness is no longer in the direct application in criticality evaluations via the resonance escape probability term in the simple factorized criticality formulation using a resonance integral, Eq. (5.101). Today, rather, these simple treatments of resonance effects in cells are employed primarily as corrections in the generation of "effective" cross sections for use in multigroup calculations (see Chapter 7), employing multigroup versions of the self-shielding factor Eq. (5.102). Many computer codes illustrate the application of the various heterogeneous corrections. The SPHINX code[22] and the MICROX code[23] are two recent examples that should be consulted for the application details.

5-7 Temperature-Dependent Resonance Absorption

5-7A General Discussion

The shape of the resonances in nuclear cross sections was given by the Breit-Wigner formula in Eq. (5.3). This is the so-called "natural line shape," which is the shape the resonances would have in the idealized case of neutron interaction with nuclei at rest [or in the center-of-mass system (CMS)]. However, in all practical cases, nuclei are in thermal motion. What one needs for practical calculations is a resonance cross section for the absorber material at rest. Within the lump of material, the nuclei have the velocity distribution of their thermal motion. The cross section for the material at rest, in which proper account is taken of the thermal agitation, exhibits strongly broadened resonances as compared to their natural shape (Doppler broadening).

The broadened resonances cause lesser spectral depressions. This results in a reduced self-shielding and in turn in increased resonance absorption. Increasing resonance absorption (which is primarily capture) with increasing temperature leads to a decrease in reactivity, and thus to an inherent negative feedback effect (Doppler feedback). The corresponding reactivity reduction, which is as automatic as the temperature increase itself, plays a vital role in safety considerations. If, in a power transient, the temperature should increase, the inherent action of the Doppler feedback quickly reduces the reactivity and the power. Subsequent power changes are slow enough that man-made equipment (scram rods) can finally terminate the transient.

The negative feedback response has been investigated in many sample experiments, especially for fast reactors for which the negative Doppler feedback is vital.[24–27] But in order to demonstrate the inherent and absolutely fail-safe action and the magnitude of the Doppler feedback, a special reactor had to be built and operated,[28, 29] the Southwest Experimental Fast Oxide Reactor (SEFOR). A special attachment allowed the reactor to become critical very rapidly, which results in a rapid power transient. In one of the key experiments, the power increased quickly from its nominal level of 10 MW(thermal) to ~23 000 MW(thermal), from where the Doppler feedback reduced it equally rapidly back to values near the operational power.[30] Subsequent scram terminated the transient.

Only the most basic concepts are discussed in this section. For a more detailed presentation the reader is referred to specialized or higher level books.[2, 6, 14]

5-7B Temperature-Dependent Resonance Shape

The natural line shape of a single resonance is given by the Breit-Wigner formula, Eq. (5.3). It is written here in terms of its maximum value (σ_m^R) and a shape function, $\psi_n(E)$, where m refers to the maximum value and n to the natural line shape:

$$\sigma^{\text{res}}(E) = \sigma_m^R \cdot \psi_n(E) \quad . \tag{5.104}$$

The shape function is of the form

$$\psi_n(E) = \frac{\left(\frac{\Gamma}{2}\right)^2}{\left(\frac{\Gamma}{2}\right)^2 + (E - E_R)^2} \quad ; \tag{5.105}$$

it is often called a "Lorentzian." The maximum value, σ_m^R, is the cross section at $E = E_R$; Γ is the total width of the resonance peak at half of its maximum value:

$$\psi_n\left(E = E_R \pm \frac{\Gamma}{2}\right) = \frac{1}{2}\psi_n(E = E_R) \quad . \tag{5.106}$$

See Eq. (5.3) for the value of σ_m^R.

Because of the thermal motion of the resonance nuclei, neutrons with energy greater or less than the resonance energy can react with the resonance, depending on the relative motion between neutron and nucleus. This clearly leads to a "broadening" of the energy range of neutrons that can react with the resonance. It is the same phenomenon originally discovered in acoustics by the Austrian physicist Doppler who found that the apparent pitch of a sound wave changes depending on the relative motion of sound emitter and receiver. This basic phenomenon is called the "Doppler effect." Practically, this is taken into account by averaging over the entire thermal motion. The resulting cross section describes the neutron reaction with material in which the individual nuclei are in thermal motion.

The resonance width is broadened and the height is reduced due to the thermal motions of the resonance nuclei. The resonance width is enlarged and approaches with increasing temperature a quantity Δ, called the "Doppler width," which is a function of the thermal velocity of the resonance nuclei.

Collisions between incident neutrons and a nucleus were discussed in Sec. 4-2. The relative energy between neutron and nucleus is proportional to:

$$(\boldsymbol{v}-\boldsymbol{V})^2 = v^2 - 2vV\cos\vartheta + V^2 \quad . \tag{5.107}$$

The term V^2 can be dropped as negligible compared to $2vV$. The second term on the right side of Eq. (5.107) can vary from $+2vV$ to $-2vV$, depending on the relative direction of motion. The relative neutron energy then has an average spread proportional to vV, called "Doppler width," Δ:

$$\Delta \propto vV \propto \sqrt{E}\sqrt{kT} \quad . \tag{5.108}$$

The quantitative evaluation yields

$$\Delta = \left(\frac{4kTE}{A}\right)^{1/2} \quad . \tag{5.109}$$

The Doppler-broadened resonance shape is obtained by the following computation.

1. It is observed that the Breit-Wigner formula and thus the natural line shape holds at first in the CMS of the neutron and nucleus. Let E' be the relative energy that results from Eq. (5.107) through multiplication with $m/2$ (neglecting V^2, and setting $V\cos\vartheta = V_z$):

$$\frac{m}{2}(\boldsymbol{v}-\boldsymbol{V})^2 = E' = \frac{1}{2}mv^2 - mvV_z \tag{5.110}$$

and

$$E' = E - (2mE)^{1/2} V_z \quad .$$

Without loss of generality, it could be assumed in Eq. (5.110) that the z axis points in the direction of the neutron motion.

2. The nuclear thermal motion is exhibited in a Maxwellian velocity distribution:

probability for V_z:

$$M(V_z) = \text{Maxwellian} \quad . \tag{5.111}$$

3. Inserting E' from Eq. (5.110) in $\psi_n(E')$, multiplying with the probability for V_z, and integrating over V_z gives the *Doppler-broadened resonance shape:*

$$\psi_D(E,T) = \int_{V_z} \psi_n[E'(V_z)] M(V_z)\, dV_z \quad . \tag{5.112}$$

The integral in Eq. (5.112) and the resulting Doppler-broadened shape are normally expressed in relative variables:

$$x = \frac{E - E_R}{\Gamma/2}, \quad y = \frac{E' - E_R}{\Gamma/2}, \quad \theta = \frac{\Delta}{\Gamma} (\propto \sqrt{T}) \quad . \tag{5.113}$$

In these variables, Eq. (5.112) is appearing as

$$\psi(x,\theta) = \frac{1}{2\theta\sqrt{\pi}} \int_{-\infty}^{\infty} \frac{\exp - \left(\dfrac{x-y}{2\theta}\right)^2}{1 + y^2}\, dy \quad . \tag{5.114}$$

The Doppler-broadened shape is often called the "Voigt profile," since it was derived by Voigt in 1912.

The averaging over the Maxwell distribution in Eq. (5.112) does *not* change the integral under the resonance. It amounts only to a redistribution of the resonance cross section. Then, the integral over the resonance is independent of the broadening and thus of temperature:

$$\int_E \psi_D(E,T)\, dE = \pi \frac{2}{\Gamma}, \text{ independent of } T \quad . \tag{5.115}$$

The fact that the integral over the resonance does not change when the Doppler broadening is evaluated can be qualitatively understood in the following way. Consider neutrons of a certain direction of energy E_R reacting with nuclei at rest. The reaction rate is proportional to $\psi_m(E = E_R)$. However, if the nuclei are in thermal motion, only very few neutrons collide with nuclei near that energy, i.e., with nuclei moving perpendicular to the direction of the neutron motion. Thus, the reaction rate at $E = E_R$ is much smaller than for nuclei at rest. Most neutrons find the nuclei moving away from or toward them. They then can only react with the wings of the resonance. The corresponding reduction of ψ_D near E_R is compensated by increased reactions on the wings.

TABLE 5-III
Resonance and Doppler Widths for ^{238}U

T (K)	kT (eV)		10 eV	10^3 eV	10^5 eV
360	0.03	$\Delta(E,kT)$	0.07	0.70	7.0
1440	0.12	$\Delta(E,kT)$	0.14	1.40	14.0
		Γ	0.027	0.040	0.175

In Table 5-III typical resonance widths for ^{238}U are listed for various energies and temperatures. This illustrates the broadening of the resonance peaks due to temperature increase. It is seen that it is the Doppler width that—at higher temperatures and energies—completely determines the width of the resonance in reactor material. Therefore, the practical width evaluation presented in Sec. 5-4 would be better based on Doppler-broadened rather than natural line shape resonances. But since Γ_p is only applied in a qualitative sense, no serious deficiency arises from this inconsistency.

Figure 5-9 shows the Doppler-broadened line shape $\psi(x,\theta)$ on a linear energy scale compared to the corresponding natural line shape. The strong broadening and the associated reduction in height are quite apparent. Figure 5-9 also shows that the width of the broadened resonance is much larger than $\Gamma + 2kT$.

5-7C Temperature-Dependent Reaction Rates

The contribution of a single resonance to the microscopic reaction rate density is given by the following integral:

$$r^{\text{res}} = \int_{\text{res}} \sigma^{\text{res}}(E,T)\varphi(E)\, dE \quad , \tag{5.116}$$

with

$$\sigma^{\text{res}}(E,T) = \sigma_m^R \psi_D(E,T) \quad . \tag{5.117}$$

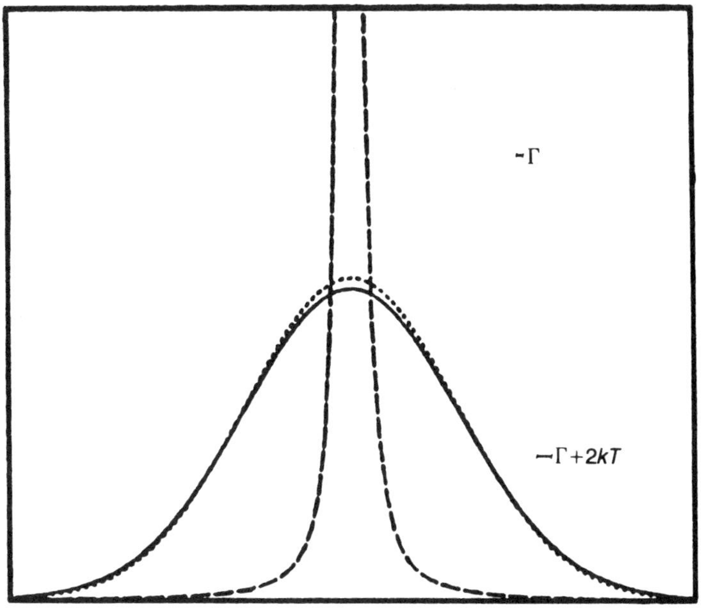

Fig. 5-9. Comparison of a Doppler-broadened resonance (solid line) with the natural shape (dashed line), depicted on a linear scale. Only about the lower half of the natural line shape is shown. Also depicted are Γ and Γ + 2*kT*. The dotted line shows a frequently used Gaussian approximation to the ψ function.

Suppose $\varphi(E)$ were virtually constant over the small width of the resonance, i.e., the resonance absorber were present in such minute amounts (infinite dilution) that it does not cause a spectral depression. Then $\varphi(E)$ can be taken out of the integral with its value at E_R:

$$r^{\mathrm{res}} \text{ (infinite dilution) } = \varphi(E_R) \cdot \int_{\mathrm{res}} \sigma^{\mathrm{res}}(E,T)\, dE \quad . \tag{5.118}$$

Observing Eq. (5.115) shows that r^{res} (infinite dilution) is independent of T.

For moderate concentrations, the macroscopic resonance cross section becomes comparable with or even much larger than the background scattering cross section. Then the resonance causes a spectral depression. If the resonance width, i.e., the broadened resonance width, is small compared to the energy loss through elastic scattering on all nuclei (NR approximation), the spectral depression is described by:

$$E\varphi_{\mathrm{NR}}(E) = \frac{\Sigma_p}{\Sigma_p + \Sigma_t^{\mathrm{res}}(E,T)} \quad . \tag{5.119}$$

The total resonance cross section is given by:

$$\Sigma_t^{res}(E,T) = \Sigma_t^{Rm} \cdot \psi_D(E,T) \quad . \tag{5.120}$$

Inserting Eqs. (5.119) and (5.120) into the reaction rate of Eq. (5.116) and approximating $1/E$ by $1/E_R$ gives:

$$r^{res} = \frac{1}{E_R} \int_{res} \frac{\Sigma_p \sigma_m^R \cdot \psi_D(E,T)}{\Sigma_p + \Sigma_t^{Rm} \psi_D(E,T)} \, dE \quad . \tag{5.121}$$

Dividing by Σ_t^{Rm} and denoting

$$\beta = \frac{\Sigma_p}{\Sigma_t^{Rm}} \tag{5.122}$$

gives

$$r^{res}(\beta,\theta) = \sigma_m^R \cdot \frac{\beta}{E_R} \cdot \int_{res} \frac{\psi_D(E,T)}{\beta + \psi_D(E,T)} \, dE \quad . \tag{5.123}$$

The integral in Eq. (5.123) is proportional to the J function:

$$J(\beta,\theta) = \frac{1}{2} \int_{res} \frac{\psi_D(E,T)}{\beta + \psi_D(E,T)} \frac{dE}{\Gamma/2} \quad . \tag{5.124}$$

The most important point is that $r^{res}(\beta,\theta)$ *always* increases with T. This is because, with increasing T, the resonances always broaden; then, the self-shielding is always decreased, which of necessity increases the reaction rate.

If, as it is in all commercial power reactors and in FBRs, *resonance capture* strongly dominates *resonance fission,* then an increase in the resonance absorption rate is always accompanied by a net increase in neutron losses. This leads directly, inherently, and instantaneously to a decrease in reactivity and thus an associated reduction in power.

Therefore, the Doppler effect provides a physically inherent and prompt negative (reactivity) feedback effect, which is very important for the safety of nuclear reactors.

5-8 Summarizing Remarks

The sharp resonances that occur in heavy nuclei within the neutron slowing down energy range between ~100 keV and a few electron volts are quite important in reactor design evaluations. Resonance absorption in ^{238}U in water-cooled reactors and in ^{232}Th in gas-cooled reactors leads to ^{239}Pu and ^{233}U, respectively; it determines, therefore, to a large extent the core reactivity and burnup characteristics. In thermal reactors as well as in fast reactors, the Doppler effect produced by these resonances has a strong influence on the dynamic and safety characteristics of the reactor.

The calculation of the resonance effect is in practice complicated by the extremely complex nature of both the number and sharpness of the resonance peaks and the heterogeneity effects of reactor fuel geometries. To account for these effects, transport theory (see Chapter 6) methods are often employed. Both the Boltzmann equation as well as the integral transport equation, the latter most often in the form of the so-called "collision probability method," have been used in the detailed evaluation of the large resonance cross sections. Because of their strong energy dependency, cross sections in this resonance range are needed in full detail in reactor calculations with thousands of discrete resonances.

Although these complex calculations are capable of producing a very accurate representation of the resonance absorption effect, they are time consuming even using large-scale computers. Detailed transport theory methods have been largely relegated to a limited role because of the cost of time-consuming computer computations. For reactor evaluations, computer models employing a number of basic approximations and simplifications are in general use.[31–33]

To obtain a first-order approximation of the resonance absorption effect on the slowing down spectrum, it is generally assumed that the slowing down region containing resonances is below the fission spectrum. This allows the neutron source to be due entirely to slowing down neutrons. Futhermore, this slowing down region is assumed to be above the thermal region since no neutron upscattering is considered. The problem of the heterogeneous structure is simplified by considering the geometry as comprised of only fuel and moderator with the resonances assumed to be sufficiently separated so that each resonance can be treated individually. This separation assumption allows a $1/E$ asymptotic flux dependency to develop between resonances.

Simplified resonance absorption models have been derived for scoping-type calculations, which generally cover the slowing down energy range. For high energies, the resonances are structured such that the NR approximation is applicable. In this NR approach, the resonance is considered to be so narrow that any scattering event within the resonance removes the neutron from the energy range of the resonance.

The low-energy range contains some rather broad resonances, and for these the NRIM approximation works pretty well. In this approximation, the slowing down due to collisions with heavy absorber nuclei can be neglected. The moderator is again treated as in the NR approach. This resonance region is important in water reactor design.

The third energy region is between the above two and is sometimes referred to as the intermediate energy range. Neither of the above two simplifications hold in this energy range, so special techniques (e.g., the Goldstein-Cohen method) are applied. In addition, improved approximations (e.g., the Greuling-Goertzel method) are available for use when hydrogen is present.

Very detailed experiments on resonance absorption have been performed to determine resonance integrals, geometry and temperature dependence, and reactivity effects of resonance absorption. Most experiments addressed heterogeneous configurations, ranging from thin wires to thick rods, from bundles of rods to subcritical and eventually critical (zero-power) configurations. Much has been learned in these experiments to improve and support the theoretical description so that one can be fully confident of the results for practical applications. Eventually, the successful operation of many power reactors provided final proof tests.

Homework Problems

1a. Derive the energy-dependence of $\sigma_c(E)$ and $\sigma_f(E)$ for small E (limit $E \rightarrow 0$) from the Breit-Wigner formula for the lowest s-wave resonance. Note, Γ_γ and Γ_f are independent of E.

b. Express the result in terms of the neutron velocity.

c. In the same way, find the energy dependence of $\sigma_s(E)$ for small E.

2. Apply the limit formula for $\sigma_c(E)$ derived in problem 1a to determine approximately the value of Γ_n^0 for the first s-wave resonance in ^{238}U. Use $E_{R=1} = 6.67$ eV; $\Gamma_\gamma = 26$ meV; $g = 1$; and $\sigma_c(0.025 \text{ eV}) = 2.7$ b. Also find $\sigma_t(E_R)$. Interpret the result. Note that the contributions of higher resonances to the thermal cross section are neglected here.

3. Explain qualitatively the low-energy cross sections of ^{135}Xe and ^{113}Cd in terms of position and height of the lowest E_R resonance and the low energy behavior of $\sigma_c(E)$ derived in problem 1a.

4. Plot on top of each other a scattering, a capture, and a total resonance shape according to the single-level Breit-Wigner formula with $\Gamma = 1.0$ eV; $\Gamma_\gamma = 0.7$ eV, and $\Gamma_n = 0.3$ eV. Draw a curve connecting the half-height values located on the upper and lower wings of the resonances (six points) and discuss its significance.

5a. Calculate the J function for the natural line shape.

b. On the basis of problem 5a, calculate the ratio of the capture reaction rates with and without self-shielding for the first resonance in ^{238}U, assuming uranium metal. Use $\sigma_p = 9$ b; $\Gamma_\gamma = 26$ meV; $\Gamma_n = 5.0$ meV; and $E_R = 6.67$ eV.

c. Calculate as problem 5b but for ^{238}U in a mixture in which ^{238}U contributes only 11.1% to the potential cross section.

6. An approximate expression for the resonance integral in a homogeneous reactor is given in Ref. 15 as $I_{eff} = 3.8\ (\Sigma_s/N_U)^{0.42}$ b with Σ_s as the total macroscopic scattering cross section and N_U as the number density of uranium atoms (use cm in Σ_s and N_U). Calculate the resonance escape probability for:

a. a natural-uranium-fueled graphite reactor where the number of carbon atoms (N_C) to uranium (N_U) is 500

b. a natural uranium light water reactor where the number of water to uranium atoms is 5.

Use $\sigma_s{}^U = 8.3$ b; $\sigma_s{}^C = 4.7$ b; and $\sigma_s{}^{H_2O} = 44.8$ b.

7. Using the expression for the homogeneous effective resonance integral given in problem 6, show by concentrating all of the uranium into one half of a reactor's volume that the resonance absorption is reduced to ~75% of the value for a uniformly distributed homogeneous reactor. (Hint: Assume that most of the scattering is due to the moderator even in the volume that contains the uranium.) In addition to the physical impracticability, why is this example, although illustrative of the effect of "lumping," grossly inaccurate?

8. For widely spaced absorber elements in a heterogeneous system, the resonance integral has been shown to approximate the following geometric dependency:

$$I = a + b(S/M)^{1/2} \quad ,$$

where a and b are constants depending on the absorber, S is the surface area in cm^2, and M is the specific mass of the absorber in g. Using the following integral data, calculate the resonance integrals for typical high-temperature gas-cooled reactor (ThO_2) and pressurized water reactor (UO_2) fuel S/M ratios:

$$ThO_2\text{: } a = 3.4,\ b = 17.3$$

and

$$UO_2\text{: } a = 3.0,\ b = 28.0 \quad .$$

Discuss the corrections needed and the implications of assuming closely packed lattices on the above widely spaced lattice results.

9. Show that the area under a Doppler-broadened resonance is the same regardless of temperature.

10. Show for any normalized Gaussian function, $\psi(x)$, that the integral over the square, i.e.,

$$\int_{-\infty}^{\infty} \psi^2(x)\, dx$$

always decreases with increasing width, s. Suggestion: Use $\psi(x) \propto \exp(-x^2/s^2)$.

11. Prove for Σ_t resonances that are small compared to Σ_p that the resonance reaction rate always increases with increasing temperature. Use a Gaussian resonance shape and the theorem found in problem 10. Assume the width, s, proportional to $\sqrt{T}$. Explain the consequences of this fact for the Doppler coefficient of reactors.

Review Questions

1. What is potential scattering?
2. Describe the reaction mechanism that involves a compound nucleus. Comment on resonances in the cross section and on inelastic scattering thresholds.

3a. How is the finite width of resonance cross sections related to the lifetime of the respective states?
b. What can be described by the partial widths Γ_j?
c. What is the sum of all Γ_j?

4. Sketch two typical resonance cross sections: a capture cross section and a total cross section (e.g., of ^{238}U between 1 and 100 eV).

5a. Give the formula for the natural line shape.
b. Give the energy-dependent factors in the Breit-Wigner formula.

6a. Derive the $1/v$ cross section for $\sigma_c(E)$ and $\sigma_f(E)$ at small E.
b. Why does $\sigma_s(E)$ not have a $1/v$ dependence for small E?

7a. Between which cross section does interference happen?
b. How does interference change the shape of a resonance cross section?
c. Sketch the scattering cross section of ^{56}Fe between 10 and 40 keV and comment on the effect of this special behavior.

8a. Why are resonance cross sections of heavy nuclei not experimentally resolved at higher energies?
b. Do unresolved resonances have to be included in the calculation? Why?

9. Give the slowing down spectrum for hydrogen and for nonhydrogenous materials.

10a. Decompose the slowing down spectrum in the resonance range into its two essential parts and discuss the behavior of both components as a function of E.

b. Discuss the meaning of the loss-free spectrum.

11a. Derive the exponential function expression for $p(E)$.

b. Write $p(E)$ in terms of the loss-free spectrum.

c. Derive $p(k)$ for a sequence of resonances.

12a. Explain qualitatively the NR and NRIM approximations.

b. Give the loss-free spectrum in the NR approximation.

c. Give the loss-free spectrum in the NRIM approximation.

13. Discuss the energy dependency of the applicability of the NR and NRIM approximations.

14a. Give the formula for p and convert it into the formula based on the resonance integrals.

b. Give the formula for the resonance integral.

c. Give the infinite dilute resonance integral.

d. Why is the resonance integral calculated with a $1/E$ spectrum?

15a. What is the resonance self-shielding? Explain how it occurs.

b. What is its physical significance?

16a. Give the background cross section, σ_b.

b. Discuss its influence on I with varying concentrations of the resonance absorber.

c. Why is resonance self-shielding not important for fission products?

17. Give the energy-dependent self-shielding factor, $f(E)$.

18. Describe the essence of the following approaches: (a) Fermi age theory in an homogeneous medium, (b) Selengut-Goertzel, (c) Greuling-Goertzel, and (d) Goldstein-Cohen.

19. Give the formula for $\varphi(E)$ in Fermi age theory and discuss its suitability for the treatment of resonance absorption.

20. Describe the approach applied for the treatment of resonance absorption in heterogeneous media, especially in a fuel-moderator cell.

21a. Give Wigner's approximation for $P_{f\to m}$.

b. Describe the concept of and give a formula for the average chord length, $\bar{l}$.

22. Describe the idea of the Dancoff correction, and present the way it is applied.

23a. Explain qualitatively why neutrons with energies below and above a resonance still can react with it.

b. How is this effect incorporated into the treatment?

24a. Derive the Doppler width, Δ.

b. What is the effect of the Doppler width?

c. Give the basic dependencies of Δ.

d. Compare typical values of Δ with Γ of ^{238}U.

25a. Sketch the derivation of the Doppler-broadened line shape.

b. What is $\psi(x,\theta)$ sometimes called?

c. Does Doppler broadening change the integral over a resonance?

26a. What is the effect of Doppler broadening on resonance self-shielding?

b. Explain why the Doppler reactivity effect in dilute fuel reactors is negative.

REFERENCES

1. H. A. Bethe, "Theoretical Analysis of Neutron Resonances in Fissile Materials," in *Physics and Mathematics,* Vol. I of *Progress in Nuclear Energy Series,* Pergamon Press, New York (1956).
2. L. Dresner, *Resonance Absorption in Nuclear Reactors,* Pergamon Press, New York (1960).
3. D. J. Hughes, *Neutron Cross Sections,* Pergamon Press, New York (1957).
4. J. J. Schmidt, "Resonance Properties of the Main Fertile and Fissile Nuclei," in *Resonance Absorption,* Vol. II, p. 223, of *Reactor Physics in the Resonance and Thermal Regions,* The MIT Press, Cambridge, Massachusetts (1966).
5. M. S. Moore and O. D. Simpson, "Measurements and Analysis of Cross Sections of Fissile Nuclides," in *Resonance Absorption,* Vol. II, p. 261, of *Reactor Physics in the Resonance and Thermal Regions,* The MIT Press, Cambridge, Massachusetts (1966).
6. Harry H. Hummel and David Okrent, *Reactivity Coefficients in Large Fast Power Reactors,* American Nuclear Society, La Grange Park, Illinois (1978).
7. D. Selengut and G. Goertzel, unpublished work; also see J. H. Ferziger and P. F. Zweifel, *The Theory of Neutron Slowing Down in Nuclear Reactors,* p. 153, The MIT Press, Cambridge, Massachusetts (1966).
8. E. Greuling and G. Goertzel, "An Approximate Method for Treating Neutron Slowing Down," *Nucl. Sci. Eng.,* **7,** 69 (1960).
9. R. Goldstein and E. R. Cohen, "Theory of Resonance Absorption of Neutrons," *Nucl. Sci. Eng.,* **13,** 132 (1962).
10. R. E. Hellens, R. W. Long, and B. N. Mount, "Multigroup Transform Calculation Description of MUFT-III Code," WAPD-TM-4, Westinghouse Electric Corp. (July 1956); also see H. Bohl, Jr., E. M. Gelbard, and G. H. Ryan, "MUFT-4: Fast Neutron Spectrum Code," WAPD-TM-22, Westinghouse Electric Corp. (1957).
11. J. E. Suich and H. C. Honeck, "The HAMMER System: Heterogeneous Analysis by Multigroup Methods of Exponentials and Reactors," DP-1064, Savannah River Lab. (Jan. 1967).
12. L. W. Nordheim, "The Contribution of the Individual Resonances to the Resonance Integrals in Uranium and Thorium," GA-2563, General Atomic (1961); also see L. W. Nordheim, "Resonance Absorption," GA-3973, General Atomic (1963).

13. M. M. R. Williams, *Mathematical Methods in Particle Transport Theory*, Chapter 9, Butterworth and Co., Canada (1971).
14. G. I. Bell and S. Glasstone, *Nuclear Reactor Theory*, Chapter 8, Van Nostrand Reinhold Co., New York (1970).
15. A. M. Weinberg and E. P. Wigner, *The Physical Theory of Neutron Chain Reactors*, p. 312, University of Chicago Press (1958).
16. A. F. Henry, *Nuclear Reactor Analysis*, The MIT Press, Cambridge, Massachusetts (1975).
17. K. M. Case, F. DeHoffman, and G. Placzek, "Introduction to the Theory of Neutron Diffusion," Los Alamos National Laboratory (1953); also see A. Sauer, "Approximate Escape Probabilities," *Nucl. Sci. Eng.*, **16,** 329 (1963).
18. S. M. Dancoff and M. Ginsburg, "Surface Resonance Absorption in a Close-Packed Lattice," USAEC report CP-2157, U.S. Atomic Energy Commission (1944).
19. R. D. Richtmyer, "Resonance Capture Calculations for Lattices by the Monte Carlo Method," *Proc. Brookhaven Conf. Resonance Absorption of Neutrons in Nuclear Reactors*, Upton, New York, September 24–25, 1956, BNL 433, p. 82, Brookhaven National Lab. (1957).
20. G. I. Bell, "A Simple Treatment for Effective Resonance Absorption Cross Sections in Dense Lattices," *Nucl. Sci. Eng.*, **5,** 138 (1959).
21. G. I. Bell, "Theory of Effective Cross Section," LA-2322, Los Alamos National Laboratory (1959).
22. W. J. Davis, M. B. Varbrough, A. B. Bortz, "SPHINX: A One-Dimensional Diffusion and Transport Nuclear Cross Section Processing Code," WARD-XS-3045-17, Westinghouse Electric Corp. (Aug. 1977).
23. P. Walti and P. Koch, "MICROX—A Two Region Spectrum Code for the Efficient Calculation of Group Cross Sections," GA-10826, General Atomic (1972).
24. P. Blomberg, E. Hellstrand, and S. Horner, Paper A/Conf/P/150, *Proc. 2nd Geneva Conf. Peaceful Uses of Atomic Energy* (1958).
25. G. J. Fisher, D. A. Meneley, R. N. Hwang, E. F. Groh, and C. E. Till, "Doppler Effect Measurements in Plutonium-Fueled Fast Power Breeder Reactor Spectra," *Nucl. Sci. Eng.*, **25,** 37 (May 1966); also see G. J. Fisher, D. A. Meneley, R. N. Hwang, E. F. Groh, and C. E. Till, "Measurement and Analyses of Doppler Effect in Plutonium-Fueled Fast Reactor Assemblies," *Proc. Conf. Safety, Fuels, and Core Design in Large Fast Power Reactors*, Argonne, Illinois, October 11–14, 1965, ANL 7120, p. 603, Argonne National Laboratory (1966).
26. G. J. Fisher, H. H. Hummel, J. R. Folkrod, D. A. Meneley, "Doppler Coefficient Measurements for ^{238}U in Fast Reactor Spectra," *Nucl. Sci. Eng.*, **18,** 290 (1964).
27. W. Häfele, K. Ott, L. Caldarola, W. Schikarski, K. P. Cohen, B. Wolfe, P. Greebler, and A. B. Reynolds, "Static and Dynamic Measurements on the Doppler Effect in an Experimental Fast Reactor," *Proc. 3rd Int. Conf. Peaceful Uses of Atomic Energy*, Conf. 28, Vol. 6, paper 644 (1964).
28. R. A. Meyer, A. B. Reynolds, S. L. Steward, M. L. Johnson, and E. R. Craig,

"Design and Analysis of SEFOR Core 1," GEAP-13598, General Electric Co. (June 1970).
29. L. D. Noble, G. Kussmaul, and G. R. Pflasterer, "Sub-Prompt Critical Transients in SEFOR," *Trans. Am. Nucl. Soc.*, **14,** 741 (1971).
30. G. Kussmaul, L. D. Noble, and G. R. Pflasterer, "Super-Prompt Critical Transients in SEFOR," *Trans. Am. Nucl. Soc.*, **15,** 339 (1972).
31. C. A. Stevens and C. V. Smith, "GAROL—A Computer Program for Evaluating Resonance Absorption Including Resonance Overlap," GA-6637, General Atomic (Aug. 24, 1965).
32. P. H. Kier, "A Method of Computing Resonance Integrals in Multiregion Reactor Cells," paper presented at the Int. Conf. Research Reactor Utilization and Reactor Mathematics, Mexico City, Mexico, May 2–4, 1967.
33. P. H. Kier, "A Program for Computation of Resonance Integrals in a Two-Region Cell," ANL-7033, Argonne National Laboratory (Aug. 1965).

Six

NEUTRON TRANSPORT

6-1 Introduction

The previous chapters have dealt with the overall spatial and energy dependencies of the neutron flux. The remaining variables on which the stationary neutron flux depends—the angular coordinates of the neutron velocity vector—are considered in this chapter. As was derived and discussed in Chapter 2, the bases for this detailed angular description of neutron motion are the neutron transport equations—the Boltzmann equation or the integral transport equation. The general subject of the more detailed description of the angular dependencies is known as "transport theory."

It is not the intent of this chapter to present in depth the wealth of mathematical theory and numerical methods that comprises the subject of transport theory. The literature contains several extensive and excellent treatises on the mathematical derivations[1–4] as well as accounts of the numerical methods[5] involved in neutron transport theory. The major purpose here is to provide the essential elements of transport theory, to outline the basic approaches, and to discuss explicitly a first-order correction that is required to overcome one of the shortcomings of the original diffusion theory.

In the next section, the necessity of considering the angular dependencies, both in the descriptions of the neutron flux as well as in the scattering processes, is discussed. This material will identify the problem areas involved in applying diffusion theory.

In Sec. 6-3, a survey of the methods for the solution of the transport equations is presented. The material will be introductory, scoping in nature, and without derivation. The major purpose of this section is to acquaint the reader with the various approaches and methods available to solve angular-dependent neutronics problems.

Section 6-4 contains a derivation of the so-called "P_1 theory." This derivation also yields the "transport correction," an angular correction for use in the diffusion model. From this P_1 development, the relationships between diffusion theory and transport theory will be clarified. In particular, a mathematical justification is given for invoking Fick's Law in neutron diffusion theory.

In this chapter, the energy dependency of the neutron flux is ignored, i.e., the angular dependencies are discussed in the framework of a one-group model. The extension of transport theory methods to incorporate neutron energy dependency is similar to the multigroup methods presented for diffusion theory in Chapter 7.

6-2 Basic Considerations

6-2A Fundamental Inaccuracies of Diffusion Theory

In the first chapter, it was shown that the neutron flux is defined as the integral of the angular flux with respect to the solid angle:

$$\phi(\boldsymbol{r},E) = \int_{\Omega} \phi(\boldsymbol{r},E,\boldsymbol{\Omega})\, d\Omega \quad . \tag{6.1}$$

Thus, in theory at least, any rigorous calculation of the neutron flux requires at first the calculation of the angular flux. One can, however, calculate the flux *approximately* without finding the angular flux first (for example, by diffusion theory). These approximate theories require simplifying assumptions about the angular flux. It is shown in Secs. 6-3B and 6-4 (and it has been stated and used above) that diffusion theory is based on the assumption (in plane geometry):

$$\phi(x,E,\mu) = \frac{1}{2}[\phi(x,E) + 3J(x,E)\mu] \quad , \tag{6.2}$$

with $\mu = \cos\vartheta$ as illustrated in Fig. 6-1.

Therefore, in any case in which the simplifying assumption on the angular distribution is not valid, it must be expected that the flux calculated from the corresponding model is inaccurate. This direct relation of the simplification of the angular distribution [for example, Eq. (6.2)] and inaccuracy of the model for the flux calculation (from diffusion theory) helps to identify cases or geometrical areas in which the diffusion theory is inaccurate. For these cases or areas, one then needs a more sophisticated theory that is based on a more accurate description of the angular distribution.

The following typical cases are examples in which the simple angular distribution given by Eq. (6.2) quite obviously breaks down:

1. Near the interface to a vacuum, the angular distribution, Eq. (6.2), fails to portray the actual situation.

This effect was introduced in Sec. 2-2E in conjunction with evalu-

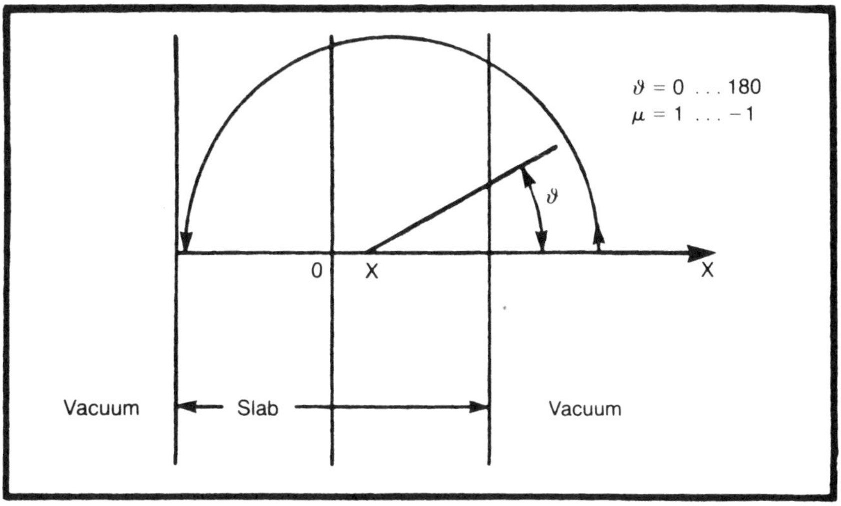

Fig. 6-1. Illustration of the angular coordinates in a plane geometry model where $\mu = \cos\vartheta$.

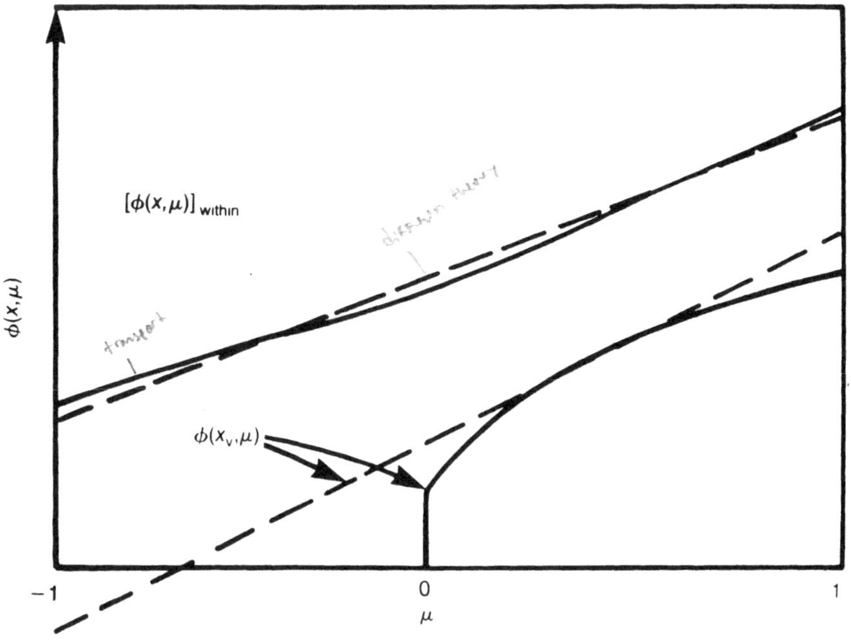

Fig. 6-2. Illustration of angular fluxes in transport (solid lines) and diffusion theory (dashed lines) at a vacuum boundary (x_v) and within a slab. The angular domain from $\mu = -1$ to 0 contains the neutrons that come from the vacuum.

ation of the boundary conditions for the transport and diffusion equations. Figure 6-2 illustrates the type of inaccuracy or discrepancy introduced by diffusion theory (dashed lines) and transport theory (solid lines). The diffusion theory or P_1 angular flux curve is that of Eq. (6.2).

This discrepancy is gradually reduced with x moving away from the interface, with the result that two or three mean-free-paths away from the interface, the "constant plus cosine" angular distribution becomes a good approximation.

2. The error of the angular distribution in the artificial situation of a core interface to vacuum will, in a milder form, occur at all interfaces between media of significantly different absorption and neutron source characteristics, for example, between:

a. core/reflector interface
b. core/blanket interface
c. internal interfaces between media of different enrichments
d. control rod or plate interfaces.

Figure 6-3 illustrates the situation near a strongly neutron-absorbing control plate. It is obvious that any angular distribution $\phi(x,\mu)$, which

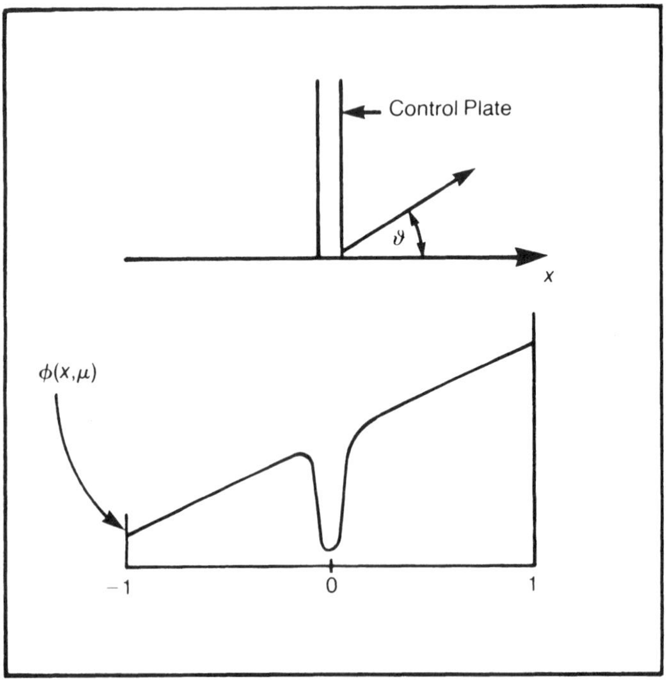

Fig. 6-3. Illustration of the angular distribution at the center of a thin control plate.

has a dip in a small angular interval, cannot be described by the straight line of Eq. (6.2).

3. The third special case of practical importance is the fuel coolant cell. In this case, diffusion theory is inadequate because of the small dimensions of all regions; thus, many boundaries are involved, as shown in Fig. 6-4. The thermal neutron flux in the fuel rod is smaller than in the moderator due to the concentration of the absorption in the fuel. This phenomenon is frequently expressed in terms of a "disadvantage factor," which reduces the flux in the fuel compared to the average flux in the cell. Diffusion theory yields an inaccurate value of a disadvantage factor, and more sophisticated methods are needed and routinely applied.

In all of these cases, the diffusion theory will be in error. Therefore, to calculate accurate flux distributions in geometries of the types discussed here, one needs more accurate methods than those of diffusion theory. All methods will, in some form, have to be based on more accurate angular distributions than the one used in diffusion theory.

6-2B Anisotropic Scattering

The approximation equation [Eq. (6.2)] of the angular flux is not the only assumption on which diffusion theory is based. The second assumption was introduced in the first chapter: namely, the calculation of the current from the gradient of the flux, i.e., Fick's Law. Fick's Law can be derived from P_1 theory under certain assumptions about the angular distribution of the neutron source (see Sec. 6-4). In this section,

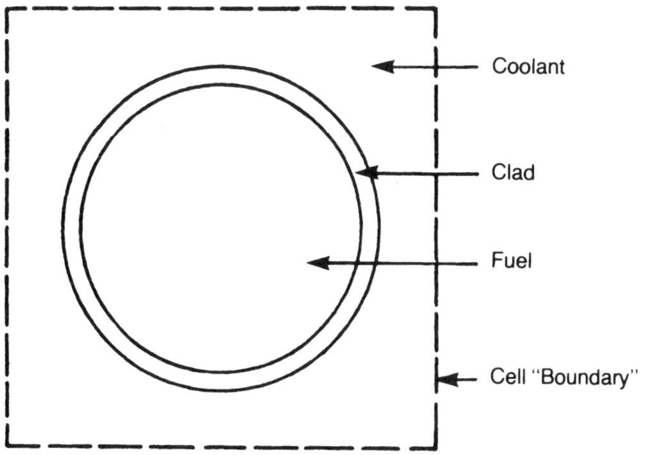

Fig. 6-4. Typical unit cell.

the angular distributions of elastic and inelastic scattering are briefly discussed; they affect the proportionality factor between current and flux gradient.

The transport of neutrons is obviously influenced by the angular distribution of scattering in the laboratory system (LS). If scattering caused only a minimal angular deflection, then it would have hardly any diffusing effect and neutrons could quickly leak out of a reactor. If the angular distribution of scattered neutrons were isotropic in the LS, then scattering would have a maximum diffusing effect, inhibiting the leakage of neutrons.

Of the two types of scattering processes, elastic and inelastic, anisotropic behavior is predominant in the elastic scattering phenomena. Inelastically scattered neutrons are emitted largely isotropically in the center-of-mass system (CMS). This holds also in the LS because inelastic events predominantly occur with neutrons interacting with heavy nuclei, for which the anisotropy introduced in the conversion to the LS is small. Therefore, the inelastic process can be treated as being isotropic. Actually, inelastic scattering has a slight bias both forward and backward, which—in a sense—is a second-order effect.

The angular distribution of elastically scattered neutrons was investigated in Sec. 4-2. Elastic scattering is isotropic in the CMS over most of the energy scale. The conversion of an elastic scattering event from the CMS into the LS introduces a degree of anisotropy for light nuclei. Then scattering, which is isotropic in the CMS, assumes a forward bias in the LS with an average cosine of the scattering angle being $\bar{\mu}_s = 2/3A$. Hence, the lighter the scattering nuclei, the larger the anisotropy. Whereas with heavy nuclei, little anisotropy is introduced due to this coordinate conversion.

Additional anisotropy in elastic scattering appears at higher energies. Conceptually, as discussed in Chapter 5, nuclear reactions, elastic scattering among them, can be considered as the interaction of neutron waves with the target nucleus—with the higher energy neutrons having the smaller wave lengths, λ ($\lambda = h/mv$). When the neutron wave length is larger than the nuclear size, the scattering directions are poorly defined and the scattering event in the CMS becomes isotropic. This is the case for *s*-wave neutrons. Conversely, if the neutron wave length is equal to or smaller than the scattering nucleus, the scattering event becomes anisotropic in the CMS. Then, *p*- and *d*-wave neutrons contribute significantly to scattering. In summary then, anisotropic scattering in the CMS is expected for all nuclei at higher neutron energies, for heavy nuclei at lower energies than for light nuclei due to the larger size of the heavy nuclei; i.e., the wave length that equals the nucleus radius corresponds to a lower energy in heavy rather than in light nuclei.

6-3 Survey Discussion of Transport Theoretical Methods

6-3A The Basic Scheme of Methods

The survey of the various transport theoretical methods, as discussed here, is illustrated by Fig. 6-5, which schematically presents transport theoretical processes and methods.

Neutron production, migration, and absorption is a *stochastic* process. Thus, during their flight through the medium, the individual neutrons are randomly subjected to scattering, absorption, and fission events. Then, the neutron motion has the character of a "random walk," with probability distributions for individual aspects such as point of interaction in space, type of interaction, angular deflection, energy change, and number of fission neutrons emitted. The transport equations describe

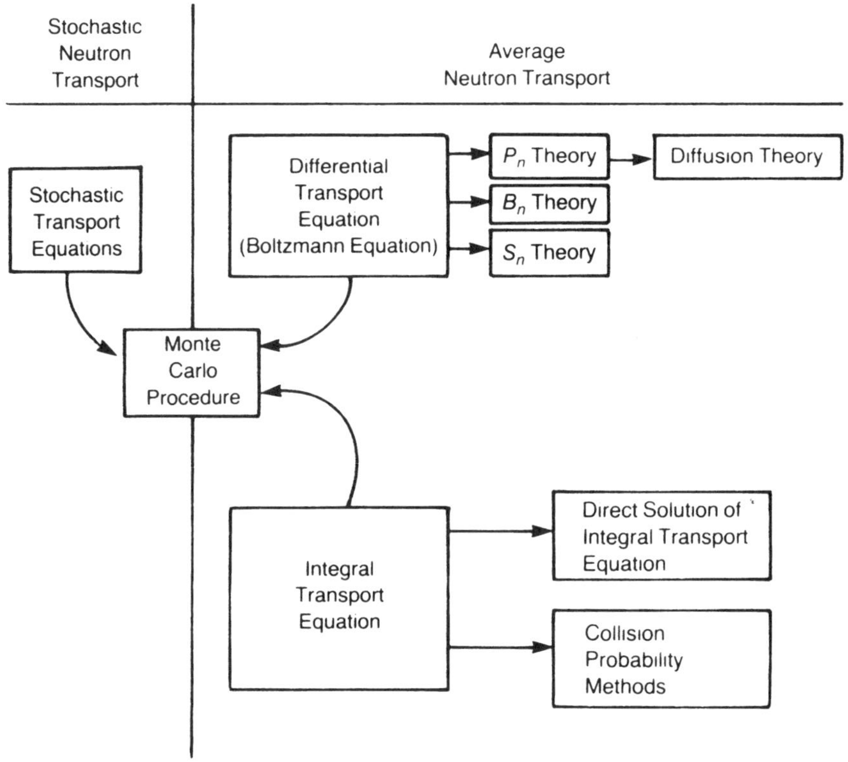

Fig. 6-5. Schematic relation of transport theoretical methods.

only the average neutron transport. Thus, a subdivision in these two categories (stochastic and average transport) is indicated in Fig. 6-5.

The Monte Carlo procedure was named after the gambling casino at Monte Carlo. The reason for this choice of name is the fact that the Monte Carlo procedure basically consists of:

1. translating a physical or mathematical problem into a game (for an inherently stochastic phenomenon, this "translation" into a game may not be necessary)
2. playing this game very often
3. forming the average value of the results.

The translation is performed in such a way that this average value yields asymptotically the solution of the stated problem.

The Monte Carlo procedure can be applied to treat directly the *stochastic* neutron transport. It can also be applied to treat the *average* neutron transport, that is, to *solve* the Boltzmann equation or the integral neutron transport equation. In this case, a translation into a game is required, it does not have to agree, however, with the original stochastic phenomenon. The only goal is to obtain, on the average, the solution of the respective transport equation.

The Boltzmann and the integral transport equation represent the general description of the average neutron transport. A solution procedure of these equations can be devised in the form of a Monte Carlo procedure. The translation of these mathematical problems in a Monte Carlo procedure is often performed in such a way that the solution can be obtained quicker than with a straightforward Monte Carlo procedure, which simulates the actual physical stochastic process (for example, by employing an average ν rather than the probability distribution of ν, which ranges from about one to seven neutrons per fission). The disadvantage of the Monte Carlo method is its poor convergence, i.e., the error is proportional to the inverse square root of the number of neutron histories investigated.

In cases of practical interest, the Boltzmann and the integral transport equations cannot be solved directly. One must devise approximate solutions procedures for these equations. The most common solution procedures are indicated in Fig. 6-5; they comprise P_n or S_n methods as procedures to solve the Boltzmann equation approximately or the numerical integration and multiple collision methods for approximately solving the integral transport equation.

The Boltzmann equation is to be solved approximately by the P_n or S_n methods. For one energy group and three space dimensions, the Boltzmann equation is given by:

Boltzmann

$$\mathbf{\Omega}\cdot\nabla\phi(\mathbf{r},\mathbf{\Omega})+\Sigma_t(\mathbf{r})\phi(\mathbf{r},\mathbf{\Omega})$$
$$=\int_{\Omega}\Sigma_s(\mathbf{r},\mathbf{\Omega}'\rightarrow\mathbf{\Omega})\phi(\mathbf{r},\mathbf{\Omega}')\,d\Omega'+\frac{1}{4\pi}S_f(\mathbf{r})+\mathbf{S}(\mathbf{r},\mathbf{\Omega}) \quad . \tag{6.3}$$

Note that the fission source is isotropic.

Only the one-dimensional transport problem is discussed explicitly below. The one-dimensional Boltzmann equation can be deduced from Eq. (6.3) as (see also Sec. 2-2):

$$\mu\frac{d}{dx}\phi(x,\mu)+\Sigma_t(x)\phi(x,\mu)$$
$$=\int_{\Omega'}\Sigma_s(x,\mathbf{\Omega}'\rightarrow\mathbf{\Omega})\phi(x,\mathbf{\Omega}')\,d\Omega'+\frac{1}{2}S_f(x)+S(x,\mu) \quad . \tag{6.4}$$

The scattering kernel transfer cross section, $\Sigma_s(x,\mathbf{\Omega}'\rightarrow\mathbf{\Omega})$, is expressed in Eq. (6.4) in terms of three-dimensional angular vectors. For further discussion, see Eqs. (6.16), (6.17), and (6.18).

6-3B Spherical Harmonics Approach

The spherical harmonics or P_n approach in solving Eq. (6.3) or Eq. (6.4) consists of expanding the angular flux distribution into a complete set of orthogonal functions, the "spherical harmonics." If only one angle is required for the description, as is the case for one-dimensional problems in plane and spherical geometry, then the spherical harmonics are just the "Legendre polynomials" of the cosine, μ, of this one angle, ϑ; $P_n(\mu)$ denotes the Legendre polynomials. Specifically, the first three $P_n(\mu)$'s are given by:

$$P_0(\mu)=1 \quad ;$$
$$P_1(\mu)=\mu \quad ;$$

and

$$P_2(\mu)=\frac{1}{2}(3\mu^2-1),\ \ldots \quad . \tag{6.5}$$

The $P_n(\mu)$ are normalized and orthogonal in the following sense:

$$\int_{-1}^{1}P_m(\mu)P_n(\mu)\,d\mu=\begin{cases}2/(2n+1) & \text{for } m=n\\ 0 & \text{for } m\neq n\end{cases} \quad . \tag{6.6}$$

If two angles are required for the description, as is generally the case, the complete orthogonal expansion system of the angular dependence consists of the surface (spherical) harmonics $Y_{n,m}$ (μ, α). The surface harmonics are composed of the associated Legendre functions $(0 \leq m \leq n)$:

$$P_n^m(\mu) = [(1-\mu^2)^{1/2}]^m \frac{d^m}{d\mu^m} P_n(\mu) = P_n^{-m}(\mu) \quad , \tag{6.7}$$

and $\sin(m\alpha)$ and $\cos(m\alpha)$. Since P_n is a polynomial of the n'th order, all derivatives of higher order than n are zero.

The associated Legendre functions for $n = 0$ and 1 are:

$$P_0^0(\mu) = P_0(\mu) = 1 \quad ,$$

$$P_1^0(\mu) = P_1(\mu) = \mu = \cos\vartheta \quad ,$$

and

$$P_1^{\pm 1}(\mu) = (1-\mu^2)^{1/2} = \sin\vartheta \quad . \tag{6.8}$$

The surface harmonics $Y_{n,m}(\mu,\alpha)$ are often defined as complex functions:

$$Y_{n,m}(\mu,\alpha) = P_n^m(\mu)\exp(im\alpha) \quad . \tag{6.9}$$

Then the expansion coefficients become complex also. The result certainly is real and is identical with the result obtained by using a real definition.

The nomenclature and even the definition of the surface harmonics are not unique: for example, some authors include normalization constants in the polynomial, while others do not.

Of special interest is $Y_{1,m}$; its three components are given by:

$$Y_{1,m}(\mu,\alpha) = \begin{cases} \sin\vartheta\cos\alpha & \text{for } m = +1 \\ \sin\vartheta\sin\alpha & \text{for } m = -1 \\ \cos\vartheta & \text{for } m = 0 \end{cases} \quad . \tag{6.10}$$

The three components of $Y_{1,m}$ are the x, y, and z components of the solid angle:

$$\mathbf{\Omega} = \begin{pmatrix} \Omega_x \\ \Omega_y \\ \Omega_z \end{pmatrix} = \begin{pmatrix} \sin\vartheta\cos\alpha \\ \sin\vartheta\sin\alpha \\ \cos\vartheta \end{pmatrix} ; \tag{6.11}$$

thus,

$$\mathbf{\Omega} = \begin{pmatrix} Y_{1,1} \\ Y_{1,-1} \\ Y_{1,0} \end{pmatrix} \quad . \tag{6.12}$$

The derivations of the P_n equations are sketched in the following, discussing only plane geometry for simplicity:

1. The angular flux is expanded into Legendre polynomials

$$\phi(x,\mu) = \sum_{n=0}^{\infty} \frac{2n+1}{2} \phi_n(x) P_n(\mu) \tag{6.13}$$

where formally

$$\phi_n(x) = \int_{-1}^{1} \phi(x,\mu) P_n(\mu) \, d\mu \quad . \tag{6.14}$$

2. If there is an independent source, it is expanded in the same way:

$$S(x,\mu) = \sum_{n=0}^{\infty} \frac{2n+1}{2} S_n(x) P_n(\mu) \quad , \tag{6.15a}$$

with

$$S_n(x) = \int_{-1}^{1} S(x,\mu) P_n(\mu) \, d\mu \quad . \tag{6.15b}$$

3. It was mentioned in Sec. 6-2B that the second feature that has to be treated in improving the diffusion approximation is the anisotropy of the scattering process. The angular dependence of the scattering kernel can be expressed in terms of the angle between $\mathbf{\Omega}'$ and $\mathbf{\Omega}$:

$$\Sigma_s(\mathbf{\Omega}' \rightarrow \mathbf{\Omega}) = \Sigma_s(\mathbf{\Omega}' \cdot \mathbf{\Omega}) = \Sigma_s(\mu_s) \quad , \tag{6.16}$$

with

$$\mu_s = \mathbf{\Omega}' \cdot \mathbf{\Omega} = \cos(\text{angle between } \mathbf{\Omega}' \text{ and } \mathbf{\Omega}) \quad . \tag{6.17}$$

This scattering angle is not just a function of μ and μ', since the azimuthal angle α is also changed in scattering, i.e., the scattering cross section depends on the change in direction in three dimensions. This fact complicates the derivation as μ_s depends on all angles:

$$\mu_s = \mu_s(\mu,\mu',\alpha,\alpha') \quad .$$

The angular dependence of elastic scattering has been briefly discussed in Sec. 6-2B. It is very significant for scattering on lighter elements. The forward bias increases at large energies, a fact that is not specifically addressed in this chapter. Generally, for scattering that is isotropic in the CMS, the angular distribution $\Sigma_s(\mu_s)$ can be readily calculated (see Sec. 4-2). The added forward bias at larger energies is determined experimentally. The resulting angular distribution, comprising all types of anisotropies, is expressed in terms of an expansion in $P_n(\mu_s)$:

$$\Sigma_s(\mu_s) = \sum_{n=0}^{\infty} \frac{2n+1}{2} \Sigma_{s,n} P_n(\mu_s) \quad . \tag{6.18}$$

Equations for the coefficients $\phi_n(x)$ in Eq. (6.13), which are only formally defined by Eq. (6.14), are obtained by (a) inserting the expansion equations, Eqs. (6.13), (6.15a), and (6.18), into the Boltzmann equation and (b) multiplying with $P_{n'}(\mu)$, and (c) integrating with respect to μ and α. The fact that the Legendre polynomials are orthogonal greatly simplifies the equations resulting from that integration.

This procedure yields an infinite set of equations that is truncated in a practical application. There are good reasons to truncate the set of equations at odd numbers.[a] This then leads to the so-called "P_1, P_3, P_5, ... theories."

The application of the P_n method clearly allows a more detailed description of the angular distribution of the angular flux. An example is given in Fig. 6-6 for the angular flux near a vacuum interface. It becomes apparent from Fig. 6-6 that a larger number of spherical harmonics are required to approximate the angular distribution near a vacuum interface.

6-3C Double P_n Theory

The P_n theory has the disadvantage of not being able to accurately satisfy the vacuum boundary condition with a small number of terms, as apparent from Fig. 6-6. To improve angular distributions of the type shown in Fig. 6-6, the P_n expansion can be carried out for both half-spaces separately. This theory is called the "double P_n theory." It allows the exact representation of a vacuum boundary condition in plane geometry and it is generally applicable in cases of clearly identifiable half-spaces, as in plane geometry problems.

The application of double P_n theory is less advantageous if it is not a given half-space in which the angular distribution is substantially different than in the other one. Consider various points around a control blade. The highly absorbing control blade and the associated dip in the angular flux appears in different angular intervals if viewed from different spatial positions (compare Fig. 6-7). Thus, there are no two half-spaces for which separate P_n expansions would be appropriate.

[a]For example, Weinberg and Wigner discuss this problem in Ref. 1.

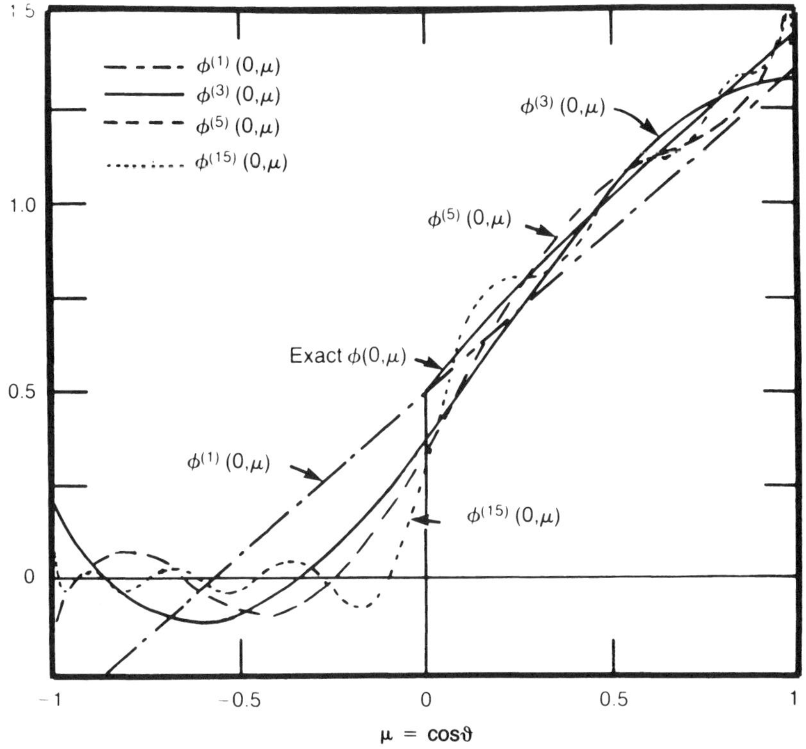

Fig. 6-6. Angular flux at the boundary in various spherical harmonics approximations compared with the exact $\phi(0,\mu)$ (from Ref. 1).

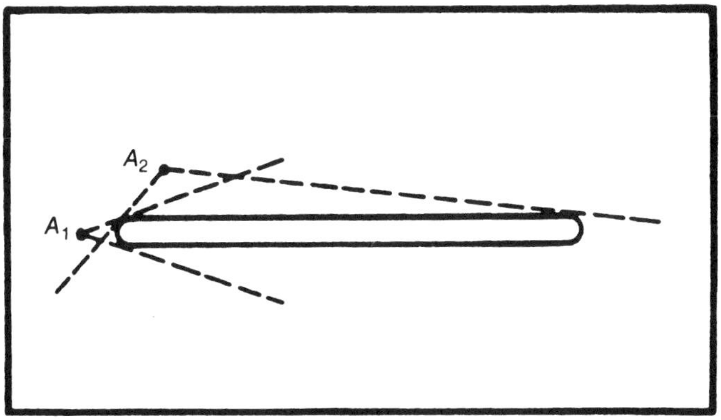

Fig. 6-7. Illustrations of the different viewing angles of a control blade from two positions, A_1 and A_2.

6-3D The S_n Theory

In the S_n approach, the angular interval is subdivided into n segments. Within each segment, the angular flux is approximated as

$$\phi(x,\mu) \simeq c_{0n} + c_{1n}\mu \quad , \tag{6.19}$$

in plane or spherical geometries.

These pieces of the angular distribution are continuously connected at the ends of the intervals. Examples are shown in Fig. 6-8.

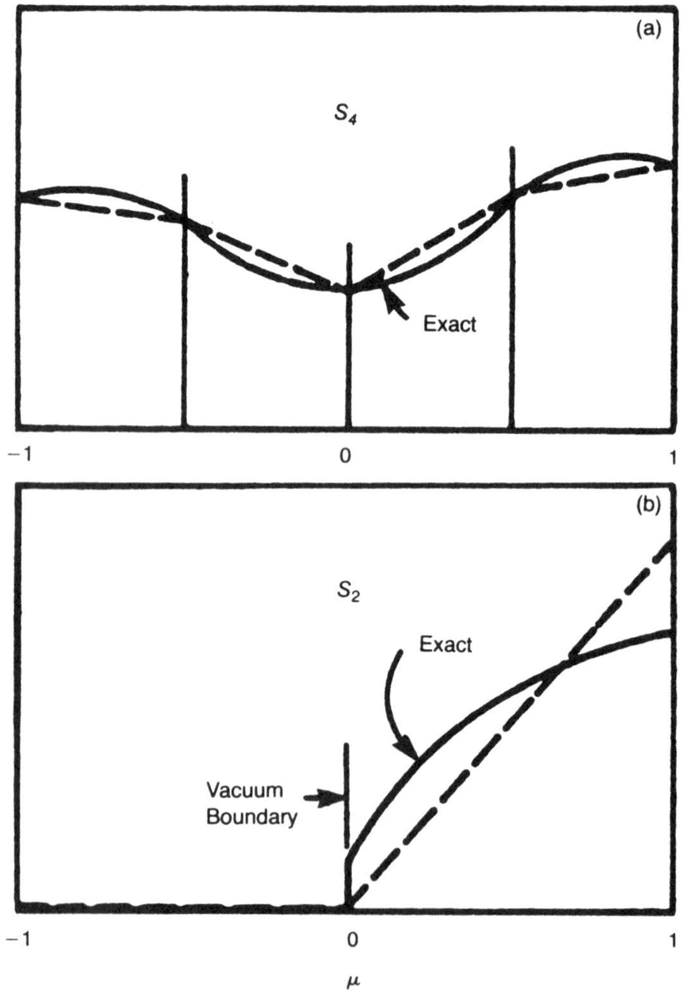

Fig. 6-8. Examples of S_n angular flux approximations: (a) S_4 near a control blade and (b) S_2 at the vacuum boundary.

The example in Fig. 6-8b shows that S_2 can already exactly satisfy the vacuum boundary condition:

$$\phi(x_b,\mu) = 0 \text{ for } \mu < 0 \quad . \tag{6.20}$$

The original S_n method was developed at Los Alamos National Laboratory by B. G. Carlson. It is, therefore, often called "Carlson's S_n method."[7,8] Many currently used computer codes have evolved from this method. They are also called "discrete ordinates" programs.

The S_n computer programs are generally devised for an even number of intervals S_2, S_4, S_8, ... Older programs use equidistant angular intervals and provide only the doubling of the intervals if more detailed angular segmentation is required. This leads to S_2, S_4, S_8, S_{16}, S_{32}, ... theories. Modern programs are more flexible in their angular segmentation and allow intervals of different sizes. Therefore, small intervals can be used when the angular distribution varies sharply and large segments for slow variation. Figure 6-9 shows an example.

This variable segmentation looks straightforward and plausible. Practical situations, however, may not be that simple. Apparently, an angular distribution near the edge of a control blade needs the angular detail in a different angular range than a point on the side (see Fig. 6-7). However, one cannot vary the structure of the angular segmentation as a function of space. Therefore, segmentation into equal intervals may often be the only reasonable approach.

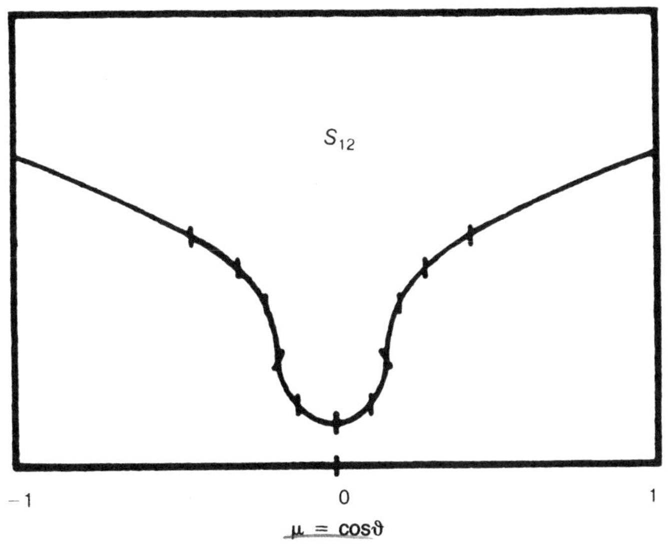

Fig. 6-9. Illustration of variable angular segmentation for the S_n method.

6-3E One-Group Integral Transport Theory

The integral transport equation for one group of neutrons was derived in Chapter 3. In general terms, the angular flux was obtained by adding the respective contributions:

$$\phi(\boldsymbol{r},E,\boldsymbol{\Omega}) = \begin{cases} \text{all contributions} \\ \text{arriving at } \boldsymbol{r} \text{ through} \\ \text{free flight.} \end{cases} \tag{6.21}$$

In the main practically useful application of the integral transport equation, the integral in Eq. (6.21) can at first be carried out over a set of spatial domains; subsequently, one sums over all domains. In addition, the result is integrated over $\boldsymbol{\Omega}$ and over the space of a given domain. Instead of $\phi(\boldsymbol{r})$, one calculates in this fashion $\hat{\phi}_d$, the flux integral in the domain "d." The terms on the right side, which describe the contributions of all domains, are also expressed in terms of the integrated sources, $\hat{S}_{td}$. This then gives, instead of Eq. (6.21), a discrete form of the integral transport equation, a multiple-collision approach:

$$\hat{\phi}_d = \sum_{d'} P_{d'd}\hat{S}_{td'} \quad . \tag{6.22}$$

The quantity $P_{d'd}$ describes the transfer of neutrons from domain d' to domain d by free flight.

6-4 P_1 Theory, Diffusion Theory, and Transport Correction

6-4A Derivation of the One-Group P_1 Equations

The discussion of the surface spherical harmonics in Sec. 6-3B showed that the first two expansion functions of the angular flux are given by one and $\boldsymbol{\Omega}$. Terminating the expansion with P_1 leads to the P_1 approximation:

$$\phi(\boldsymbol{r},\boldsymbol{\Omega}) = \frac{1}{4\pi}[\phi(\boldsymbol{r}) + 3\boldsymbol{\Omega}\cdot\boldsymbol{J}(\boldsymbol{r})] \tag{6.23}$$

or in one dimension as

$$\phi(x,\mu) = \frac{1}{2}[\phi(x) + 3\mu J(x)] \quad . \tag{6.24}$$

In P_n theory, *all* coefficients are to be treated as unknown functions. Thus, in P_1 theory, the current $\boldsymbol{J}(\boldsymbol{r})$ is treated as an unknown function in the same way as the flux ϕ. Therefore, two equations, the P_1 equations, must be found for $\phi(\boldsymbol{r})$ and $\boldsymbol{J}(\boldsymbol{r})$.

The basis for the derivation of the one-group P_1 equations is the one-group Boltzmann equation:

$$\mathbf{M\Phi} = S_t(\boldsymbol{r},\boldsymbol{\Omega}) \quad , \tag{6.25}$$

with

$$\mathbf{M\Phi} = \boldsymbol{\Omega}\cdot\nabla\phi(\boldsymbol{r},\boldsymbol{\Omega}) + \Sigma_t(\boldsymbol{r})\phi(\boldsymbol{r},\boldsymbol{\Omega}) - \int \Sigma_s(\boldsymbol{r},\boldsymbol{\Omega}'\cdot\boldsymbol{\Omega})\phi(\boldsymbol{r},\boldsymbol{\Omega}')\,d\Omega' \quad . \tag{6.26}$$

The total source, S_t, may be different for different one-group problems; for example, S_t may equal:

$$S_t(\boldsymbol{r},\boldsymbol{\Omega}) = \lambda S_f(\boldsymbol{r},\boldsymbol{\Omega}) \tag{6.27}$$

in a one-group theory λ eigenvalue problem for *all* neutrons or S_t may describe the scattering into the thermal group in a one-group theory for thermal neutrons. In both cases, there may also be an independent source present.

According to the general prescription to find the P_1 equations, one multiplies Eq. (6.26) with each of the expansion functions and integrates with respect to $\boldsymbol{\Omega}$. Thus, the one-group P_1 equations result from:

$$\int_\Omega \mathbf{M\Phi}\,d\Omega = \int_\Omega S_t(\boldsymbol{r},\Omega)\,d\Omega \tag{6.28}$$

and

$$\int_\Omega \boldsymbol{\Omega}\mathbf{M\Phi}\,d\Omega = \int_\Omega \boldsymbol{\Omega} S_t(\boldsymbol{r},\boldsymbol{\Omega})\,d\Omega \quad . \tag{6.29}$$

The integrations are carried out term by term; use is made of the approximation, Eq. (6.23), only when needed. The spatial argument is omitted during the derivation. The interchangeability of the vectors in the dot product allows the following evaluation of the first term in Eq. (6.28) [see Eq. (6.26)] (space dependence is suppressed):

$$\int_\Omega \boldsymbol{\Omega}\cdot\nabla\phi(\boldsymbol{\Omega})\,d\Omega = \nabla\cdot\int_\Omega \boldsymbol{\Omega}\phi(\boldsymbol{\Omega})\,d\Omega = \nabla\cdot\boldsymbol{J} \quad . \tag{6.30}$$

The second term in Eq. (6.28) [see Eq. (6.26)] is simply given by

$$\int_\Omega \Sigma_t\phi(\boldsymbol{\Omega})\,d\Omega = \Sigma_t\phi \quad . \tag{6.31}$$

In the third term in Eq. (6.28) [see Eq. (6.26)], the order of the two $\boldsymbol{\Omega}$ integrations is interchanged, which then simply leads to $\Sigma_s\phi$:

$$\int_{\Omega}\int_{\Omega'} \Sigma_s(\boldsymbol{\Omega}'\rightarrow\boldsymbol{\Omega})\phi(\boldsymbol{\Omega}')\; d\Omega'\; d\Omega = \int_{\Omega'}\int_{\Omega} \Sigma_s(\boldsymbol{\Omega}'\cdot\boldsymbol{\Omega})\; d\Omega\phi(\boldsymbol{\Omega}')\; d\Omega'$$

$$= \Sigma_s \int_{\Omega'} \phi(\boldsymbol{\Omega}')\; d\Omega' = \Sigma_s\phi \quad . \tag{6.32}$$

On the right side of Eq. (6.28) one obtains

$$\int_{\Omega} S_t(\boldsymbol{r},\boldsymbol{\Omega})\; d\Omega = S_t(\boldsymbol{r}) = \text{total source density (fission plus independent source)} \quad . \tag{6.33}$$

Note that it was not necessary to use any assumption on the angular distribution of the angular flux to derive the first of the P_1 equations. Since this is also the first of all P_n equations, the result is quite general.

Adding all terms derived above gives the first of the one-group P_1 equations:

$$\nabla\cdot\boldsymbol{J}(\boldsymbol{r}) + \Sigma_a(\boldsymbol{r})\phi(\boldsymbol{r}) = S_t(\boldsymbol{r}) \quad . \tag{6.34}$$

The vector gradient $\nabla\cdot\boldsymbol{J}$ is often called "divergence" and is denoted by "div":

$$\nabla\cdot\boldsymbol{J}(\boldsymbol{r}) = \text{div}\boldsymbol{J}(\boldsymbol{r}) \quad . \tag{6.35}$$

The analogous derivation for the one-dimensional case is obtained readily. The Boltzmann equation and the angular distribution are given by Eqs. (6.4) and (6.24), respectively. The integration of the individual terms yields:

$$\int_{\mu} \mu \frac{d}{dx}\phi(x,\mu)\; d\mu = \frac{d}{dx}\int_{\mu} \mu\phi(x,\mu)\; d\mu = \frac{d}{dx}J(x) \quad , \tag{6.36}$$

$$\int_{\mu} \Sigma_t\phi(x,\mu)\; d\mu = \Sigma_t\phi(x) \quad , \tag{6.37}$$

and

$$\int_{\mu} S_t(x,\mu)\; d\mu = S_t(x) \quad . \tag{6.38}$$

Only the integration of the scattering integral represents a complication in one dimension, since the scattering leads inherently to a change in the angular direction in *three* dimensions. This problem can be avoided by carrying out the angular integration in three dimensions, which simply gives $\Sigma_s\phi(x)$, according to Eq. (6.32). Collecting all terms gives the first of the P_1 equations (or P_n equations) for plane geometry:

$$\frac{d}{dx}J(x) + \Sigma_a(x)\phi(x) = S(x) \quad . \tag{6.39}$$

The second of the P_1 equations is derived by carrying out the integrations in Eq. (6.29) or in its one-dimensional analog. The approximation for the angular distribution Eq. (6.23) is only used when necessary.

To maintain clarity throughout the derivation, the three types of integrals that appear in these integrations are treated first:

1. *Integrals over one angular vector (or cosine):*

$$\int_\Omega \mathbf{\Omega}\, d\Omega = 0 \text{ or } \int_\mu \mu\, d\mu = 0 \quad . \tag{6.40}$$

In Eq. (6.40), $\mathbf{\Omega}$ may be a row or a column vector.

2. *Integrals over two angular vectors:* Also needed is the integral over the $\mathbf{\Omega\Omega}$, appearing in reverse order compared to the dot product. Recall that the dot product is equal to unity:

Dot product:

$$\mathbf{\Omega} \cdot \mathbf{\Omega} = (\Omega_x\ \Omega_y\ \Omega_z)\begin{pmatrix}\Omega_x \\ \Omega_y \\ \Omega_z\end{pmatrix} = 1 \quad . \tag{6.41}$$

If the sequence of row and column vectors appears in reverse order from the dot product, a matrix is obtained—the so-called "dyad," which is defined as:

$$\mathbf{\Omega\Omega} = \begin{pmatrix}\Omega_x \\ \Omega_y \\ \Omega_z\end{pmatrix}(\Omega_x\ \Omega_y\ \Omega_z) \equiv \begin{pmatrix}\Omega_x\Omega_x & \Omega_x\Omega_y & \Omega_x\Omega_z \\ \Omega_y\Omega_x & \Omega_y\Omega_y & \Omega_y\Omega_z \\ \Omega_z\Omega_x & \Omega_z\Omega_y & \Omega_z\Omega_z\end{pmatrix} \quad . \tag{6.42}$$

It can be easily verified that the integral over this dyad equals $\frac{4\pi}{3}$ times the unit matrix:

$$\frac{3}{4\pi}\int_\Omega \begin{pmatrix}\Omega_x \\ \Omega_y \\ \Omega_z\end{pmatrix}(\Omega_x\ \Omega_y\ \Omega_z)\, d\Omega = \begin{pmatrix}1 & 0 & 0 \\ 0 & 1 & 0 \\ 0 & 0 & 1\end{pmatrix} \quad . \tag{6.43}$$

All off-diagonal terms are zero since at least one sine or cosine function appears with an odd power. The integrals over the diagonal terms yield $\frac{4\pi}{3}$.

In the one-dimensional case, one needs, instead of the dyad, merely μ^2 with the integral

$$\frac{3}{2}\int_\mu \mu^2 \, d\mu = 1 \quad . \tag{6.44}$$

3. *Integrals over three angular vectors:* The following integral over three angular vectors is needed:

$$\int_\Omega \mathbf{\Omega\Omega\Omega} \, d\Omega = 0 \quad . \tag{6.45}$$

The following terms appear in the one-dimensional plane geometry case, Eq. (6.4):

Leakage term:

$$\int_\mu \mu^2 \frac{d}{dx}\left[\frac{1}{2}\phi(x) + \frac{3}{2}\mu J(x)\right] d\mu = \frac{1}{3}\frac{d}{dx}\phi(x) \quad .$$

Total loss term:

$$\Sigma_t(x)\int_\mu \mu\phi(x,\mu) \, d\mu = \Sigma_t(x) J(x) \quad . \tag{6.46}$$

Scattering integral term:

$$= \bar{\mu}_s \Sigma_s(x) J(x) \quad .$$

[For the derivation see the general case, Eqs. (6.47) through (6.59).]

Source term (right side):

$$\int_\mu \mu S_t(x,\mu) \, d\mu = S_t^{(1)}(x)[\text{right side}] \quad .$$

The quantity $S_t^{(1)}(x)$ describes the first-order anisotropy of the angular distribution of the source; $S_t^{(1)}$ disappears if the source is isotropic, as the fission source.

The P_1 approximation of the angular flux was only needed in the leakage term. All other terms on the left side could be integrated by employing the definitions of the flux and the current as an integral over the angular flux. The same holds for the general case presented below.

In the three-dimensional Eq. (6.3), one has instead of Eq. (6.46) the following terms (the space dependence is omitted in the notation):

Leakage term:

$$\int_\Omega \boldsymbol{\Omega\Omega}\cdot\nabla\phi(\boldsymbol{r},\boldsymbol{\Omega})\,d\Omega=\frac{1}{4\pi}\int_\Omega \boldsymbol{\Omega\Omega}\cdot\nabla\phi\,d\Omega+\frac{3}{4\pi}\int_\Omega \boldsymbol{\Omega\Omega}\cdot\nabla\boldsymbol{\Omega}\cdot\boldsymbol{J}\,d\Omega$$

$$=\frac{1}{3}\begin{pmatrix}1&0&0\\0&1&0\\0&0&1\end{pmatrix}\nabla\phi=\frac{1}{3}\nabla\phi\quad .$$

Total loss term:

$$\Sigma_t(\boldsymbol{r})\int_\Omega \boldsymbol{\Omega}\phi(\boldsymbol{r},\boldsymbol{\Omega})\,d\Omega=\Sigma_t(\boldsymbol{r})\boldsymbol{J}(\boldsymbol{r})\quad .\tag{6.47}$$

Scattering integral term:

$$\int_\Omega \boldsymbol{\Omega}\int_{\Omega'}\Sigma_s(\boldsymbol{\Omega}'\cdot\boldsymbol{\Omega})\phi(\boldsymbol{\Omega}')\,d\Omega'\,d\Omega=\boldsymbol{I}_{\Omega'\Omega}\quad .$$

[See Eq. (6.57).]

Source term (right side):

$$=\int_\Omega \boldsymbol{\Omega}S_t(\boldsymbol{r},\boldsymbol{\Omega})\,d\Omega=S_t^{(1)}(\boldsymbol{r})\text{ [right side]}\quad .$$

In calculating $\boldsymbol{I}_{\Omega'\Omega}$, it is easiest to carry out first the integration with respect to $\boldsymbol{\Omega}$:

$$\boldsymbol{I}_{\Omega'\Omega}=\int_{\Omega'}\int_\Omega \boldsymbol{\Omega}\Sigma_s(\boldsymbol{\Omega}'\cdot\boldsymbol{\Omega})\,d\Omega\phi(\boldsymbol{\Omega}')\,d\Omega'=\int_{\Omega'}\boldsymbol{I}_{\Omega'}\phi(\boldsymbol{\Omega}')\,d\Omega'\quad .\tag{6.48}$$

The scattering transfer cross section depends on the angle between $\boldsymbol{\Omega}'$ and $\boldsymbol{\Omega}$:

$$\Sigma_s(\boldsymbol{\Omega}'\cdot\boldsymbol{\Omega})=\frac{1}{2\pi}\Sigma_s(\mu_s)\quad .\tag{6.49}$$

It is advisable, and at the same time simple, to express $\boldsymbol{\Omega}$ in components around $\boldsymbol{\Omega}'$ for this particular integration, i.e., in terms of the scattering angle:

$$\boldsymbol{\Omega}=\begin{pmatrix}\sin\vartheta_s\ \cos\alpha_s\\ \sin\vartheta_s\ \sin\alpha_s\\ \cos\vartheta_s\end{pmatrix}\quad .\tag{6.50}$$

The differential is then given by

$$d\Omega=d\mu_s\,d\alpha_s\quad .\tag{6.51}$$

Note that the integral over Ω equals 4π. The Ω integral in Eq. (6.48) (i.e., $\boldsymbol{I}_{\Omega'}$) then becomes:

$$\boldsymbol{I}_{\Omega'} = \frac{1}{2\pi}\int_{\mu_s}\int_{\alpha_s}\begin{bmatrix}(1-\mu_s^2)^{1/2}\cos\alpha_s\\(1-\mu_s^2)^{1/2}\sin\alpha_s\\\mu_s\end{bmatrix} d\alpha_s\,\Sigma_s(\mu_s)\,d\mu_s \quad . \tag{6.52a}$$

Carrying out the α_s integration first yields

$$\boldsymbol{I}_{\Omega'} = \begin{pmatrix}0\\0\\1\end{pmatrix}_{\Omega'}\int_{-1}^{1}\mu_s\Sigma_s(\mu_s)\,d\mu_s \quad . \tag{6.52b}$$

The μ_s integration introduces the average cosine $\bar{\mu}_s$:

$$\boldsymbol{I}_{\Omega'} = \begin{pmatrix}0\\0\\1\end{pmatrix}_{\Omega'}\bar{\mu}_s\Sigma_s \quad , \tag{6.53}$$

with

$$\Sigma_s = \int_{-1}^{1}\Sigma_s(\mu_s)\,d\mu_s \tag{6.54}$$

and

$$\bar{\mu}_s = \frac{1}{\Sigma_s}\int_{-1}^{1}\mu_s\Sigma_s(\mu_s)\,d\mu_s \quad . \tag{6.55}$$

Equation (6.53) shows that $\boldsymbol{I}_{\Omega'}$ is a vector pointing in the direction of the z axis used in the representation of $\boldsymbol{\Omega}$, i.e., in the direction of $\boldsymbol{\Omega}'$. Thus, for the subsequent integration, Eq. (6.53) is to be written as

$$\boldsymbol{I}_{\Omega'} = \boldsymbol{\Omega}'\bar{\mu}_s\Sigma_s \quad . \tag{6.56}$$

With $\boldsymbol{I}_{\Omega'}$ from Eq. (6.56), the integral, Eq. (6.48), has the following form:

$$\boldsymbol{I}_{\Omega'\Omega} = \bar{\mu}_s\Sigma_s\int_{\Omega'}\boldsymbol{\Omega}'\phi(\boldsymbol{\Omega}')\,d\Omega' \quad . \tag{6.57}$$

Thus,

$$\boldsymbol{I}_{\Omega'\Omega} = \bar{\mu}_s\Sigma_s\boldsymbol{J} \quad . \tag{6.58}$$

The index s, which distinguishes the cosine between $\boldsymbol{\Omega}$ and $\boldsymbol{\Omega}'$ from the cosine in the angular coordinate system, is commonly omitted in $\bar{\mu}$; $\bar{\mu}$ is understood as the average cosine of the deflection angle.

Adding all terms of Eqs. (6.46) and (6.47), respectively, yields the second of the one-group P_1 equations, given here in one planar dimension [see Eq. (6.62) for the three-dimensional form]:

$$\frac{1}{3}\frac{d\phi(x)}{dx} + \Sigma_t(x)J(x) - \bar{\mu}_s\Sigma_s(x)J(x) = S_t^{(1)}(x) \quad . \tag{6.59}$$

The two terms proportional to J are combined. The new coefficient of J is called the "transport cross section":

$$\Sigma_{tr} = \Sigma_t - \bar{\mu}_s\Sigma_s = \Sigma_a + (1 - \bar{\mu}_s)\Sigma_s \quad . \tag{6.60}$$

The term $\bar{\mu}_s\Sigma_s$ is called the "transport correction."

As the final result, the one-group P_1 equations in one planar dimension are obtained:

$$\frac{dJ(x)}{dx} + \Sigma_a(x)\phi(x) = S_t(x)$$

and

$$\Sigma_{tr}(x)J(x) + \frac{1}{3}\frac{d\phi(x)}{dx} = S_t^{(1)}(x) \quad , \tag{6.61}$$

and in three dimensions

$$\nabla\cdot\boldsymbol{J}(\boldsymbol{r}) + \Sigma_a(\boldsymbol{r})\phi(\boldsymbol{r}) = S_t(\boldsymbol{r})$$

and

$$\Sigma_{tr}(\boldsymbol{r})\boldsymbol{J}(\boldsymbol{r}) + \frac{1}{3}\nabla\phi(\boldsymbol{r}) = \boldsymbol{S}_t^{(1)}(\boldsymbol{r}) \quad . \tag{6.62}$$

The second of the P_1 equations is the replacement of Fick's Law of The diffusion theory. The $\boldsymbol{J}$ is generally not proportional to the gradient of the flux; rather it is to be composed of $\nabla\phi$ *and* the anisotropic component of the neutron source.

6-4B Derivation of the One-Group Diffusion Equation

Inspection of Eq. (6.62) shows that an angular dependency of the neutron source perturbs the proportionality of $\nabla\phi$ and $\boldsymbol{J}$. Only if the source is isotropic, i.e., only if $\boldsymbol{S}_t^{(1)}(\boldsymbol{r})\equiv 0$, can one express $\boldsymbol{J}$ by $\nabla\phi$:

$$\boldsymbol{J}(\boldsymbol{r}) = -\frac{1}{3\Sigma_{tr}(\boldsymbol{r})}\nabla\phi(\boldsymbol{r}) = -D(\boldsymbol{r})\nabla\phi(\boldsymbol{r}) \quad . \tag{6.63}$$

This now gives the definition of the diffusion constant:

$$D(\boldsymbol{r}) = \frac{1}{3\Sigma_{tr}(\boldsymbol{r})} = \frac{\lambda_{tr}}{3} \quad , \tag{6.64}$$

with λ_{tr} being the so-called "transport mean-free-path." Note that up to this point the diffusion constant was used as a given quantity. Now the one-group diffusion constant is derived from the second P_1 equation.

Inserting Eqs. (6.63) and (6.64) into the first of the P_1 equations gives the one-group diffusion equation:

$$-\nabla \cdot D(\boldsymbol{r})\nabla\phi(\boldsymbol{r}) + \Sigma_a(\boldsymbol{r})\phi(\boldsymbol{r}) = S_t(\boldsymbol{r}) \quad . \tag{6.65}$$

In Eq. (6.65), $S_t(\boldsymbol{r})$ may be any one-group neutron source, particularly the corresponding source of either of the two types of one-group theories discussed above:

$$S_t(\boldsymbol{r}) = \lambda S_f(\boldsymbol{r}) \quad , \tag{6.66a}$$

with S_f being the fission source in one-group theory for all neutrons or

$$S_t(\boldsymbol{r}) = S_s(\boldsymbol{r}) \tag{6.66b}$$

as the scattering source into the thermal group in one-group theory for thermal neutrons.

The fission source is isotropic, i.e., the one-group P_1 theory and the one-group diffusion theory for *all* neutrons are equivalent. The P_1 theory provides the appropriate formula for the one-group diffusion constant.

In the one-group theory for thermal neutrons only, the neutrons are provided by scattering from the epithermal range into the thermal group. This scattering source is generally anisotropic, particularly if hydrogen is the dominating moderator. Thus, the accuracy of the one-group theory of thermal neutrons may be significantly hampered by the neglect of the anisotropy of the scattering source.

The original Fick's Law was postulated for the diffusion of gases, i.e., for a case with no absorption and isotropic scattering. The corresponding diffusion constant is

$$D = \frac{1}{3\Sigma_s} \tag{6.67}$$

without absorption and with isotropic scattering. Inclusion of absorption and anisotropic scattering requires a new derivation, which the P_1 theory provides.

As long as the scattering is still isotropic, by including the absorption, the following is obtained:

$$D = \frac{1}{3\Sigma_t} \quad . \tag{6.68}$$

By including also the forward bias of the scattering, one then obtains the more general formula

$$D = \frac{1}{3\Sigma_{tr}} = \frac{1}{3(\Sigma_t - \bar{\mu}_s \Sigma_s)} \quad . \tag{6.69}$$

With increasing forward bias (increasing $\bar{\mu}_s$), the current increases for a given flux gradient:

$$J = \frac{-1}{3(\Sigma_t - \bar{\mu}_s \Sigma_s)} \nabla \phi \quad . \tag{6.70}$$

As a consequence, the leakage, such as that approximately described by

$$DB^2 = \frac{B^2}{3(\Sigma_t - \bar{\mu}_s \Sigma_s)} \quad , \tag{6.71}$$

increases with increasing forward bias of the elastic scattering. This is particularly important at higher neutron energies.

6-5 Summarizing Remarks

The contents of this chapter on the angular dependency of the neutron flux constitute only the barest of introductions. In closing, the importance of incorporating the angular neutron flux effects for certain reactor situations must be emphasized again. Typical examples of areas where the diffusion theory neutron flux model is inappropriate include: near the interface to a vacuum, at interfaces between media, and in the small regions associated with fuel coolant cells or control rods. In these situations, use of diffusion theory will be in error and one will have to turn to the more accurate transport theory model, which incorporates the angular distribution of the flux into its formulations.

While the details of the various transport models were left for more advanced texts, a derivation of the one-group P_1 equations in one planar dimension was provided. The solution of this transport model provides a more accurate formula for the one-group diffusion constant. This transport-theory-derived diffusion constant expression includes the effects of neutron absorption and first-order anisotropic scattering and, therefore, with these corrections, the range of diffusion theory application has been extended. (Note that the original Fick's Law definition applies only to the case of no absorption and isotropic scattering).

Homework Problems

1. Derive the one-group integral transport equation for the angular flux, $\phi(x,\mathbf{\Omega})$, for a purely x-dependent geometry consisting of three slabs of scattering and absorbing material only, with vacuum outside. Assume isotropic scattering and uniformly distributed independent sources.
2. Derive the one-group integral transport equation for the angular flux, $\phi(r,\mathbf{\Omega})$, for a spherical geometry consisting of two media (a center sphere in a spherical shell with a vacuum outside); apply the assumptions of problem 1.
3. Modify the x-dependent transport equation derived in problem 1 to describe a symmetrical three-region slab problem, with a fuel plate (d_f) in the center of a large moderator region with a vacuum outside (thickness d_m on both sides of the fuel plate).
4. Assuming the flux $\phi(x)$ is known, apply the integral transport equation for the angular flux $\phi(x,\mathbf{\Omega})$, obtained in problem 3, to find the integral equation for the flux $\phi(x)$.
5. Investigate the iterative calculational procedure for finding the flux, $\phi(r)$, in the spherical problem addressed in problem 2. Assume the inner sphere of radius $R = 2$ cm is black; take $\Sigma_s = 0.2$/cm and $\Sigma_a = 0$ in the outer medium, assumed to be infinitely large, with $\phi(r\rightarrow\infty) = \phi_\infty$.
   a. Calculate the first iterate, $\phi^{(1)}(r)$, from an assumed flat flux distribution as the "zeroth iterate," $\phi^{(0)} = \phi_\infty$, in the outer medium.
   b. Plot $\phi^{(1)}(r)/\phi_\infty$ and discuss the result.
   c. Discuss qualitatively the changes to be expected in the next iterate, $\phi^{(2)}(r)$, as well as in the converged result.
   d. Solve the one-group diffusion equation for this case; plot, compare, and discuss the results.
6. Prove that the integral over the dyad matrix, Eq. (6.42), equals $4\pi/3$.

7a. Give the dyad product $\mathbf{\Omega}\cdot\mathbf{\Omega}$ for the two-dimensional case, say, in the x-y plane.

b. Find its integral over Ω.

8a. Calculate the average scattering angle, $\overline{\mu}_s$, for an angular distribution with an assumed anisotropy in the laboratory system represented by the first three Legendre polynomials, $P_n(\mu)$:

$$p(\mu) = (a_0P_0 + a_1P_1 + a_2P_2) \quad .$$

b. Find the maximum value of $\overline{\mu}_s$ obtainable with such an angular distribution (note that $p(\mu)\geq 0$).

9. Find the average scattering angle, $\bar{\mu}_s$, for inelastic scattering with an anisotropy contribution proportional to the second Legendre polynomial, $p(\mu) = a_0P_0 + a_2P_2$, with $a_2/a_0 = 0.25$.

Review Questions

1a. What is the angular distribution of $\phi(x,\mu)$, which is assumed in diffusion theory?
b. Describe four typical cases in which this angular distribution is not a good approximation.
2. Explain why the neutron population in a reactor is fluctuating statistically.
3. Which method can be applied to treat stochastic phenomena?
4. Explain the Monte Carlo procedure (a) in general, (b) in its application to stochastic processes, and (c) in its application to non-stochastic equations (Boltzmann equation). How is ν treated in the latter case?
5a. Give the basic dependencies of the (a) Legendre polynomials, $P_n(\mu)$, (b) the associated Legendre polynomials $P_n^m(\mu)$, and (c) the surface harmonics, $Y_{n,m}(\mu,\alpha)$.
b. Give the first two (n = 0 and 1) P_n and $Y_{n,m}$.
c. Give the orthogonality relation of the P_n's.
6. Explain the idea of the P_n theory and sketch the derivation.
7. Present the idea of the double P_n method and discuss its applicability.
8. Present the S_n approach and explain its key advantage compared to the P_n method.
9. In which regionally discrete form is the integral transport theory practically applied?
10. Sketch the derivation of the one-group P_1 equation.
11a. Give the one-group P_1 equations.
b. Assume $S^{(1)} = 0$ and derive the diffusion equation.
12. Give the one-group macroscopic transport cross section and explain how an increasing forward bias of the elastic scattering affects the current and the leakage (DB^2).
13. What is a dyad product? Give two examples ($\mathbf{\Omega\Omega}$ and $\mathbf{F}$).
14. What is the correct substitute for $\bar{\mu}_s$ in the multigroup treatment?

REFERENCES

1. A. M. Weinberg and E. P. Wigner, *The Physical Theory of Neutron Chain Reactors*, University of Chicago Press (1958).
2. M. M. R. Williams, *Mathematical Methods in Particle Transport Theory*, Butterworth and Co., Canada (1971).
3. B. Davison, *Neutron Transport Theory*, Oxford University Press (1957).
4. G. I. Bell and S. Glasstone, *Nuclear Reactor Theory*, Van Nostrand Reinhold Co., New York (1970).
5. H. Greenspan, C. N. Kelber, and D. Okrent, Eds., *Computing Methods in Reactor Physics*, Gordon & Breach Science Publishers, New York (1968).
6. K. D. Lathrop, "Discrete-Ordinates Method for the Numerical Solution of the Transport Equation," *Reactor Technol.*, **15,** No. 2 (Summer 1972).
7. B. G. Carlson, "Solutions of the Transport Equation by S_n Approximations," LA-1599, Los Alamos National Lab. (1953).
8. B. G. Carlson and K. D. Lathrop, Chapter 3 in *Computing Methods in Reactor Physics*, H. Greenspan, C. N. Kelber, and D. Okrent, Eds., Gordon & Breach Science Publishers, New York (1968).

Seven

MULTIGROUP DIFFUSION THEORY

7-1 Introduction

In the previous chapters, the analysis and form of the individual components of the neutron flux were studied and discussed extensively. The overall spatial characteristics of the total flux were examined in Chapter 3 for several simple reactor geometries. For some standard unreflected reactor configurations (for example, bare sphere), it was shown that a straightforward analytical expression for the flux shape could be obtained rather easily. The elimination of the energy dependencies, through separation or integration, which yielded the spatial neutron flux, was valid only for one-region systems. Although this approach cannot be applied to complex multiregion geometries and nonuniform conditions (for example, variation in the enrichment) present in actual reactors, it is useful to describe approximately the leakage in a transverse one-region dimension in terms of a DB^2 contribution to Σ_a.

Similarly, the forms of the neutron spectra, i.e., the pure energy dependency of the neutron flux, in Chapters 4 and 5 were treated based on the assumption of flux separation. It was shown that an exact analytical treatment can only be obtained for hydrogen as an elastic scatterer, and the inclusion of any other scatterer required approximations in the treatment. Nevertheless, a good representation of the neutron spectrum in the slowing down range was possible by properly piecing together information from various simpler cases. Also in Chapter 4, the relative magnitude of the slowing down spectrum and the thermal spectrum was found by using a global balance consideration. In this slowing down treatment, all results were found for the space-independent spectrum, which was obtained by eliminating the space dependence. The process of elimination of the space dependence showed the assumptions required to obtain an equation that contains neither space dependence nor leakage.

Recall that for both of these space- and energy-independent investigations, the basic assumption employed in isolating either one of the two dependencies was flux separability:

$$\phi(x,E) = \phi(x) \cdot \varphi(E) \quad . \tag{7.1}$$

Substitution of this flux expression into the diffusion equation, Eq. (2.28), resulted in the required two working equations, Eqs. (3.1) and (3.2).

In the treatment of the energy dependency after the elimination of the space dependence, the leakage appeared in the form of $D(E)B^2$, added to $\Sigma_t(E)$:

$$\Sigma_{tl}(E) = \Sigma_t(E) + D(E)B^2 \quad . \tag{7.2}$$

For large reactors, one has

$$D(E)B^2 << \Sigma_t(E) \quad ; \tag{7.3}$$

then, the slowing down spectrum in the finite system, $\varphi(E)$, can be factored into two parts (as was shown in Chapter 4):

$$\varphi(E) = \varphi^{\infty}(E) \cdot P_f(E) \quad , \tag{7.4}$$

with $\varphi^{\infty}(E)$ being the spectrum in the infinite medium and $P_f(E)$ the nonleakage probability down to energy E. The factorization, Eq. (7.4), is normally applied for the treatment of resonance absorption.

In the thermal range, the spectrum is closely coupled through frequent up and down scattering. In the case of a separable flux, i.e., for a single large region, the leakage is largely described by an energy-independent factor P_{th}.

Unfortunately, in most actual problems, as in the calculation of the flux or determination of the conditions for criticality in commercial reactors, the flux is not separable. For the treatment of these actual problems, a different calculational approach must be applied that retains both the space and energy dependencies. The normally applied analysis procedure is the "multigroup method," which was introduced in a preliminary form in Sec. 2-2D. In this chapter, a more detailed description of the multigroup method is presented, including the definition and generation of the group constants and a brief outline of the numerical solution procedure. The emphasis is on diffusion theory, which represents the model most extensively used in actual production reactor design calculations. It should be pointed out, however, that the same multigroup methods are applied also to the "transport theory" models outlined in Chapter 6.

The information, derived above, on neutron spectra is explicitly used in the multigroup theory as weighting spectra in the process of generating the group constants. These weighting spectra in the slowing down range are most often calculated for homogeneous media, but heterogeneity corrections are always applied. A most useful concept in this context is the collision probability method (compare Sec. 6-5).

7-2 The Multigroup Approach

7-2A General Discussion of Approaches

If the flux is separable, the separation techniques and procedures yield energy and space dependencies. If the flux is not separable, it may be concluded that $\phi(x,E)$ must be calculated directly. One therefore seems to have two extremes:

Calculation of a separable flux:

$$\phi(x)\varphi(E)$$

and

direct calculation of the nonseparable flux:

$$\phi(x,E) \quad .$$

In most practical cases, there is room for a compromise, in the form of approaches between the two extremes, approaches based on a partial separability. This is practically assumed in the multigroup approach.

The separability is assumed to be valid only over parts of the energy scale, i.e., in the simplest case of two parts (for example, a "thermal" and a "fast," or "above-thermal," part):

$$\Phi(x,E) = \begin{bmatrix} \phi_f(x) \cdot \varphi_f(E) \\ \phi_{th}(x) \cdot \varphi_{th}(E) \end{bmatrix} \quad , \tag{7.5}$$

with $\phi_{th}(x) \neq \phi_f(x)$ since thermal and the fast neutrons normally have different spatial dependencies. The assumption [Eq. (7.5)] leads to the two-group theory. Its generalization,

$$\Phi(x,E) = \begin{bmatrix} \phi_1(x) \cdot \varphi_1(E) \\ \cdot \quad \cdot \\ \cdot \quad \cdot \\ \cdot \quad \cdot \\ \phi_g(x) \cdot \varphi_g(E) \\ \cdot \quad \cdot \\ \cdot \quad \cdot \\ \cdot \quad \cdot \\ \phi_G(x) \cdot \varphi_G(E) \end{bmatrix} \quad , \tag{7.6}$$

leads to a "multigroup theory." For simplicity, $\varphi_g(E)$ may be normalized to unity,

$$\int_g \varphi_g(E)\, dE = 1 \quad , \tag{7.7}$$

so that $\phi_g(x)$ represents the group integral. The "weighting spectra," $\varphi_g(E)$, are used to condense the energy-dependent cross sections into "group constants." Equations (7.5) and (7.6) are applied in most practical cases, although the formal derivation of the multigroup equations is based on the general flux $\phi(x,E)$.

There is another logical possibility to express a partial separability. In Eq. (7.5), one has weakened the strong separability assumption by applying it only over parts of the *energy* scale. One can weaken it also by applying it only over part of the *space* domain:

$$\phi(x,E) = [\phi_1(x)\varphi_1(E)]_{\text{reg. 1}} \quad \text{and} \quad [\phi_2(x)\varphi_2(E)]_{\text{reg. 2}} \quad . \tag{7.8}$$

This flux representation is rarely applied in the form of Eq. (7.8). It is usually combined with the assumption, Eq. (7.6); it then leads to a multigroup theory with region-dependent group constants (e.g., different group constants for core and reflector). A second form of applying the idea expressed in Eq. (7.8) is to blend the two components and use both in the entire space. Then the factor in front of each spectrum is not the full flux in this region since there is a contribution from the other terms. Therefore, the notation is changed to:

$$\phi(x,E) = g_1(x)\varphi_1(E) + g_2(x)\varphi_2(E) \quad , \tag{7.9a}$$

or in general form:

$$\phi(x,E) = \sum_m g_m(x)\varphi_m(E) \quad . \tag{7.9b}$$

This method is called "space-energy synthesis," "the overlapping groups method," or the "multimode theory."

The multigroup theory has advantages over the space-energy synthesis; therefore, it is the method generally used. The multigroup diffusion approach is the method that has attained almost universal usage in the nuclear reactor design field. A major reason for its acceptance is that the numerical techniques used for its evaluation are uniquely suited to efficient solution by high-speed computers. Briefly, the multigroup diffusion method consists of converting the continuous diffusion equation into a discrete structure in both space and neutron energy. The spatial discretization is affected by overlaying a pointwise mesh or grid structure over the region. The energy domain is subdivided into a discrete group flux structure. For each energy group, the required nuclear parameters of "group constants" are determined and the diffusion equation solved, usually applying finite difference techniques. In the following sections, the group structure and multigroup diffusion equations are presented with the important topic of group constant definitions presented in Sec. 7-3.

7-2B Group Intervals and Group Fluxes

The energy scale is divided into "group" intervals, beginning with the first interval at the highest energy according to the standard notation. The basis for the definition of the group structure is usually the lethargy scale. Equal lethargy increments are often used in the slowing down range (see Fig. 7-1).

The neutron flux is integrated over each group interval to form the "group flux," which represents a "neutron group":

$$\phi_g(x) = \int_{E_g}^{E_{g-1}} \phi(x,E)\, dE = \int_g \phi(x,E)\, dE \quad . \tag{7.10}$$

Group 1 contains the highest energy neutrons. For example, in a two-group model for a thermal reactor, the fast group flux is $\phi_1(x)$, and the thermal flux is $\phi_2(x)$.

In most thermal reactor design analysis, 2 to 4 energy groups have proven to be sufficient. Fast reactors, on the other hand, normally require from 10 to 100 separate group fluxes for reactor evaluation studies, because of their more complicated energy dependencies.

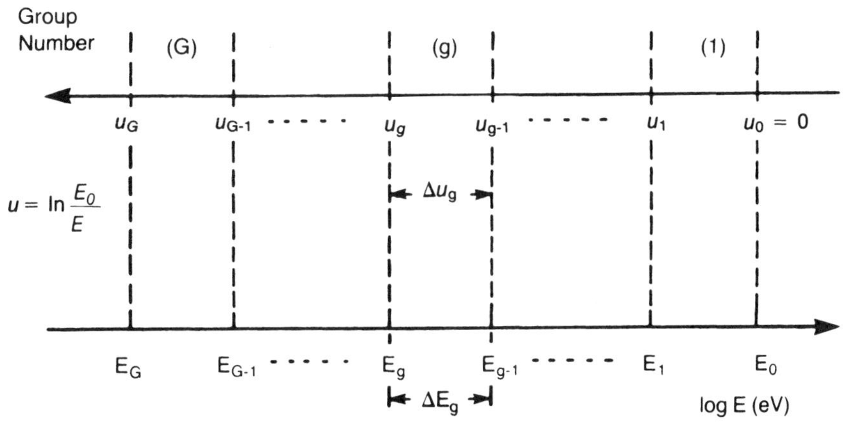

Fig. 7-1. The segmentation of the lethargy and energy scales for the multigroup model.

7-2C Formal Derivation of the Multigroup Diffusion Equations

The basis for the derivation of the multigroup diffusion equations is the energy-dependent diffusion equation [see Eq. (2.20)], which will be given here for regionwise constant composition in one space dimension:

$$-\frac{d}{dx}D(E)\frac{d}{dx}\phi(x,E)+\Sigma_t(E)\phi(x,E)-\int_0^\infty \Sigma_s(E'\rightarrow E)\phi(x,E')\,dE'$$
$$-\lambda\chi(E)\int_0^\infty \nu(E')\Sigma_f(E')\phi(x,E')\,dE'=0 \quad . \qquad (7.11)$$

In addition to Eq. (7.11), boundary and interface conditions, which impose the continuity of flux and current, are required at each region boundary as well as at the outer surface of the reactor for completion of the problem definition (see Secs. 2-2E and 7-5A).

Each term in Eq. (7.11) is a reaction rate density. The definition of the multigroup approach is based on the principle of formal conservation of the groupwise reaction rate densities. This leads, in the same way as for one group (see Chapter 2), to the group constants as average values, after the group fluxes have already been defined as integrals [Eq. (7.10)].

The implementation of this principle first requires that one produce the groupwise reaction rates, i.e., Eq. (7.11) is integrated individually over each group interval. To simplify the presentation, use is made of the operator notation:

$$\left.\begin{aligned}&\int_1(\mathbf{M}-\lambda\mathbf{F})\Phi\,dE=0\\ &\quad\vdots\\ &\int_g(\mathbf{M}-\lambda\mathbf{F})\Phi\,dE=0\\ &\quad\vdots\\ &\int_G(\mathbf{M}-\lambda\mathbf{F})\Phi\,dE=0\end{aligned}\right\}\quad G \text{ equations for the } G \text{ unknown functions } \phi_1(x),\ldots,\phi_G(x) \quad . \qquad (7.12)$$

Note that, in the formal derivation of the multigroup equations, one did *not* have to make use of a groupwise separability nor of any other approximations, i.e., the formal multigroup equations are as exact as the original equation. It is in the generation of group constants where approximations are applied.

Deferring temporarily (until the next section) the definitions of these group constants, the integration of Eq. (7.12) yields the "multigroup diffusion equation" (in one dimension with regionwise constant compositions):

$$-D_g \frac{d^2}{dx^2}\phi_g(x) + \Sigma_{tg}\phi_g(x) - \sum_{g'}\Sigma_{sg'g}\phi_{g'}(x) = \lambda\chi_g\sum_{g'}\nu\Sigma_{fg'}\phi_{g'}(x) \quad . \tag{7.13}$$

The multigroup versions of the operators **M**, **F**, and Φ can be readily extracted from Eq. (7.13):

$$\Phi = \begin{bmatrix} \phi_1(x) \\ \cdot \\ \cdot \\ \cdot \\ \phi_G(x) \end{bmatrix} , \tag{7.14}$$

$$\mathbf{M} = \begin{pmatrix} -D_1 \frac{d^2}{dx^2} + \Sigma_{r1} & -\Sigma_{s21} & \ldots & -\Sigma_{sG1} \\ -\Sigma_{s12} & -D_2 \frac{d^2}{dx^2} + \Sigma_{r2} & \ldots & -\Sigma_{sG2} \\ \cdot & \cdot & & \cdot \\ \cdot & \cdot & & \cdot \\ \cdot & \cdot & & \cdot \\ -\Sigma_{s1G} & -\Sigma_{s2G} & \ldots & -D_G \frac{d^2}{dx^2} + \Sigma_{rG} \end{pmatrix} , \tag{7.15}$$

and

$$\mathbf{F} = \begin{pmatrix} \chi_1\nu\Sigma_{f1} & \cdots & \chi_1\nu\Sigma_{fG} \\ \cdot & & \cdot \\ \cdot & & \cdot \\ \cdot & & \cdot \\ \chi_G\nu\Sigma_{f1} & \cdots & \chi_G\nu\Sigma_{fG} \end{pmatrix}$$

$$= \begin{pmatrix} \chi_1 \\ \cdot \\ \cdot \\ \cdot \\ \chi_G \end{pmatrix} (\nu\Sigma_{f1} \cdots \nu\Sigma_{fG}) \quad . \tag{7.16}$$

The multigroup flux, Φ, as given by Eq. (7.14), is called the "flux vector." In Eq. (7.16), the fission matrix actually is the dyad product of the two vectors χ and $\nu\Sigma_f$ (see Sec. 6-4).

The multigroup equation, Eq. (7.13), then assumes the familiar form:

$$(\mathbf{M} - \lambda\mathbf{F})\Phi = 0 \quad , \tag{7.17}$$

with **M, F,** and Φ given by Eqs. (7.14), (7.15), and (7.16).

Remaining now for completion of Eq. (7.13) are the definitions of the group constants. These are provided in the next section.

7-3 Definition of the Group Constants

7-3A Flux-Weighted Group Constants

The implementation of the formal derivation of the multigroup equations leads to the definition of the group constants. The condensation of the individual terms of the general equation, Eq (7.11), is discussed in detail in the following. The simplest case is the treatment of the total reaction rate:

$$\int_g \Sigma_t(E)\phi(x,E)\, dE \quad . \tag{7.18}$$

One has to introduce the group quantities such that the total reaction rate is conserved. This is done by setting this integral equal to the group flux times the average $\Sigma_t(E)$ in group g:

$$\int_g \Sigma_t(E)\phi(x,E)\, dE = \overline{\Sigma}_t^g \cdot \phi_g(x) = \Sigma_{tg}\phi_g(x) \quad . \tag{7.19}$$

This procedure formally preserves the reaction rates and leads to the definition of group constants as flux-weighted average values:

$$\Sigma_{tg}(x) = \frac{\int_g \Sigma_t(E)\phi(x,E)\, dE}{\int_g \phi(x,E)\, dE} = \overline{\Sigma}_t^g(x) \quad . \tag{7.20}$$

The definition, Eq. (7.20), leads to a group constant depending on space if $\phi(x,E)$ is not separable. In practically all applications, the space-dependent group constants are approximated by space-independent values; either because the information is not available to calculate space-dependent group constants or because one wants to avoid the complications resulting from their space dependency. Space independency of group constants is equivalent to assuming separability within group g:

$$[\phi(x,E)]_{\text{group } g} \simeq \phi_g(x) \cdot \varphi_g(E) \quad . \tag{7.21}$$

This obviously leads to space-independent group constants:

$$\Sigma_{tg} = \frac{\int_g \Sigma_t(E)\varphi_g(E)\, dE}{\int_g \varphi_g(E)\, dE} \quad . \tag{7.22}$$

The derivation yields at first macroscopic group constants, which can be separated in their isotopic components:

$$\Sigma_{tg} = \sum_i N_i \sigma_{tg}^i \quad , \tag{7.23}$$

with

$$\sigma_{tg}^i = \frac{\int_g \sigma_t^i(E)\varphi_g(E)\, dE}{\int_g \varphi_g(E)\, dE} \quad . \tag{7.24}$$

Basic group constant sets are tabulated and stored as microscopic group constants from which macroscopic group constants can be calculated for different compositions of materials.

If different weighting spectra are used for different regions in Eq. (7.24), region-dependent sets of group constants are obtained. This is often done in more sophisticated calculations.

The total cross section is composed of capture, scattering, and possibly fission. Thus, Eq. (7.22) also defines the corresponding group constants, i.e.:

$$\Sigma_{tg} = \Sigma_{cg} + \Sigma_{fg} + \Sigma_{sg} + \Sigma_{ing} \quad , \tag{7.25}$$

where all of the elastic and inelastic scattering events, which occur within group g, are included in the Σ_{sg} and Σ_{ing} terms, respectively. Furthermore, the group-to-group scattering or transfer cross sections are components of these group-scattering cross sections:

$$\Sigma_{sg} = \sum_{g''} \Sigma_{sgg''} \quad ; \tag{7.26}$$

and for Σ_{ing}, where one has only down scattering:

$$\Sigma_{ing} = \Sigma_{ingg} + \Sigma_{ingg+1} + \ldots + \Sigma_{ingG} \quad . \tag{7.27}$$

Equations (7.26) and (7.27) are the multigroup versions of the corresponding integrals, which relate the scattering cross sections with the corresponding transfer cross sections:

$$\Sigma_s(E) = \int_0^\infty \Sigma_s(E \rightarrow E'')\, dE'' \tag{7.28}$$

and

$$\Sigma_{in}(E) = \int_0^E \Sigma_{in}(E \rightarrow E'')\, dE'' \quad . \tag{7.29}$$

In the fission source, the integration for the derivation of the multigroup equations applies to $\chi(E)$ only:

$$\chi_g = \int_g \chi(E)\, dE, \tag{7.30}$$

which is the fraction of fission neutrons born into group g. Thus, there is no flux weighting in the χ_g group constants.

One must also express the integral in the fission source in terms of group fluxes:

$$S_f(x) = \int_0^\infty \nu\Sigma_f(E')\phi(x,E')\, dE' = \sum_g \int_g \nu\Sigma_f(E')\phi(x,E')\, dE'$$

and

$$S_f(x) = \sum_g \nu\Sigma_{fg}\phi_g(x) \quad . \tag{7.31}$$

Applying the same conservation principle as for other reaction rates leads to

$$\frac{\int_g \nu(E)\sigma_f(E)\varphi(E)\, dE}{\int_g \varphi(E)\, dE} = \nu\sigma_{fg} \quad . \tag{7.32}$$

Note that $\nu\sigma_{fg}$ is considered to be *one* quantity. Since σ_{fg} has already been defined along with σ_{tg}, one can obtain a group value for ν as

$$\nu_g = \frac{\nu\sigma_{fg}}{\sigma_{fg}} = \frac{(\nu\sigma_{fg})}{\sigma_{fg}} \quad , \tag{7.33}$$

where in the numerator on the right side σ_{fg} normally cannot be canceled. If the energy range of a group is very small (for example, a thermal group), $\nu(E)$ is then virtually independent of E in this group, and σ_{fg} cancels in Eq. (7.33). Example:

$$\nu_{th}(^{235}\text{U}) = 2.43 \quad .$$

The scattering integral (both up and down scattering) is separated into group contributions in the same manner as the fission source energy integral:

$$\int_g \int_0^\infty \Sigma_s(E' \to E)\phi(x,E')\, dE'\, dE = \int_g \sum_{g'} \int_{g'} \Sigma_s(E' \to E)\phi(x,E')\, dE'\, dE$$

$$= \sum_{g'} \Sigma_{sg'g}\phi_{g'}(x) \quad . \tag{7.34}$$

The macroscopic scattering matrix is composed of the following microscopic contributions

$$\sigma_{sg'g} = \frac{\int_g \int_{g'} \sigma_s(E' \to E)\varphi(E')\, dE'\, dE}{\int_{g'} \varphi(E')\, dE'} \quad . \tag{7.35}$$

The double integral is not discussed in detail, although its calculation is straightforward. However, it has different integration limits (because of different ΔE_{max}) for different isotopes.

The scattering rate within a group appears twice in the balance equation (in Σ_{tg} and in the scattering source). It is common to cancel this term and to introduce the "removal" group constant:

$$\Sigma_{rg} = \Sigma_{tg} - \Sigma_{sgg} = \Sigma_{ag} + \sum_{g' \neq g} \Sigma_{gg'} \quad . \tag{7.36}$$

In doing so, the diagonal terms in the scattering matrix disappear. If one has merely down scattering, Eq. (7.36) is simply given by:

$$\Sigma_{rg} = \Sigma_{ag} + \sum_{g' > g} \Sigma_{gg'} \quad . \tag{7.37}$$

7-3B The Diffusion Group Constant

The most complicated group constant is the "diffusion group constant." Taking its formal definition out of Eq. (7.12), one obtains at first

$$\text{leakage rate} = -\int_g \frac{d}{dx} D(E) \frac{d}{dx}\phi(x,E)\, dE \quad . \tag{7.38}$$

If one inserts the groupwise separability, one obtains simply [$\varphi(E) = \varphi_g(E)$]:

$$\text{leakage rate} = -\frac{d^2}{dx^2}\phi_g(x) \cdot \int_g D(E)\varphi_g(E)\, dE \quad . \tag{7.39}$$

Equation (7.39) yields, as a definition of the diffusion group constant,

$$D_g = \frac{\int_g D(E)\varphi_g(E)\, dE}{\int_g \varphi_g(E)\, dE} \quad . \tag{7.40}$$

This procedure of a definition of D_g appears to be fairly simple, but one should keep in mind that $\phi(x,E)$ is actually nonseparable. This is particularly true at regional interfaces. Then one should actually weight $D(E)$ with the gradient of $\phi(x,E)$. The separated flux is a poor approximation around interfaces; then the actual gradient can be quite different from the separated value, since differentiation generally increases inaccuracies. Thus, D_g as calculated from groupwise flux separation is likely to be more inaccurate than other group constants.

There is another complication in the definition of the diffusion group constant; it contains a macroscopic cross section in the denominator:

$$D(E) = \frac{1}{3\Sigma_{tr}(E)} = \frac{1}{3\sum_i N_i \sigma^i_{tr}(E)} \quad . \tag{7.41}$$

Equation (7.40) defines a diffusion group constant for the given composition represented by the N_i's in the denominator of Eq. (7.41). It is impossible to correctly break up the resulting diffusion group constant into individual isotopic contributions or components.

If a microscopic transport group constant, σ^i_{trg}, has been calculated, the corresponding macroscopic group constant can be formed by:

$$\Sigma_{trg} = \sum_i N_i \sigma^i_{trg} \quad , \tag{7.42}$$

similar to the previous homogenization process used for flux-weighted group constants. Then,

$$D_g = \frac{1}{3\Sigma_{trg}} \quad . \tag{7.43}$$

The simplest prescription for a calculation of σ_{trg} consists of flux weighting of $\sigma_{tr}(E)$:

$$\sigma_{trg} = \frac{1}{\varphi_g} \int \sigma_{tr}(E)\varphi(E)\, dE \quad . \tag{7.44}$$

A more sophisticated procedure is based on a current weighting:

$$\sigma_{trg} = \frac{1}{J_g} \int_g \sigma_{tr}(E) J(E)\, dE \quad . \tag{7.45}$$

Information on separated currents can be obtained, along with separated fluxes, by solving the P_1 equations.

The detailed discussion of the various calculational procedures for finding a value of σ^i_{trg} is beyond the scope of this text.

7-4 Explicit Few-Group Theories

In the previous sections, the formal definitions of the multigroup equations and constants were given, which defined the multigroup calculational scheme. The treatment of the detailed energy dependence requires a very large number of groups. The numerical solution of the set of multigroup diffusion equations can only be performed on large computers. This is particularly true when the system is multidimensional. To hold down the computation time, a large number of groups are employed with one-dimensional calculations. In two- or three-dimensional treatments, relatively "few" groups are employed. The method of reducing the neutron energy group structure from a large number of groups to a relatively small number is called "group collapsing." If only a fairly approximate solution is desired for a thermal reactor problem, few-group constants can be obtained directly, without an explicit group collapsing procedure.

The procedures employed in group collapsing are described in Sec. 7-5, where the emphasis is on the structure of the lowest level of the few-group methods, namely one- and two-group methods. One- and two-group models provide insight into the workings of higher order multigroup models; a model with a larger number of groups is, in principle, merely an extension of the two-group model. Few-group models are also used in scoping- and survey-type thermal reactor calculations. In fact, in the light water reactor field, many burnup codes and reactor simulator models are successfully structured around two-group models. Fast reactor calculations, because of their more detailed energy treatment requirements, need a larger number of groups (>10 groups) for use in quantitative criticality studies.

In addition to the formal few-group models, the more traditional formulations of one- and two-group models are also presented. These procedures, which are structured around four-factor-type parameters, are in little actual use today. Nevertheless, they are presented for background and completeness.

In this treatment, not much emphasis is placed on the numerical aspects of the solution of these models. Since it is the physics that is of concern, students interested in the numerical mathematical aspects of the solution should consult the specialized textbooks and computer program descriptions that deal extensively with this topic.[1,2] The basic aspects of the numerical approaches are presented in Sec. 7-5.

7-4A One-Group Theory for All Neutrons

If *all* neutrons are collected into a single group by forming the total flux,

$$\phi(\boldsymbol{r}) = \int_0^\infty \phi(\boldsymbol{r},E)\, dE \quad , \tag{7.46a}$$

obviously a "one-group theory" is obtained. In this one-group theory for all neutrons, the thermal, intermediate, and fast ranges are combined in the same "group" interval. This one-group theory was derived in Chapter 2 by integration of the energy-dependent balance equation:

$$-\nabla \cdot D(E)\nabla\phi(\boldsymbol{r},E) + \Sigma_t(E)\phi(\boldsymbol{r},E) - \int_{E'} \Sigma_s(E'\rightarrow E)\phi(\boldsymbol{r},E')\, dE' = \lambda \int_{E'} \nu\Sigma_f(E')\phi(\boldsymbol{r},E')\, dE' \quad . \tag{7.46b}$$

The scattering terms are then canceled, yielding the one-group equation:

$$-\nabla \cdot D\nabla\phi(\boldsymbol{r}) + \Sigma_a\phi(\boldsymbol{r}) = \lambda\nu\Sigma_f\phi(\boldsymbol{r}) \quad . \tag{7.47a}$$

The one-group constants appearing in this equation are averaged over the entire energy range as described in Sec. 7-3. For a separated neutron flux, the following is obtained:

$$D = \frac{\int_0^\infty D(E)\varphi(E)\, dE}{\int_0^\infty \varphi(E)\, dE} \quad ,$$

$$\Sigma_t = \frac{\int_0^\infty \Sigma_t(E)\varphi(E)\, dE}{\int_0^\infty \varphi(E)\, dE} \quad ,$$

$$\nu\Sigma_f = \frac{\int_0^\infty \nu\Sigma_f(E)\varphi(E)\, dE}{\int_0^\infty \varphi(E)\, dE} \quad ,$$

and

$$\Sigma_a = \Sigma_t - \Sigma_s \quad . \tag{7.47b}$$

The energy spectrum, $\varphi(E)$, needed to evaluate the above reactor constants, is obtained by first solving a proper energy-dependent equation. Commensurate with the limited accuracy of the one-group model, even in a multiregion system, only one weighting spectrum is determined, i.e., for a one-region system with an average composition.

Completing this one-group formulation are the boundary and interface conditions as described in Sec. 2-2E.

The one-group theory for *all* neutrons is helpful for qualitative considerations in fast reactors. In thermal reactors, one generally prefers to treat the thermal and epithermal neutrons in separated groups (compare Sec. 7-4B).

The one-group formulation, expressed by Eq. (7.47a), is sometimes called the "one-speed" balance, since it can be obtained by assuming that all the neutrons have the same speed. The difference, however, between the one-group and the one-speed approaches is more than merely a problem of semantics; in the one-group model, the constants are well-defined average values given by Eq. (7.47b). The one-speed model has no such precise formalism.

7-4B One-Group Theory for Thermal Neutrons

The simplest treatment of a thermal reactor problem consists of a one-group theory for just the thermal neutrons. The neutrons above the thermal group (epithermal neutrons) are treated separately, often by a multigroup theory. The thermal neutron group is then the lowest energy group, i.e., group *G*:

$$\phi_{th}(\boldsymbol{r}) = \int_G \phi(\boldsymbol{r},E)\, dE = \phi_G(\boldsymbol{r}) \quad . \tag{7.48}$$

Since no fission neutrons are emitted directly into the thermal group, the balance equation (for regionwise constant cross sections) is given by:

$$-D_{th}\nabla^2\phi_{th}(\boldsymbol{r}) + \Sigma_{a,th}(\boldsymbol{r})\phi_{th}(\boldsymbol{r}) - \int_0^{E_{th}}\int_{E_{th}}^{\infty} \Sigma_s(\boldsymbol{r},E'\rightarrow E)\phi(\boldsymbol{r},E')\, dE'\, dE = 0 \quad . \tag{7.49}$$

Note that in the scattering integral the E' integration is carried out from E_{th} to ∞, since the within-group scattering element, i.e., the integral from 0 to E_{th}, has been canceled in Eq. (7.49) with the scattering term in $\Sigma_t\phi_{th}$. The λ eigenvalue does not appear explicitly in Eq. (7.49); it is implicitly contained in the scattering source.

In multigroup theory, the scattering term in Eq. (7.49) is just calculated from the higher fluxes:

$$S_{sG} = \sum_{g'<G} \Sigma_{sg'G}\phi_{g'} \propto s_0 \quad , \tag{7.50}$$

with s_0 being λ times the fission source.

If one does not want to apply multigroup theory explicitly, one has to find an estimate of this scattering source. This requires an investigation of the scattering of epithermal neutrons into the thermal group as presented below.

The considerations involved in choosing the upper limit of the thermal energy range, E_{th}, which is often taken between 0.4 and 0.65 eV, were presented in Sec. 4-7.

In the range between E_{th} and the 6.7-eV resonance of ^{238}U, there is little absorption. Therefore, recall from Chapter 4, the spectrum between resonances in a separable large system is given by the loss-free spectrum times the two reduction factors for resonance absorption and epithermal leakage, respectively:

$$\varphi(E) = \frac{s_0}{E \cdot \overline{\xi\Sigma_s}} p \cdot P_f \tag{7.51}$$

and

$$p = p(E_{th});\ P_f = P_f(E_{th}) \quad ,$$

where

s_0 = neutron source
$\overline{\xi\Sigma_s}$ = average slowing down power
p = resonance escape probability
P_f = fast nonleakage probability.

In a range of small absorption, the scattering integral, which is equal to the slowing down density, $q(E)$, is given by (see, for example, Sec. 4-7):

$$q(E) = E\varphi(E)\overline{\xi\Sigma_s} \quad . \tag{7.52}$$

Applying Eqs. (7.51) and (7.52) to E_{th} gives:

$$q(E_{th}) = E_{th} \cdot \varphi(E_{th})\overline{\xi\Sigma_s} \quad . \tag{7.53}$$

Inserting Eq. (7.51) gives:

$$q_{th} = q(E_{th}) = s_0 \cdot p \cdot P_f \quad , \tag{7.54}$$

i.e., the number of neutrons scattered into the thermal group equals the number of source neutrons reduced by resonance absorption and fast leakage.

Relating s_0 to the fission source and the eigenvalue λ results in:

$$q_{th} = \lambda \varepsilon \nu \Sigma_{fth} \varphi_{th} p \cdot P_f \quad , \tag{7.55}$$

where ε is the fast fission factor also defined earlier in Chapter 2. Then, by using the approximate assumption that Eq. (7.55), which is derived for the separable large system, holds for all points in space, $\phi_{th}(\boldsymbol{r})$ can be substituted for φ_{th}:

$$q_{th}(\boldsymbol{r}) = \lambda \varepsilon p P_f \nu \Sigma_{fth} \phi_{th}(\boldsymbol{r}) \quad . \tag{7.56}$$

Inserting Eq. (7.56) into Eq. (7.49) gives the balance equation (the indices on the group constants are often omitted in this one-group theory):

$$-D_{th}\nabla^2\phi_{th}(\boldsymbol{r}) + \Sigma_{ath}\phi_{th}(\boldsymbol{r}) - \lambda\nu\Sigma_{fth}\varepsilon p P_f \phi_{th}(\boldsymbol{r}) = 0 \quad . \tag{7.57}$$

Introducing ηf (compare Chapter 2),

$$\frac{\nu\Sigma_{fth}}{\Sigma_{ath}} = \frac{\nu\Sigma_{fth}}{\Sigma_{ath}^{\text{fuel}}} \cdot \frac{\Sigma_{ath}^{\text{fuel}}}{\Sigma_{ath}} = \eta \cdot f \quad , \tag{7.58}$$

brings Eq. (7.57) to the historically used form of the one-group equation for thermal neutrons:

$$-D_{th}\nabla^2\phi_{th}(\boldsymbol{r}) + (1 - \lambda k_\infty P_f)\Sigma_{ath}\phi_{th}(\boldsymbol{r}) = 0 \tag{7.59a}$$

or

$$-\nabla^2\phi_{th}(\boldsymbol{r}) + \frac{(1 - \lambda k_\infty P_f)}{L_{th}^2}\phi_{th}(\boldsymbol{r}) = 0 \quad , \tag{7.59b}$$

with

$$L_{th}^2 = \frac{D_{th}}{\Sigma_{ath}} \quad . \tag{7.60}$$

The factor in the second term is B_λ^2 (the λ-modified material buckling defined in Sec. 3-5A). The equation is then of the familiar form of the Helmholtz equation:

$$\nabla^2\phi(\boldsymbol{r}) + B_\lambda^2\phi(\boldsymbol{r}) = 0 \quad . \tag{7.61}$$

For solutions of this equation, see Chapters 2 and 3.

7-4C Two-Group Theory

The general multigroup theory gives the balance relationships for the two-group theory in multigroup notation (see also Sec. 2-2D):

$$-D_1\nabla^2\phi_1+\Sigma_{r1}\phi_1-\lambda(\nu\Sigma_{f1}\phi_1+\nu\Sigma_{f2}\phi_2)=0 \tag{7.62a}$$

and

$$-D_2\nabla^2\phi_2+\Sigma_{a2}\phi_2-\Sigma_{12}\phi_1=0 \quad , \tag{7.62b}$$

with

$$\Sigma_{r1}=\Sigma_{a1}+\Sigma_{12} \quad , \tag{7.63a}$$

$$\Sigma_{r2}=\Sigma_{a2} \text{ (no removal through scattering)} \quad , \tag{7.63b}$$

and

$$\left.\begin{matrix}\chi_1=1\\ \chi_2=0\end{matrix}\right\} \text{ all fission neutrons appear in the first group.} \tag{7.64}$$

Consistent with standard designations, the subscript 1 denotes the fast group whereas the 2 refers to the thermal energy group. With this symbolism and these approximations, Eq. (7.62a) is the balance relationship for the fast group with Eq. (7.62b) providing the description of the thermal group. The energy ranges for the two groups are again determined by the considerations presented in Sec. 4-7 for E_{th}. The groups fluxes are then defined as:

$$\phi_1(\boldsymbol{r})=\int_{E_{th}}^{\infty}\phi(\boldsymbol{r},E)\,dE \tag{7.65a}$$

and

$$\phi_2(\boldsymbol{r})=\int_{0}^{E_{th}}\phi(\boldsymbol{r},E)\,dE \quad . \tag{7.65b}$$

The group constant sets for the two groups are integrated average values over their respective energy ranges, as prescribed by the formulas given in Sec. 7-3.

In all computer programs used for two-group calculations in practical applications, the two-group equations are used as expressed by Eqs. (7.62). The group constants are found by weighting with detailed spectra. For example, the HAMMER program,[3] which is applied in reactor calculations, employs 38 fine groups in the thermal range and 54 groups in the epithermal range. Again, finite difference techniques are employed to solve the space-dependent system of equations, Eqs. (7.62), in an actual reactor situation. Continuity relationships and boundary conditions must be satisfied in the solution procedure (see Sec. 7-5 for a general finite difference solution approach).

The two-group model, Eqs. (7.62), is routinely solved with present-day computers. Historically, however, before the advent of high-speed computers, various analytical two-group formulations were developed, based on the classical "lumped" reactor parameters. A typical model was approximated in the following way.

The relatively small contribution of fast fissions is combined with the thermal fissions by introducing the "fast fission" factor, ε:

$$\nu\Sigma_{f1}\phi_1 + \nu\Sigma_{f2}\phi_2 = \left(\frac{\nu\Sigma_{f1}\phi_1}{\nu\Sigma_{f2}\phi_2} + 1\right)\nu\Sigma_{f2}\phi_2 = \varepsilon''\nu\Sigma_{f2}\phi_2 \quad . \tag{7.66}$$

The parentheses define the fast fission factor, ε'', which is somewhat different from the previously defined factors ε and ε' (see Sec. 2-5). The $\phi_1(\boldsymbol{r})$ and $\phi_2(\boldsymbol{r})$ are assumed proportional to each other, so that ε may become independent of $\boldsymbol{r}$.

The transfer group constant, Σ_{12}, is evaluated based on its definition, Eq. (7.35). Since most of the transfer of neutrons into the thermal group comes from a small epithermal range above E_{th}, $p(E)$, which is a factor in the slowing down spectrum, can be taken out of the integral in the numerator with its value at E_{th}. Further, $p(E)$ has minimal impact on the denominator integral. Applying both simplifications allows the expression of Σ_{12} in terms of the loss-free spectrum $\varphi_0(E)$:

$$\Sigma_{12} \simeq p \cdot \frac{\displaystyle\int_0^{E_{th}}\int_{E_{th}}^{E_0} \Sigma_s(E'\rightarrow E)\varphi_0(E')\, dE'\, dE}{\displaystyle\int_{E_{th}}^{E_0} \varphi_0(E')\, dE'} \quad . \tag{7.67}$$

Equation (7.67) is often further simplified by approximating $\varphi_0(E)$ with a $1/E$ spectrum:

$$\Sigma_{12} = p \cdot \frac{\displaystyle\int_0^{E_{th}}\int_{E_{th}}^{E_0} \Sigma_s(E'\rightarrow E)\, \frac{dE'}{E'}\, dE}{\displaystyle\int_{E_{th}}^{E_0} \frac{dE'}{E'}}$$

$$= \frac{p}{u_{th}}\int_{u_{th}}^{\infty}\int_0^{u_{th}} \Sigma_s(u'\rightarrow u)\, du'\, du \quad , \tag{7.68}$$

with u_{th} being the lethargy at E_{th}. Thus the transfer group cross section is approximated as:

$$\Sigma_{12} = p \cdot \Sigma_{12}^0 \quad . \tag{7.69}$$

The physical interpretation of Σ_{12}^0 is the transfer group constant, calculated with the loss-free slowing down spectrum; Σ_{12}^0 is often denoted by Σ_L.

Based on this interpretation of the transfer group constant, one can say that in an infinite medium

$$\Sigma_{12}^0 \phi_1 \simeq s_0 \tag{7.70}$$

and

$$\Sigma_{12}\phi_1 \simeq p \cdot s_0 \quad , \tag{7.71a}$$

i.e., $p \cdot s_0$ neutrons are transferred into the thermal group. The transfer reaction rate in Eq. (7.62b) then becomes:

$$\Sigma_{12}\phi_1 = p \cdot \Sigma_{12}^0 \phi_1 \quad . \tag{7.71b}$$

The absorption of neutrons is estimated as:

$$\Sigma_{a1}\phi_1 \simeq (1-p)s_0 \simeq (1-p)\Sigma_{12}^0\phi_1 \quad . \tag{7.72}$$

For the group removal rate out of group 1, this gives:

$$\Sigma_{r1}\phi_1 = \Sigma_{a1}\phi_1 + \Sigma_{12}\phi_1 \simeq (1-p)\Sigma_{12}^0\phi_1 + p\Sigma_{12}^0\phi_1 = \Sigma_{12}^0\phi_1 \quad . \tag{7.73}$$

Furthermore, the fission source is often written in the following expanded fashion[4]:

$$\varepsilon\nu\Sigma_{f2} = \varepsilon\frac{\nu\Sigma_{f2}}{\Sigma_{a2}^{\text{fuel}}} \cdot \Sigma_{a2}^{\text{fuel}} = \varepsilon\eta_2\Sigma_{a2}^{\text{fuel}} \quad , \tag{7.74}$$

or it was further expanded as[5]:

$$\varepsilon\nu\Sigma_{f2} = \varepsilon\eta_2\Sigma_{a2}^{\text{fuel}} = \varepsilon\eta_2\frac{\Sigma_{a2}^{\text{fuel}}}{\Sigma_{a2}}\Sigma_{a2} = \varepsilon\eta_2 f_2\Sigma_{a2} = \frac{k_\infty}{p}\Sigma_{a2} \quad . \tag{7.75}$$

Inserting Eqs. (7.71b), (7.73), and (7.75) into Eqs. (7.62) brings the two-group equations into the approximate form:

$$-D_1\nabla^2\phi_1 + \Sigma_{12}^0\phi_1 - \lambda\frac{k_\infty}{p}\Sigma_{a2}\phi_2 = 0$$

and

$$-D_2\nabla^2\phi_2 + \Sigma_{a2}\phi_2 - p\Sigma_{12}^0\phi_1 = 0 \quad . \tag{7.76}$$

The modified form of the two-group equations is difficult to interpret.

One must go back through the derivation in order to realize the precise physical nature of some of the terms in Eq. (7.76). One now applies mostly the standard form of the two-group equations given by Eq. (2.48). The analytical solution procedure, presented in Sec. 2-7, is repeated here to include the historical concepts.

This set of homogeneous differential equations [Eq. (7.76)] for the two-group neutron model describing a bare core through flux separation leads to the wave equation for the spatial factor

$$\nabla^2\phi(\boldsymbol{r}) + B^2\phi(\boldsymbol{r}) = 0 \quad , \tag{7.77}$$

and the following two-group equations for the energy dependence:

$$-(D_1B^2 + \Sigma_{12}^0)\varphi_1 + \lambda\left(\frac{k_\infty}{p}\Sigma_{a2}\right)\varphi_2 = 0$$

and

$$(p\Sigma_{12}^0)\varphi_1 - (D_2B^2 + \Sigma_{a2})\varphi_2 = 0 \quad . \tag{7.78}$$

The general results of Sec. 2-6 as well as the procedure of Sec. 2-7 apply directly to the solutions of Eqs. (7.78). If $\lambda = 1$, Eqs. (7.78) represent an eigenvalue problem for $B^2 = B_m^2$, the material buckling. For a given geometry, B^2 has to be determined from Eqs. (7.77) and the boundary conditions; then B_{geo}^2 is obtained. Inserting this in Eqs. (7.78) requires one to retain λ and solve the resulting equations as an eigenvalue problem in λ.

In the evaluation of this set of simultaneous equations for the nontrivial solution, Cramer's Rule is invoked, which requires the determinant of the coefficient matrix to be zero. Applying this condition to Eq. (7.78), with $\lambda = 1$, results in the two-group equation for the material buckling, given here in its historical form:

$$(D_1B^2 + \Sigma_{12}^0)\,(D_2B^2 + \Sigma_{a2}) - k_\infty\Sigma_{a2}\Sigma_{12}^0 = 0 \quad . \tag{7.79}$$

Solving Eq. (7.79) gives two values for B^2; the larger one is the material buckling B_m^2.

The criticality expression, Eq. (7.79), can be rearranged into a more compact form by dividing through by $\Sigma_{a2}\Sigma_{12}^0$ yielding:

$$\left(\frac{D_1B_m^2}{\Sigma_{12}^0} + 1\right)\left(\frac{D_2B_m^2}{\Sigma_{a2}} + 1\right) = k_\infty \quad . \tag{7.80}$$

By recognizing that D_2/Σ_{a2} is the square of the diffusion length, L_2^2, for thermal neutrons and that D_1/Σ_{12}^0 is a type of pseudo-diffusion length, L_1^2, for fast neutrons, Eq. (7.80) becomes:

$$1 = \frac{k_\infty}{(1 + L_1^2 B_m^2)(1 + L_2^2 B_m^2)} \quad . \tag{7.81}$$

In the context of this derivation where Σ_{12}^0 is a transfer or slowing down cross section, L_1^2 can be interpreted as equal to τ or the Fermi age parameter. Thus, L_1^2 could be replaced by τ in Eq. (7.81).

It should be noted again that Eqs. (7.79), (7.80), and (7.81) are equations for the material buckling, from which the critical size can be inferred.

Retaining the eigenvalue λ $(=\frac{1}{k})$ in the balance equation, Eq. (7.78), and assuming B^2 to be known $(B^2 = B_{\text{geo}}^2)$ yields instead of Eq. (7.81) the respective equation for λ or $k = k_{\text{eff}}$:

$$k = k_{\text{eff}} = \frac{k_\infty}{(1 + L_1^2 B^2)(1 + L_2^2 B^2)} \quad . \tag{7.82}$$

Other analytical two-group models exist that attempt to describe the slowing down and absorption processes more accurately. One of these is referred to as the "modified two-group theory." It is claimed[6] that this model, which is modified to include the effects of resonance processes, is equivalent to a three-group model. In this approach, although only two flux groups are employed, the fast group source term is redefined to include resonance capture and epithermal fissions. With the exception of this source term, the two-group flux balance equations are quite similar to those presented previously for the standard two-group model [see Eq. (7.78)]. The form of the criticality expression is also identical, with the exception that a modified age, τ', which incorporates the resonance absorption processes, is employed.

7-4D Two-Group Core Reflector Model

The formal set of two-group equations, Eqs. (7.62), presented in the previous section is quite general and can be solved both for bare and multiregion problems using standard finite difference techniques. For a reactor that includes a reflector, the dissimilar core reflector regions would automatically be considered in the region-dependent group constant set in which the reflector region material compositions would not contain any fuel material. The numerical solution then for the discrete two-group fluxes would yield the desired spatial flux distributions as well as the critical mass. While this method and solution using a computer is the present-day "modus operandi" in the reactor design field, it does not provide insight into the physics of a reflected reactor system. Therefore, in this section, the analytical two-group model of a core reflector system is presented. For comparison, the one-group core reflector problem discussion in Chapter 3 should be reviewed (see Sec. 3-5).

The analytical two-group core reflector model is derived from neutron balances for the thermal and fast flux written over both the core and reflector control volumes (see Fig. 7-2). For the core region, the two-group flux relationships are those derived in the previous section for the bare core, namely, Eq. (7.77):

$$-D_{1,c}\nabla^2\phi_{1,c} + \Sigma^0_{12,c}\phi_{1,c} - \lambda\frac{k_\infty}{p_c}\Sigma_{a2,c}\phi_{2,c} = 0$$

and

$$-p_c\Sigma^0_{12,c}\phi_{1,c} - D_{2,c}\nabla^2\phi_{2,c} + \Sigma_{a2,c}\phi_{2,c} = 0 \quad , \tag{7.83}$$

where the subscript c has been added to denote the core region. In the reflector region, the steady-state diffusion balance relationships for the fast and thermal fluxes are merely:

$$-D_{1,r}\nabla^2\phi_{1,r} + \Sigma_{12,r}\phi_{1,r} = 0$$

and

$$-D_{2,r}\nabla^2\phi_{2,r} + \Sigma_{a2,r}\phi_{2,r} - \Sigma_{12,r}\phi_{1,r} = 0 \quad , \tag{7.84}$$

where again the subscript 1 denotes the fast flux, 2 the thermal flux, and r the reflector region. The transfer group cross section, $\Sigma_{12,r}$, in the fast flux balance equation is defined similar to that for the core, for example, Eq. (7.83). However, since there are no heavy elements in the

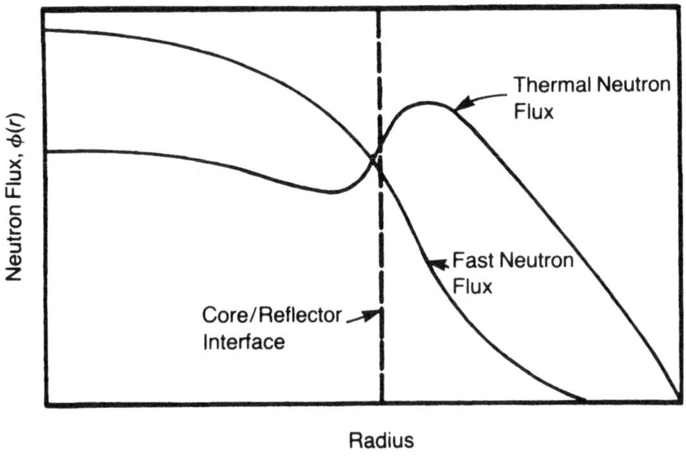

Fig. 7-2. Typical two-group neutron flux distributions in a core reflector geometry.

reflector, there is no correction required for resonance absorption as was the case in the core. Note also that in the fast group there is no fission source term as the reflector does not contain any fissionable materials. In the thermal flux balance expression, $\Sigma_{a2,r}$ is a true absorption cross section. These constants along with the other parameters appearing in Eqs. (7.83) and (7.84) are again integrated average quantities over their respective energy ranges.

The method of solution for the set of four equations posed by this core reflector problem is not described here. Suffice it to say that boundary and interface conditions are required for each of the two-group fluxes. In addition, the standard requirements that the fluxes be finite and nonnegative are imposed. Analytical expressions for two-group fluxes for the common one-dimensional symmetric reactor geometries (e.g., slab, sphere, cylinder) similar to those studied in Chapter 3 are available for this two-region problem. Furthermore, a trial-and-error method is applied for determining the core's required critical radius (or critical mass for a given reflector thickness). These procedures are rather involved and require a considerable amount of algebra; therefore, they are not presented here. The interested reader should consult texts in which two-group theory is presented in more detail, for example, Refs. 4, 5, and 7, for the criticality condition as well as the flux distribution.

As mentioned in Chapter 3, the one-group diffusion model cannot describe one of the important physical features of a reflector surrounding a thermal core region. This effect is the rather marked peaking of the thermal flux in the reflector region near the core reflector interface as illustrated in Fig. 7-2. There are two reasons for this increase in the thermal flux: There are more neutrons thermalized in the reflector, and considerably fewer neutrons are absorbed due to the absence of fuel in this region. This effect also has a beneficial influence on the thermal flux distribution in the core region for, as Fig. 7-2 indicates, it tends to flatten the thermal flux in the vicinity of the reflector. Therefore, the presence of a reflector improves the important flux peaking situation in the core as well as reduces the critical mass of the reactor.

7-5 The Multigroup Model

The formal definitions of the multigroup model were presented in Sec. 7-2C. In this section, this set of equations is examined in more detail primarily from the standpoint of their application in solving actual reactor problems. Since the numerical techniques used in their solution are quite involved, only the general approach is presented.

As outlined in Sec. 7-2C, the basis for the multigroup formalism is the subdivision of the total energy spectrum into discrete segments.

Within each energy segment, ΔE, or group g, it is assumed that the group constants are available, obtained as average values using the flux spectrum appropriate to that energy group. In the case of thermal water reactors, normally 3 or 4 groups are all that are required to obtain a realistic neutronic description. Usually the first several groups are used to describe the slowing down and resonance effects with the last group reserved for the thermal group. Gas-cooled thermal reactor studies sometimes employ more energy groups (e.g., 7 to 9) with typically finer detail in the thermalization region (e.g., 4 thermal groups). For fast reactors, the number of groups must necessarily be increased to account for the more detailed absorption processes occurring over the large energy range. To treat these effects in fast reactors, a minimum of 10 to 30 groups is normally required in reactor physics analyses.

7-5A Multigroup Equations

Reactor configurations are generally approximated by discrete zones or regions within which materials are treated as homogeneous mixtures of various elements with given amounts and compositions. In the case where a given k_{eff} is sought by searching for the required fuel concentration (fuel search) or control rod position (poison search), several calculations with given compositions are performed to obtain the desired result. Also, the regional structure is sometimes varied in a search for a flat power distribution.

The multigroup diffusion equation, Eq. (7.13), for a space point $\boldsymbol{r}$ written for energy group g for this reactor is then

$$-\nabla\cdot D_g(\boldsymbol{r})\nabla\phi_g(\boldsymbol{r})+\Sigma_{rg}(\boldsymbol{r})\phi_g(\boldsymbol{r})-\sum_{g'\neq g}\Sigma_{sg'g}(\boldsymbol{r})\phi_{g'}(\boldsymbol{r})$$

$$=\lambda\chi_g\sum_{g'}\nu\Sigma_{fg'}(\boldsymbol{r})\phi_{g'}(\boldsymbol{r})\quad (g=1,2,3,\ldots,G)\quad . \tag{7.85}$$

The eigenvalue ($\lambda = 1/k$) again is present to allow for a nontrivial solution of a homogeneous equation with given coefficients and boundary conditions.

In Eq. (7.85), the within-group scattering has been eliminated by suitable definition of the removal cross section, $\Sigma_{rg}(\boldsymbol{r})$, as discussed in Sec. 7-3E, i.e.:

$$\Sigma_{rg}(\boldsymbol{r})=\Sigma_{ag}(\boldsymbol{r})+\sum_{g'\neq g}\Sigma_{sgg'}\quad . \tag{7.86}$$

Of course, the absorption cross section, $\Sigma_{ag}(\boldsymbol{r})$, in Eq. (7.86) includes both the capture and fission cross section,

$$\Sigma_{ag}(\boldsymbol{r})=\Sigma_{cg}(\boldsymbol{r})+\Sigma_{fg}(\boldsymbol{r})\quad . \tag{7.87}$$

The parameters of group constants in Eq. (7.85) are defined and evaluated over each energy group and each region (indicated as $\boldsymbol{r}$ dependencies), as discussed in Sec. 7-3:

$$D_g(\boldsymbol{r}) = 1/3\Sigma_{trg}(\boldsymbol{r}) \quad ,$$

$$\Sigma_{ag}(\boldsymbol{r}) = \int_g \Sigma_a(\boldsymbol{r},E)\varphi_g(E)\,dE \quad ,$$

$$\chi_g = \int_g \chi(E)\,dE \quad ,$$

$$\nu\Sigma_{fg}(\boldsymbol{r}) = \int_g \nu\Sigma_f(\boldsymbol{r},E)\varphi_g(E)\,dE \quad ,$$

and

$$\Sigma_{sg'g}(\boldsymbol{r}) = \int_g\int_{g'} \Sigma_s(\boldsymbol{r},E'\rightarrow E)\varphi_g(E')\,dE'\,dE \quad . \tag{7.88}$$

In the above definitions of the group parameters, the flux spectrum for each group has been normalized, i.e.,

$$\int_g \varphi_g(E)\,dE = 1 \quad (g = 1,2,\ldots,G) \quad ; \tag{7.89}$$

thus, the denominators lacking in the expressions in Eq. (7.88) are unity.

Similar to the one- and two-group models, boundary conditions are required to solve this system of second-order differential equations. In addition, interface relationships are necessary to couple the equations between dissimilar regions (i.e., regions of differing material compositions). In the radial direction of a cylindrical system, one of the two boundary conditions is

$$\nabla\phi_g(r=0) = 0 \quad (g = 1,2,\ldots,G) \quad , \tag{7.90a}$$

which physically excludes singular contributions at $r = 0$ and accounts for the symmetry of the flux $\phi_g(r)$ at $r = 0$. The outer boundary, at $r = R$, is in general form expressed as:

$$A_1\boldsymbol{n}\cdot\nabla\phi_g(R) + A_2\phi_g(R) = A_3 \quad (g = 1,2,\ldots,G) \quad , \tag{7.90b}$$

where $\boldsymbol{n}$ is the unit normal to the external surface. This external boundary condition is the general version and incorporates the descriptions of various reactor situations. For example:

1. When $A_1 = 0, A_2 = 1$, and $A_3 = 0$, the condition that the flux $\phi_g(r)$ vanish at $r = R$ is imposed.
2. When $A_1 = 0.71\ \lambda_{trg}$ (λ_{trg} is the transport mean-free-path length in group g), $A_2 = 1$ and $A_3 = 0$, the linear extrapolation distance of $0.71\ \lambda_{trg}$ is obtained at the outer reactor boundary.
3. When $A_1 = 0$, $A_2 = 1$, and $A_3 = C_g$, the inhomogeneous external boundary condition is obtained.

There are, of course, other boundary conditions that are possible, but the three presented are the most common.

The interface conditions for each group are imposed to guarantee continuity of flux and current across adjacent dissimilar regions, j and $j+1$. These two requirements are for an interface located at $r = R_j$:

$$\phi_g(R_j^-) = \phi_g(R_j^+) \quad (g = 1,2,...,G) \tag{7.90c}$$

and

$$D_g^j \nabla\phi_g(R_j^-)\cdot \boldsymbol{n} = D^{j+1}\nabla\phi_g(R_j^+)\cdot \boldsymbol{n} \quad (g = 1,2,...,G) \quad , \tag{7.90d}$$

where $\boldsymbol{n}$ is again the unit vector normal to the interface.

If there are N regions and G energy groups, there are $2\cdot N\cdot G$ boundary and interface conditions for the $N\cdot G$ number of second-order differential equations.

In the next sections, a brief sketch of the approach and techniques employed in the solution of this "multigroup" equation set is presented. The presentation is general; for specifics of the numerical details several fine texts[1,2] are recommended. In addition, the many diffusion theory reactor computer code manuals are also excellent sources of information concerning the numerical solution of the multigroup diffusion equations.[8–10]

7-5B Overall Solution Approach

For reactor design application in the nuclear industry, a solution to the multigroup equation is generally obtained through the use of numerical techniques. Analytical methods relying on expansion functions (for example, Fourier transformations[11]) have not received much wide scale acceptance by practicing reactor designers.[a] In general, the numerical approach and, in particular, finite difference methods have gained favor because of their compatibility with the performance characteristics

[a]The reasons are quite similar to those for the preference of the numerical neutron transport S_n method to the P_n expansion method in Chapter 6.

of high-speed computers. The flexibility afforded in the assignment of both the number and location of finite difference mesh points provides the designer with the capability to describe the rather complicated reactor geometries in several dimensions. In addition, with the finite difference method, by the simple increase in the number of mesh points, a finer neutron flux definition or increased problem accuracy can be obtained. With the analytical methods, it is generally more difficult to obtain increased accuracy since additional terms must be provided to the series expansion of the flux.

The role of analytical solutions in the solution of the multigroup diffusion equation, then, is more of a supportive one. Simplified analytical methods are sometimes employed to obtain initial flux guesses for the numerical method. Also, in some numerical computer programs, analytical expansion methods are sometimes employed in conjunction with finite difference techniques. An example of this type of application would be in a numerical code containing large mesh spacing. To optimize the computer running time, the intermediate flux values between mesh points might be obtained using analytical flux expansions rather than subdividing the problem into a finer mesh structure.

The finite difference approximation to the multigroup equation is the accepted numerical solution technique and is in widespread use within the nuclear industry, hence, it is discussed in this book. It is not the intent, however, to provide a complete treatment of this method of solution (for details the reader should consult Refs. 1 and 2) nor is it implied that this finite difference technique is the only numerical approach (see Sec. 7-6 for the "nodal" method, which is now also in widespread use). Recently, the so-called "finite element" method has received analysis attention. There are indications that this method of solution of the multigroup equation set has the potential to reduce considerably the computer running time without sacrificing accuracy relative to the "finite difference" approach. The method, however, has only rather recently been applied to the solution of the neutron diffusion problem and, therefore, there does not exist the wealth of testing and application experience that is available with the finite difference method. Future acceptance of the finite element method as the standard technique in actual reactor calculations awaits development of production-type reactor computer codes. However, for a discussion of the general finite element technique, Refs. 12, 13, and 14 are recommended.

The overall solution to the multigroup diffusion equation in the finite difference scheme is conceptually very simple. The continuous variation of the energy and space dependency is made discrete. Recall that this is first accomplished by transformation of the energy variable into a discrete form by use of the energy structure, i.e.:

$$\Delta E_g = E_{g-1} - E_g \quad . \tag{7.91}$$

Upon substitution of this energy formulation into the diffusion equation, the multigroup approximation given by Eq. (7.85) results. The discretization of the space variable is accomplished by superimposing a mesh structure over the reactor volume in question and then transforming the continuous space variables into discrete values by use of their finite difference analogs. What results from these transformations are essentially two types of problems that must be solved: First, an inhomogeneous set of algebraic equations for the local discrete fluxes and then an eigenvalue problem for the criticality factor. In general, iterative methods are used for both solutions. However, the matrix techniques differ because of the radically different nature of the two problems. The iterative solution of the pointwise fluxes in space for each group are usually referred to as the "inner" iteration, whereas the search for criticality requires so-called "outer" iterations. These solution approaches are discussed in more detail in the next three sections.

7-5C Finite Difference Approximation

As mentioned in the previous section, the finite difference method consists of overlaying the reactor volume with a mesh structure. This results in the volume being subdivided into a series of cells or regions whose boundaries or extremities are defined by constant values of one of the system's coordinates. For example, in the case of *x-y* geometry, the boundaries would be either the *x* or *y* planes, whereas in *r-z* geometry, they could be either the radial coordinate *r* or the axial coordinate *z*. The mesh structure representing discrete points at which the neutron flux is calculated is usually placed at the outer edges of the cells. In some models, the mesh fluxes are centerpoint values within each interval. However, in all approaches, the group parameters in the diffusion equation do not vary within a region. These parameters may, of course, vary from region to region across the identified boundaries due to differing material compositions or conditions (e.g., temperature).

To provide an example of the finite difference technique for simplicity, consider a one-dimensional slab reactor with a mesh structure as illustrated in Fig. 7-3. The nodal mesh spacing is not necessarily uniform (i.e., spacing may vary between points). Usually, to improve the calculational accuracy, the spacings are selected closer together in regions where the neutron flux is changing rapidly as a function of space (e.g., near dissimilar region boundaries). However, the mesh points are se-

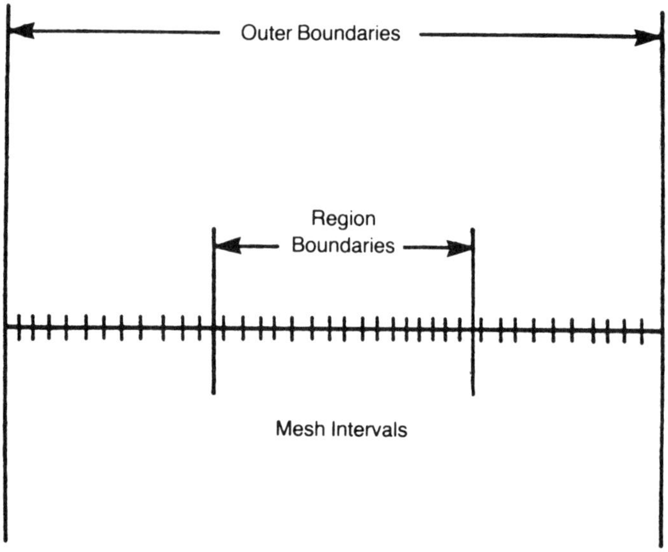

Fig. 7-3. Illustration of regional and mesh structure in a slab reactor problem. Most programs allow for a different mesh size in different regions.

lected in such a way that flux or mesh points are located at the interfaces between regions. In general, it is assumed that the same mesh structure exists for each neutron energy group. There are typically between 20 to 100 mesh intervals per dimension. Thus, two-dimensional problems may use 400 to well over 1000 mesh intervals; three-dimensional problems considerably more. For each mesh, there are the multigroup fluxes. A 20-group problem for a 2000-mesh interval structure employs 40 000 group flux values for a representation of $\phi(\mathbf{r},E)$.

The basic starting point in the finite difference approach is the previously developed group-dependent diffusion equation given by Eq. (7.85). It makes the following derivation easier to follow, however, if all source terms are combined:

$$-\nabla \cdot D_g \nabla \phi_g(x) + \Sigma_{rg}\phi_g(x) = S_{tg}(x) \quad , \tag{7.92a}$$

$$S_{tg}(x) = \sum_{g' \neq g} \Sigma_{sg'g}\phi_{g'}(x) + \lambda \chi_g S_{fg}(x) \quad , \tag{7.92b}$$

and

$$S_{fg}(x) = \sum_{g'} \nu \Sigma_{fg'}\phi_{g'}(x) \quad , \tag{7.92c}$$

where $S_{tg}(x)$ describes the group-dependent fission and scattering term, neglecting the presence of any external neutron source. The other terms are as previously defined.

Considering the one-dimensional form of the neutron leakage operator, $\nabla \cdot D_g \nabla \phi_g(x)$, for a point centered between x_m and x_{m+1}, the differential can be expressed by its finite difference analog (group indices are suppressed):

$$D \left. \frac{d\phi(x)}{dx} \right|_{x=x_m} \simeq \overline{D}_m \left(\frac{\phi_{m+1} - \phi_m}{\Delta x_m} \right) , \tag{7.93a}$$

where ϕ_m is $\phi(x_m)$ and the group subscript has been deleted for convenience. Also, in Eq.(7.93a), the spacing and diffusion coefficient are defined as:

$$\Delta x_m = x_{m+1} - x_m \tag{7.93b}$$

and

$$\overline{D}_m = \frac{D(x_m) + D(x_{m+1})}{2} . \tag{7.93c}$$

Applying this process of differencing to the entire operator, the following is obtained:

$$\frac{d}{dx} D \left. \frac{d\phi}{dx} \right|_{x=x_m} \simeq \left[\overline{D}_m \left(\frac{\phi_{m+1} - \phi_m}{\Delta x_m} \right) - \overline{D}_{m-1} \left(\frac{\phi_m - \phi_{m-1}}{\Delta x_{m-1}} \right) \right] \frac{2}{(\Delta x_m + \Delta x_{m-1})} . \tag{7.93d}$$

This expression for the leakage operator can be rewritten in a more compact form as:

$$\frac{d}{dx} D \left. \frac{d\phi}{dx} \right|_{x=x_m} \simeq a_m^+ \phi_{m+1} - a_m \phi_m + a_m^- \phi_{m-1} , \tag{7.93e}$$

where the following coefficients have been used:

$$a_m = \left(\frac{\overline{D}_m}{\Delta x_m} + \frac{\overline{D}_{m-1}}{\Delta x_{m-1}} \right) \frac{2}{(\Delta x_m + \Delta x_{m-1})} ,$$

$$a_m^- = \left(\frac{\overline{D}_{m-1}}{\Delta x_{m-1}} \right) \frac{2}{(\Delta x_m + \Delta x_{m-1})} ,$$

and

$$a_m^+ = \left(\frac{\overline{D}_m}{\Delta x_m}\right)\frac{2}{(\Delta x_m + \Delta x_{m-1})} \quad . \tag{7.93f}$$

Equation (7.93e) is the difference analog of the leakage term and is generally known as the "three-point difference formula"; it relates the flux at a given mesh point to its two neighbors. Other difference forms for the two- and three-dimensional Laplacian operator are called the "five-" and "seven-point" difference approximations, respectively.

Upon substituting Eq. (7.93f) into Eq. (7.92a), the desired difference expression is obtained ($\Sigma_m = \Sigma_{rm}$):

$$-a_m^+\phi_{m+1} - a_m^-\phi_{m-1} + (a_m + \Sigma_m)\phi_m = S_m \quad , \tag{7.94}$$

which represents a set of linear inhomogeneous equations in the unknown group fluxes. The number of algebraic equations for each group is equal to the number of inner mesh points and in order to deal with them, matrix techniques are required. In matrix notation for the g'th energy group with M number of mesh points, Eq. (7.94) becomes

$$\mathbf{A}_g\Phi_g = S_{tg} \quad , \tag{7.95a}$$

where the flux vector Φ_g and source vector S_{tg} are column vectors:

$$\Phi_g = \begin{pmatrix} \phi_1 \\ \phi_2 \\ \cdot \\ \cdot \\ \cdot \\ \phi_{M-1} \end{pmatrix} , \; S_g = \begin{pmatrix} S_{t1} \\ S_{t2} \\ \cdot \\ \cdot \\ \cdot \\ S_{tM-1} \end{pmatrix} \quad . \tag{7.95b}$$

Zero flux boundary conditions are applied for simplicity. Thus, $\phi_0 = \phi_M = 0$. The diffusion operator has the form of a band matrix defined as ($\Sigma_m = \Sigma_{rgm}$):

$$\mathbf{A}_g = \begin{pmatrix} -a_1^- & (a_1+\Sigma_1) & -a_1^+ & 0 & \ldots \\ 0 & -a_2^- & (a_2+\Sigma_2) & -a_2^+ & \ldots \\ 0 & 0 & -a_3^- & \cdot & \ldots \\ \cdot & \cdot & \cdot & \cdot & \ldots \\ \cdot & \cdot & \cdot & \cdot & \ldots \\ \cdot & \cdot & \cdot & \cdot & \ldots \end{pmatrix} \tag{7.95c}$$

To complete the formulation of Eq. (7.95), the boundary conditions are supplied for the mesh points located at the extremities of the regions (see Sec. 7-5A). The solution of Eqs. (7.95) is presented in Sec. 7-5D.

The approach just presented represents one way of obtaining the finite difference equation [i.e., Eq. (7.94)], but it is not unique. Other methods representing Eq. (7.92) by algebraic equations include the use of variational methods or a straight integration approach. In this latter method, one merely integrates Eq. (7.92) over the interval corresponding to mesh point x_m. The limits of this integration would be from $(x_m - \Delta x_{m-1}/2)$ to $(x_m + \Delta x_m/2)$ with the resulting derivative replaced by the finite difference approximation as given by Eq. (7.93a). This approach would yield a somewhat different set of coefficients in the difference equation although the calculational accuracy would be comparable.

Extensions of these differencing methods to two- and three-dimensional problems, although resulting in more complex expressions, follows almost directly from this one-dimensional example. In two dimensions, the mesh structure can be written for x-y, r-z, or $r - \vartheta$ geometries. Additional specialized mesh structures are also possible; an example being the 60-deg triangular mesh utilized in describing the hexagonal subassemblies in liquid-metal fast breeder reactors and high-temperature gas-cooled reactors.[15]

7-5D Solution of Multigroup Equations

The solution of the multigroup equation, Eq. (7.92), consists essentially of two separate and distinct operations. First a solution to the finite difference matrix expression [i.e., Eq. (7.95)] for the neutron flux values at each mesh point within each energy group is sought. For the one-dimensional model, in theory at least, a solution to Eq. (7.95) can be obtained by simply taking the inverse of the $\mathbf{A}_g$ matrix,

$$\Phi_g = \mathbf{A}_g^{-1} S_g \quad , \tag{7.96}$$

using for example direct methods such as Gaussian elimination. In practice, however, particularly in two- and three-dimensional problems, the size of this $\mathbf{A}_g$ matrix makes this inversion process impractical. Iteration schemes are generally used (e.g., successive overrelaxation) that have been called almost universally in the nuclear reactor design area "inner iterations." Elaborate numerical techniques are employed (for example, the Tschebycheff polynomial method) in some computer programs in an attempt to improve the rate of convergence. The details of these approaches can be found in Refs. 1 and 2.

The second part of the solution of the multigroup model is to solve the eigenvalue problem for the criticality factor, k, with the converged within-group spatial fluxes. This calculation also involves iterative methods where new neutron source distributions are determined based on past values. Because each of these source iterations usually requires several prior inner iterations for convergence, they are called "outer iterations." As one might expect, because of the basic difference in the nature of these two problems, the iterative techniques employed also differ significantly.

7-5E Outer Iteration

To start the calculational procedure, an initial guess is made for the spatial fission source in the group-dependent diffusion equation, Eq. (7.92):

$$\chi_g S_f(x) = \chi_g S_f^{(0)}(x) \quad , \tag{7.97a}$$

where the superscript is the iteration index. A value for the eigenvalue or criticality factor k is also assumed:

$$\lambda = \frac{1}{k} = \frac{1}{k^{(0)}} \quad , \tag{7.97b}$$

and the group fluxes can then be calculated from the diffusion equation using an iterative technique.

The initial guesses for the neutron fission source and criticality factor are substituted in Eq. (7.92), which has been written for the first energy group by setting $g = 1$ and eliminating the scatter source term:

$$D_1 \nabla^2 \phi_1^{(1)}(x) - \Sigma_{r1}(x)\phi_1^{(1)}(x) + \frac{1}{k^{(0)}} \chi_1 S_f^{(0)}(x) = 0 \quad , \tag{7.98}$$

with $\phi_1^{(1)}(x)$ determined by the finite difference methods described in Sec. 7-5F. The total neutron source for the second energy group can then be calculated:

$$S_{t2}^{(0)}(x) = \Sigma_{s12}(x)\phi_1^{(1)}(x) + \frac{1}{k^{(0)}} \chi_2 S_f^{(0)}(x) \quad . \tag{7.99}$$

Upon substitution of this source into the diffusion equation written for the second energy group ($g = 2$), the equation can then be solved for the second flux group, $\phi_2^{(1)}(x)$, by an inner iteration. Continuing this procedure through all the energy groups yields all of the first iteration group fluxes, $\phi_g^{(1)}(x)$, and a better fission source estimate:

$$S_f^{(1)}(x) = \sum_{g=1}^{G} \nu\Sigma_{fg}(x)\phi_g^{(1)}(x) \quad . \tag{7.100}$$

In addition, a new estimate of k results from the definition of k as the ratio of fission neutrons generated in two successive generations or, in this case, iterations:

$$k^{(1)} = \frac{\int S_f^{(1)}(x)\, dx}{\int \frac{S_f^{(0)}(x)\, dx}{k^{(0)}}} .$$

This general iterative process is continued until convergence is obtained. Two standard convergence criteria are usually employed. The first is:

$$\left| \frac{k^{(n)} - k^{(n-1)}}{k^{(n-1)}} \right| \lesssim \varepsilon_1 \quad , \qquad (7.101a)$$

where ε_1 is some small number. The second criterion for terminating the iteration is based on the local flux distribution:

$$\left| \frac{\phi^{(n)}(x) - \phi^{(n-1)}(x)}{\phi^{(n-1)}(x)} \right| \lesssim \varepsilon_2 \quad , \qquad (7.101b)$$

where again ε_2 is selected as a suitably small value greater than zero. Usually $k^{(n)}$ converges more rapidly than the spatial flux distribution, hence, Eq. (7.101b) would control the rate of convergence to the problem solution if ε_2 were equal to ε_1. Often, only the k convergence criterion is employed; for $\varepsilon_1 = 10^{-5}$, satisfactory flux convergence is achieved in most cases.

The number of iterations is, of course, strongly geometry dependent and varies from problem to problem. The choice of the initial fission sources also makes quite a difference in the number of iterations and ultimately in the amount of computer time required for a reactor problem calculation. Good initial guesses can be made by employing results from similar problems or, if several cases are to be run, by setting up a segment of calculations where only minor changes are made in the reactor specifications so that the results from one calculation can be used as the initial guess for the next.

It should also be evident that k iteration problems are not the only type of calculations that can be performed by this outer iteration technique. As discussed in Chapter 2, criticality searches are of interest in reactor design and they can also be performed. In these evaluations, reactor parameters other than k, such as geometry, fuel concentration, or control rod reactivity, can be varied until a desired k is achieved. Also these same types of iterative procedures are employed in the source-type problem. In this case, for example, when the flux distributions are required in a reactor during startup, the problem does not involve an eigenvalue (see Chapter 2). However, the solution to this subcritical system in the presence of an external source is in many ways similar to the one described previously for the k problem.

7-5F Inner Iteration

The solution to the finite difference diffusion equations for the spatial fluxes constitutes the inner iteration portion of the multigroup problem. The procedure is essentially the same for each group with the calculation usually starting at the highest or first energy group. For the three-point difference equation derived in Sec. 7-5C for the one-dimensional example, a straightforward solution is possible as described below.

Recall that for the one-dimensional spatial reactor model, the expression for the finite difference analog of the diffusion equation was given by Eq. (7.94). Considering the first mesh point ($m = 1$) and applying the boundary condition at $x = 0$ of $\phi_0 = 0$ (another condition could be applied if desired), the expression for ϕ_1 becomes:

$$-a_1^+\phi_2 - 0 + (a_1 + \Sigma_1)\phi_1 = S_{t1} \quad . \tag{7.102}$$

Similarly, for the second mesh point ($m = 2$), an expression can be obtained for ϕ_2 in terms of its neighboring fluxes ϕ_1 and ϕ_3:

$$-a_2^+\phi_3 - a_2^-\phi_1 + (a_2 + \Sigma_2)\phi_2 = S_{t2} \quad . \tag{7.103}$$

Substituting the expression for ϕ_1 given by Eq. (7.102) into Eq. (7.103), an expression for ϕ_2 as a function of ϕ_3 can be obtained. After repeating this substitution process throughout the mesh structure, ultimately an expression for the last mesh ($m = M - 1$) is reached

$$0 - a_{M-1}^-\phi_{M-2} + (a_{M-1} + \Sigma_{M-1})\phi_{M-1} = S_{M-1} \quad ,$$

where again the boundary condition (at $m = M$) of $\phi_M = 0$ has been imposed.

The set of finite difference equations can now be solved by the Gaussian elimination method because the ϕ_{M-2} term in Eq. (7.103) is known from the preceding equation. By calculating back in reverse order through the series of linear algebraic expressions, all values of ϕ_m can be determined.

When the finite difference reactor model consists of two or three dimensions, a direct calculational method is not possible. For these situations, an iterative procedure is employed as has been illustrated starting with Eq. (7.96).

The **A** matrix is broken into three parts:

$$\mathbf{A} = \mathbf{C} - \mathbf{U} - \mathbf{L} \quad , \tag{7.104}$$

where **C** is the matrix of all of the diagonal terms, **U** is a matrix consisting of all of the upper diagonal terms, and **L** is a matrix composed of all of the terms below the main diagonal. The matrix equation, Eq. (7.95a), upon substitution of Eq. (7.104), then becomes (g is suppressed):

$$(\mathbf{C} - \mathbf{U} - \mathbf{L})\Phi = S \quad . \tag{7.105}$$

Equation (7.105) is only partially inverted by writing:

$$\Phi = \mathbf{C}^{-1}(\mathbf{U} + \mathbf{L})\Phi + \mathbf{C}^{-1}S \quad . \tag{7.106}$$

An iterative scheme is devised by assuming that the flux on the right side comes from the previous iteration:

$$\Phi^{(n+1)} = \mathbf{C}^{-1}(\mathbf{U} + \mathbf{L})\Phi^{(n)} + \mathbf{C}^{-1}S \quad , \tag{7.107}$$

where $\Phi^{(n)}$ is the flux vector after the n'th iteration. The iteration can then begin by inserting initial guess $\Phi^{(0)}$ for each mesh point into the right side of Eq. (7.107) and computing $\Phi^{(1)}$ on the left side. This procedure is repeated until a converged solution is obtained. It is called the "simultaneous relaxation" method since the flux value at each mesh is determined or relaxed simultaneously.

A more rapidly converging technique for evaluating the finite difference equation is referred to as the "successive relaxation" or "Liebmann" method. The basic idea in this approach is to replace each $\phi_m^{(n)}$ component with the newly obtained $\phi_m^{(n+1)}$ value as soon as it is calculated throughout the inner iteration. This results in a more economical calculation than the simpler simultaneous relaxation method; it is expressed mathematically as

$$\Phi^{(n+1)} = \mathbf{C}^{-1}[\mathbf{U}\Phi^{(n)} + \mathbf{L}\Phi^{(n+1)} + S] \quad . \tag{7.108}$$

The total multigroup calculation procedure then consists of exercising the inner and outer iterative calculational procedures as outlined above. Figure 7-4 is a flow diagram of this computational process as typically employed to compute the critical size or mass of a reactor. The group constants are determined as discussed in Sec. 7-3. The weighting spectra, $\varphi_g(E)$, are either approximated by simple expressions (such as $1/E$) or they are obtained from zero-dimensional fine neutron energy group P_1 transport theory calculations (see Chapter 6). The weighting spectra in these group cross-section computations are not very sensitive to spatial dependencies; therefore, no iterations are normally employed to recalculate the cross-section library using improved weighting spectra.

7-5G Multigroup Finite Difference Computer Codes

Practical reactor design analysis can be accomplished only with the aid of computers. Therefore, a wide variety of computer codes for solving the multigroup steady-state diffusion or transport equation in one,

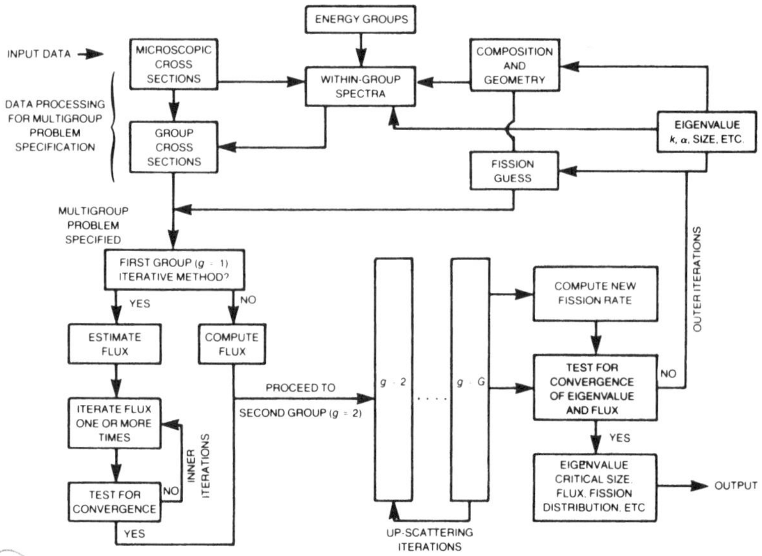

Fig. 7-4. Flow diagram for computing the critical size or mass of a reactor.

two, or three dimensions have been developed. A number of programs of varying capabilities are available for general use from the Argonne Code Center (now the National Energy Software Center)[16] and the Oak Ridge Radiation Shielding Information Center.[17] These programs are written in FORTRAN and can range from a few thousand lines up to 100,000 lines of code for the most sophisticated programs. These codes can be CPU intensive, with large memory requirements, and are typically run on large mainframe computers, e.g., supercomputers.

Finite difference, multigroup diffusion theory is the normally applied approach for flux calculations. Examples of one-dimensional codes are: 1DX (Ref. 18), DARC1D (Ref. 8), SDX (Ref. 19), and SPHINX (Ref. 20). Two-dimensional codes include 2DB (Ref. 21), EXTERMINATOR-2 (Ref. 10), and GAUGE (Ref. 15). For detailed calculations in three dimensions, PDQ-7 (Ref. 9), 3DB (Ref. 22), VENTURE (Ref. 23), and DIF3D (Ref. 24) are available.

As discussed in Chapter 6, there are certain analysis situations where transport theory instead of diffusion theory must be applied to obtain a sufficiently accurate solution. Transport theory for flux calculations generally applies the discrete ordinates (S_n) method. Examples of one-dimensional codes include DTF-IV (Ref. 25), ANISN (Ref. 26), XSDRNPM-S (Ref. 27), and ONEDANT (Ref. 28). Sophisticated two-dimensional transport codes are TWODANT (Ref. 29), DORT (Ref. 30), and TWOHEX (Ref. 31) and, in three-dimensions, THREETRAN (Ref.

32). For shielding and deep penetration problems, stochastic methods, e.g., Monte Carlo techniques as discussed in Sec. 6-3, are often applied. Some examples of Monte Carlo codes are MCNP (Ref. 33), TRIPOLI (Ref. 34), and MORSE (Ref. 35).

Although the application of multigroup theory to most reactor problems appears to be straightforward, there are many potential pitfalls for the unwary. The use of computer codes outside their area of applicability is one such pitfall. Inaccuracies can also occur if one is not careful with the problem definition, boundary conditions, and input specifications. Finally, the generation of multigroup libraries or group constant sets is not only difficult but is also a source of considerable uncertainty in the practical application of multigroup theory. These uncertainties result from differences in the treatment of weighting spectra, resonance self-shielding, heterogeneities, group collapsing, and burnup effects.

The basic "pointwise" nuclear physics data are contained in computer library files which must be processed to generate multigroup group constant sets for subsequent use in a multigroup diffusion or transport theory code. The most widely used data library in the United States is the Evaluated Nuclear Data File (ENDF) maintained by Brookhaven National Laboratory.[36] The ENDF/B is a library consisting of a "best" evaluated set of cross sections for reactor applications. It should be realized that ENDF/B represents a tremendous amount of data and very few codes can interface with ENDF directly. Most computer codes interface with detailed multigroup group constant sets that have been generated from ENDF/B data.

Codes for generating these group constant libraries from ENDF/B data for thermal reactors include: FLANGE-II (Ref. 37), ETOG-1 (Ref. 38), and XLACS (Ref. 39). Libraries of this type are then used by codes like LEOPARD (Ref. 40), LASER (Ref. 41), and EPRI-CELL (Ref. 42) to produce problem-specific group constants. These codes are based on the historical MUFT (Ref. 43) and GAM (Ref. 44) codes for generating the epithermal reactor constants and the THERMOS (Ref. 45) and SOFOCATE (Ref. 46) codes for thermal group constants. The resulting cross sections can then be used in standard one-, two-, or three-dimensional finite difference calculations. However, if subassembly averaged group constants are to be generated for use in nodal codes (Sec. 7-6), then special codes like CASMO-3 (Ref. 47) and PHOENIX (Ref. 48) must be used.

An adequate description of neutron thermalization is quite important for generating accurate thermal neutron group constants. Therefore, considerable effort has been expended in developing the THERMOS code. It is based on integral transport theory for a fuel rod/cladding/coolant cell; it uses the Nelkin kernel (Ref. 49) to explicitly represent the energy exchange between the neutron and the vibrational and rotational levels of the H_2O molecule of mass 18.

Fast reactor group constant generation requires a more detailed consideration of the energy dependence in the slowing down range. Several specialized codes have been developed applying two fundamentally different approaches: the Bondarenko self-shielding factor approach[50] is applied in MINX (Ref. 51), NJOY (Ref. 52), ETOX (Ref. 53), ENDRUN (Ref. 54), and the AMPX-II package;[55] the alternative is the MC^2–2 approach.[56]

A detailed discussion of group constant generation and how the uncertainties associated with the process are treated is beyond the scope of this text; an appreciation for the problems involved can be found in the literature.[57]

7-6 Nodal Methods

7-6A Survey of Nodal Methods

The nodal method is a popular calculational technique which eliminates the small mesh spacing limitations of the finite difference approach, producing relatively cost-effective calculational capabilities for three-dimensional power distributions. This approach is widely used in the LWR industry in reactor simulator and fuel management studies. It represents the reactor as a group of nodes where the average nodal neutron fluxes are the unknowns. In these large homogenized volume nodes, averaged group cross sections are defined using standard multigroup methods. A node is usually taken as a vertical section of a complete fuel subassembly (plus half of its surrounding gap) which, along with the axial dimension of equal size, produces a cubical volume. If the fuel assembly is large (e.g., a 21- × 21-cm PWR assembly), four nodes are sometimes used to describe the assembly laterally. Neutron balances or transport relationships are written for each node which couple the average fluxes with the total or partial neutron currents between the nodes. In this approach, each node is usually coupled with its adjacent neighbors, which requires six coupled relationships for a cubic nodal volume.

There are several types of nodal approximation schemes: response matrix, coarse mesh methods, and source-sink theory.

The basic idea of the response matrix method is to relate the average flux within each nodal volume along with the outgoing neutron currents at each surface of the nodal volume to the properties within the node and the current entering through each surface. From the incoming currents at each surface, the average flux within the node can be determined and the current leaving each surface can be calculated. The resulting flux and exiting currents as related to a set of entering currents for each volume surface are called the response matrices for that node. The matrices which describe the partial currents crossing nodal surfaces are called transmission and reflection matrices. A major feature of this approach is that the coupling coefficients which tie the nodes together depend only on the properties of a single node and are not a function of the properties of the adjacent neighbors.

With the response matrices determined, the neutron balances for each node as a function of energy group and nodal properties are developed explicitly in terms of the neutron currents at each surface. The solution of the response matrices generally makes use of diffusion theory, although transport theory can also be employed. Iterative solution methods are applied. The nodal equations are usually solved for only one to two energy groups because of the large size of a nodal representation of an entire reactor. In fact, in most nodal models, the coupling equations are solved in the epithermal groups only, as epithermal neutrons are primarily responsible for the internodal coupling effects. The thermal group fluxes are then calculated in a secondary calculation with the model termed a "one-and-a-half group model." The solution of the response matrix model provides the average energy group fluxes for each node as well as the effective multiplication factor of the reactor core. Methods have also been developed to obtain efficiently local fuel pin power production within the assembly nodal volumes. Examples of response matrix models are provided by the widely applied programs FLARE (Ref. 58), TRILUX (Ref. 59), and SIMULATE (Ref. 60).

The second nodal approach, also based on diffusion theory, is generally referred to as the "modified coarse-mesh finite difference model." In this method, the continuity of fluxes and net currents is required at the nodal surfaces. This approach also requires expressions for the transverse leakage of neutrons as functions of the nodal group fluxes at the center points of adjacent nodes. The PRESTO model[61] is an example of this approach which has found widespread application. The coupling of the volume-averaged flux is expressed as a function of the adjacent nodal volume-averaged fluxes and an adjustable parameter, which must be determined from an auxiliary calculation.

In the third type of nodal technique, called the "source-sink method," the fuel assemblies of the reactor core are each described as a source of fast neutrons and a sink for thermal neutrons. The neutron coupling relationship is between each assembly and its neighbors. Originally developed for large, thermal, natural uranium-fueled reactors, this method has been applied to an LWR system. For details of this technique as developed for a two-group LWR application, Ref. 62 should be consulted.

7-6B Two-Group Nodal Models

Some basic aspects of the nodal form of the neutron balance equation are discussed in this section. The particular method used in the popular FLARE model[58] is briefly indicated.

In most nodal models, the reactor core is subdivided into nodes of equal volume $V = \Delta X \cdot \Delta Y \cdot \Delta Z$ where ΔX, ΔY, and ΔZ are the lateral and the axial dimensions of the nodal cube (e.g., 20 cm × 20 cm × 20 cm). The basic balance equation is again the multigroup diffusion equation given by Eq. (7.85). However, the neutron current is retained in the leakage term, avoiding the application of Fick's law.

Thus, instead of Eq. (7.85), the multigroup balance equation is used in the form

$$\nabla \cdot \boldsymbol{J}_g(\boldsymbol{r}) + \Sigma_{rg}(\boldsymbol{r})\phi_g(\boldsymbol{r}) - \sum_{g \neq g'} \Sigma_{sg'g}\phi_{g'}(\boldsymbol{r}) = \lambda\chi_g \sum_{g'} \nu\Sigma_{fg'}(\boldsymbol{r})\phi_{g'}(\boldsymbol{r}) \quad . \tag{7.109}$$

The nodal mesh cube is denoted by (m). Integrating Eq. (7.109) over the volume V yields

$$\frac{1}{V}\sum_{k=1}^{6}\int_s \boldsymbol{J}_g(k) \cdot \boldsymbol{n}_k \, ds + \Sigma_{rg}(m)\phi_g(m) - \sum_{g' \neq g} \Sigma_{sg'g}(m)\phi_{g'}(m) = \lambda\chi_g \sum_{g'} \nu\Sigma_{fg'}(m)\phi_{g'}(m) \quad . \tag{7.110}$$

The leakage term is summed over the six sides of the cube; $\boldsymbol{n}$ is the normal to the surface S.

The multigroup neutron nodal balance relationship expressed by Eq. (7.110) is still in a form that is difficult to solve efficiently if many groups are required. In LWRs, two groups are usually sufficient. The further reduction to a one-and-one-half group model is suggested by the fact that the epithermal mfp is about three to four times larger than that of thermal neutrons, which in turn leads to the predominant epithermal coupling. The node-to-node coupling by thermal neutrons is therefore relatively insignificant, especially for large nodes.

The two-group version of Eq. (7.110) is given by:

$$\frac{1}{V}\sum_{k}\int_s \boldsymbol{J}_1(k) \cdot \boldsymbol{n} \, ds + \Sigma_{r1}(m)\phi_1(m) = \lambda[\nu\Sigma_{f1}(m)\phi_1(m) + \nu\Sigma_{f2}(m)\phi_2(m)] \quad , \tag{7.111a}$$

and

$$\frac{1}{V}\sum_k \int_s \boldsymbol{J}_2(k)\cdot\boldsymbol{n}\,ds + \Sigma_{a2}(m)\phi_2(m) - \Sigma_{12}(m)\phi_1(m) = 0 \quad . \tag{7.111b}$$

The reduction of these two-group equations to a 1.5 group model is achieved by approximating the thermal neutron leakage term, which normally connects a volume element with the neighboring nodes by a "local" quantity:

$$\frac{1}{V}\sum_k \int_s \boldsymbol{J}_2(k)\cdot\boldsymbol{n}\,ds = D_2(m)B_2^2(m)\phi_2(m) \quad . \tag{7.112}$$

The buckling-type quantity B_2^2 is formally defined by this equation as the "local" description of the leakage as given by the left side; as a description of the group leakage, it has the character of a geometrical buckling.

Upon substitution of Eq. (7.112) into Eq. (7.111b), ϕ_2 can be expressed in terms of ϕ_1:

$$\phi_2(m) = \frac{\Sigma_{12}(m)}{\Sigma_{a2}(m) + D_2(m)B_2^2(m)}\phi_1(m) \quad . \tag{7.113}$$

The resulting expression is used to eliminate ϕ_2 from Eq. (7.111a), giving the 1.5 group equation as:

$$\frac{1}{V}\sum_k \int_s \boldsymbol{J}_1(k)\cdot\boldsymbol{n}\,ds + \Sigma_{r1}\phi_1 = \lambda\left(\nu\Sigma_{f1} + \frac{\nu\Sigma_{f2}\Sigma_{12}}{\Sigma_{a2} + D_2B_2^2}\right)\phi_1 \quad , \tag{7.114}$$

where all quantities pertain to node m.

The epithermal leakage term is commonly treated by a coupling expression between the net or partial currents and the nodal fluxes or sources of the node and its six neighbors. Several approaches have been successful in describing this coupling, one of the oldest being the FLARE approach.[58] The literature should be consulted for the detailed model descriptions.

More recently, other response matrix methods as well as coarse-mesh nodal models have been developed which have removed some of the problems inherent in obtaining the coupling coefficients for describing the epithermal leakage in earlier nodal models. For details on these more efficient techniques for obtaining criticality and power distributions in large 3-D reactor cores, Refs. 60 and 61 should be consulted.

The nodal methods were briefly introduced in this chapter. Their importance stems from the fact that they allow a reduction by a factor of 100 in the number of mesh points required for the typical LWR core description compared to the finite difference technique. This can result in a similar reduction in the computational time and storage requirements, thereby allowing 3-D burnup calculations to be performed on a routine basis. The principal drawback of using nodal methods is that detailed information about the power shape for the heterogeneous node cannot be obtained directly as with the fine-mesh, finite difference technique. However, recent efforts[63] have shown progress in reconstructing the nodal power shape based on information available in the nodal calculation.

7-7 Summarizing Remarks

In this final chapter, the generally applied approach in solving the neutron diffusion problem, namely, multigroup theory, was presented and described in more detail than in its preliminary introduction in Chapters 2 and 4. The basis is the energy-dependent diffusion equation where each term is a reaction rate density. The multigroup equations were derived using the concept of neutron "groups" within energy group intervals, conserving the group-wise reaction rate densities. The derivation requires integrating all terms over each energy group. This results in group constants being average values with the group fluxes defined as integrals.

Although the derivation of the multigroup equations are formally exact, approximations must be made in the generation of the group constants. Generally, one approximates the space-dependent group constants by assuming space independence. This is required due to the lack of data and the complexity of carrying space-dependent cross sections throughout the calculations. Through the use of different weighting neutron flux spectra for different regions, a set of region-dependent group constants is obtained.

Basic group constant sets are tabulated as microscopic group constants from which macroscopic constants can be generated for different compositions of materials. The problems with obtaining the diffusion group constants are noted; the accuracy near interfaces suffers from the fact that the flux separation approximation is poor near interfaces and considerably more inaccurate for the gradient of the flux than for the flux itself.

This chapter illustrates the multigroup approach by starting with the one-group theory presented originally in Chapter 2. It is noted that in multidimensional calculations often a relatively small number of neutron groups are used. The process by which one obtains from a large number of groups a small number is called "group collapsing." The one-group theory leads to simple lumped parameters that describe the one-group model for thermal neutrons in the neutron balance relationship.

As more neutron groups represent a better description of the neutron balance relationships in an actual reactor, two-group theory was also presented. The criticality expression for the two-group model was developed in a form similar to the one-group expression. A two-group, two-region (core reflector) model was also presented to indicate typical effects near core/reflector interfaces.

The major emphasis in this chapter, however, was on the multigroup formulation. Discrete zones are also employed to describe the spatial dependency. In these regions, homogeneous material mixtures are assumed. The balance equations based on region-wise constant macroscopic group constants are completed with boundary and interfacial conditions as needed for a complete description of the problem.

The most widely used and accepted technique for solving the multigroup equation set is the finite difference approximation. Two distinctly different types of problems that must be solved emerge with this method. First, the inhomogeneous set of algebraic equations for the local fluxes in space are solved in an "inner iteration." Next the solution for criticality is obtained from an eigenvalue search in the "outer iteration." The finite difference formulation is sketched in one dimension using matrix notation and describing briefly the iterative methods used in their solution.

For the calculation of three-dimensional power distributions in large LWRs, and even more so for burnup and fuel management problems, detailed finite difference descriptions are infeasible. Therefore, the more coarse "nodal" methods have been developed and are in widespread use. A brief description of the nodal techniques is given in the final section of this chapter. The derivation of the widely applied one-and-a-half group model is presented as it is a key feature of most nodal methods.

Homework Problems

NOTE: For the solution of problems 3 through 11, see Tables 7-I and 7-II for macroscopic three-group constant sets for typical cores of a light water reactor (LWR) and a sodium-cooled fast breeder reactor (FBR), respectively.

1a. Calculate a three-group constant set for epithermal and fast neutrons for the following cross sections:

- $\sigma_s(E)$of hydrogen as given in Fig. 4-6
- $\sigma_s = 3.8$ b of oxygen, assumed to be constant
- $\sigma_s = 9.0$ b for actinides, assumed to be constant (use the s-wave scattering kernel in all cases)

- $\sigma_a = \sigma_{a1}/\sqrt{E}$, with E in electron volts
- $\nu\sigma_f = \nu\sigma_{f1}/\sqrt{E}$, with E in electron volts
- $\chi(E)$, assumed to all be in group 1.

Assume a $1/E$ spectrum and use the following group boundaries: $E_0 = 10$ MeV, $E_1 = 1$ MeV, $E_2 = 10$ keV, and $E_3 = 1.0$ eV.

b. Use the solid line spectrum given in Fig. 4-10 instead of the $1/E$ spectrum and discuss the differences in the resulting group constants.

2a. Take the spectrum given by the solid line in Fig. 4-10 and determine approximately the three-group fluxes (spectra) for the group structure given in problem 1a.

b. Assuming that these three-group fluxes are the result of an independent calculation, plot the corresponding spectrum as a histogram over a logarithmic energy scale, and compare it with the spectrum of Fig. 4-10.
Reminder: The spectrum in Fig. 4-10 is unrealistically high around 1 MeV because of the neglect of the inelastic scattering in that energy domain.

3. Find a three-group constant set by reducing $\nu\Sigma_f$ in the FBR group constant set given in Table 7-II by a factor of 5 and Σ_a by a factor of 2. This set is then approximately representative of an FBR blanket with a composition similar to the one in the core, except for the fissile material content.

4a. Give the three-group diffusion theory balance equation, which is the basis for the determination of the material buckling.

b. Find the material buckling and the fundamental spectra for the three media characterized by the three-group constant sets: the LWR case of Table 7-I, the FBR case of Table 7-II, and the FBR blanket developed in problem 3.

c. Calculate also the higher material bucklings and the corresponding eigenspectra and compare them with the fundamental mode results.

d. Find the critical dimension of an unreflected cylinder (height = diameter) from the two core group constant sets, the LWR case of Table 7-I and the FBR case of Table 7-II.

5a. Find the λ eigenvalue for the FBR blanket case using the group constant set developed in problem 3, assuming an infinite medium.

b. Find the B_m^2 eigenvalue for the FBR blanket case of problem 3 and the fundamental spectrum.

6. Collapse the three-group sets (the LWR case of Table 7-I, the FBR case of Table 7-II, and the FBR blanket developed in prob-

TABLE 7-I
Representative PWR Macroscopic Three-Group Constants
(cm^{-1})

Group	Energy Range	χ	Σ_t	$\nu\Sigma_f$	Σ_a	Σ_{tr}	Σ_{Sgg}	Σ_{Sgg+1}	Σ_{Sgg+2}
1	5.53 keV to 10 MeV	1.000	0.502	4.131×10^{-3}	3.056×10^{-3}	0.255	0.451	0.048	0.0
2	0.625 eV to 5.53 keV	0.000	1.070	1.223×10^{-2}	2.219×10^{-2}	0.555	0.973	0.076	—
3	0 to 0.625 eV	0.000	1.569	1.615×10^{-1}	1.003×10^{-1}	1.313	1.469	—	—

TABLE 7-II
Representative 300-MW(e) LMR Macroscopic Three-Group Constants
(cm^{-1})

Group	Energy Range (Mev)	χ	Σ_t	$\nu\Sigma_f$	Σ_a	Σ_{tr}	Σ_{Sgg}	Σ_{Sgg+1}	Σ_{Sgg+2}
1	0.82 to 14.2	0.7712	0.16705	1.453×10^{-2}	5.901×10^{-3}	1.135×10^{-1}	2.263×10^{-2}	1.965×10^{-3}	1.965×10^{-3}
2	0.16 to 0.82	0.2044	0.21219	0.515×10^{-2}	3.173×10^{-3}	1.722×10^{-1}	2.027×10^{-1}	6.306×10^{-3}	—
3	0 to 0.16	0.0244	0.30362	0.568×10^{-2}	6.476×10^{-3}	2.695×10^{-1}	2.971×10^{-1}	—	—

lem 3) to 3 two-group sets, retaining the first group.

7. Find the material bucklings and the fundamental spectra for the two-group cases and compare them with the three-group results of problem 4.

8a. Collapse the 3 three-group sets (the LWR case of Table 7-I, the FBR case of Table 7-II, and the FBR blanket developed in problem 3) to one-group sets.

b. Collapse the 3 two-group sets developed in problem 6 to one-group sets and discuss the comparison with the results of problem 8a.

9. Find the one-group material bucklings for the three-group constant sets developed in problem 8 and compare them with the two- and three-group results of problems 7 and 4.

10. Find from the two-group set the higher material buckling and the corresponding eigenspectrum for the FBR blanket case.

11a. Consider a half-space blanket, extending from $x = 0$ to ∞, with a source in the first group at $x = 0$. Calculate analytically the two-group flux distribution using the group constant set developed in problem 6. (See Sec. 3-6 for procedure.)

b. Plot the flux distribution, plot the ratio of $\phi_2(x)/\phi_1(x)$, compare this quantity with the corresponding quantity of the fundamental spectrum, and discuss all results.

Review Questions

1. How can the separability assumption be weakened without going over to the complete nonseparable case?

2a. Sketch the derivation of the multigroup diffusion equations.

b. Explain the advantage of using the formally correct flux in that derivation.

3a. Which conservation idea is applied in casting a continuous energy dependence into a multigroup structure?

b. Define the group flux, ϕ_g.

c. Derive Σ_{tg} and $\Sigma_{sg'g}$.

d. Derive the multigroup representation of $\mathbf{F}\Phi$.

4. Define the removal cross section.

5a. Derive D_g under the assumption of group-wise separability.

b. Derive Σ_{trg} from the second of the P_1 equations [with $S^{(1)} = 0$].

c. Decompose the latter formula in isotopic contributions.

d. What is the problem of formulas involving the averaging of $D(E)$ or $1/\Sigma_{tr}(E)$ with respect to isotopic decomposition?

6. Give the one-group equation for all neutrons.

7a. Give the one-group equation for thermal neutrons; also give the multigroup calculated source.
b. Where is the λ eigenvalue?
8a. Give the two-group equations in multigroup notation ($\chi_1 = 1$, $\chi_2 = 0$).
b. Derive from that the two-group formula for k_{eff} for a one-region problem (do not introduce any of the "four-factor" complications).
9a. Sketch the typical ϕ_{th} and ϕ_f behavior in a thermal core reflector geometry.
b. Explain the physical reasons for this behavior.
10a. Give the three-point formula for the leakage operator in one dimension.
b. Give the structure of the finite-differenced spatial diffusion equation for a group.
c. Give the same in operator notation (indicate only on which band the operator **A** has nonvanishing elements).
11. Explain the inner iteration in a one-dimensional problem with $\phi = 0$ at the boundaries.
12a. Explain the outer iteration for obtaining the eigenvalue λ.
b. How is the outer iteration modified for an enrichment search?
c. Give two typical convergence criteria for outer iterations.

REFERENCES

1. M. Clark, Jr. and R. J. Hansen, *Numerical Methods of Reactor Analysis,* Academic Press, New York (1964).
2. H. Greenspan, C. N. Kelber, and D. Okrent, Eds., *Computing Methods in Reactor Physics,* Gordon & Breach Science Publishers, New York (1968).
3. J. E. Suich and H. C. Honeck, "The HAMMER System—Heterogeneous Analysis by Multigroup Methods of Exponentials and Reactors," DP-1064, Savannah River Lab. (1964).
4. J. R. Lamarsh, *Introduction to Nuclear Reactor Theory,* Addison-Wesley Publishing Co., Reading, Massachusetts (1966).
5. S. Glasstone and A. Sesonske, *Nuclear Reactor Engineering,* D. Van Nostrand Co., Princeton, New Jersey (1967).
6. R. L. Murray, *Nuclear Reactor Physics,* Prentice-Hall, Inc., Englewood Cliffs, New Jersey (1957).
7. S. Glasstone and M. C. Edlund, *The Elements of Nuclear Reactor Theory,* D. Van Nostrand Co., Princeton, New Jersey (1952).
8. D. E. Neal, G. K. Leaf, and A. S. Kennedy, "The Arc System One-Dimensional Diffusion Theory Capability," ANL-7715, Argonne National Lab. (May 1971).
9. W. R. Cadwell, "PDQ-7 Reference Manual," WAPD-TM-678, Westinghouse Electric Corp. (Jan. 1967).
10. T. B. Fowler, M. L. Tobias, and D. R. Vondy, "EXTERMINATOR-2: A Fortran IV Code for Solving Multigroup Neutron Diffusion Equations in Two Dimensions," ORNL-4078, Oak Ridge National Lab. (Apr. 1967).
11. K. J. Kobayski, "Solution of Multigroup Diffusion Equation in X-Y-Z Geometry by Finite Fourier Transformation," *J. Nucl. Sci. Technol.,* p. 218 (1975).
12. A. F. Henry, *Nuclear Reactor Analysis,* The MIT Press, Cambridge, Massachusetts (1975).
13. H. G. Kaper, G. K. Leaf, and A. J. Lindeman, "A Timing Comparison Study for Some High Order Finite Element Approximation Procedures and a Low Order Finite Difference Approximation Procedure for the Numerical Solution of the Multigroup Neutron Diffusion Equation," *Nucl. Sci. Eng.,* **49,** 27 (1972).
14. K. F. Hansen and C. M. Kang, "Finite Element Methods in Reactor Physics Analysis," *Nucl. Sci. Eng.,* **51,** 456 (1973).
15. N. R. Wagner, "GAUGE: A Two-Dimensional Few Group Neutron Diffusion-Depletion Program for a Uniform Triangular Mesh," GA-8307, General Atomic (Mar. 1968).
16. M. K. Butler, N. Hollister, M. Legan, and L. Ranzini, "Argonne Code Center: Compilation of Program Abstracts," ANL-7411, Argonne National Lab. (1968); see also J. M. Brown, M. K. Butler, M. M. DeBruler, F. K. Deggs, H. S. Edwards, L. R. Eyberger, C. E. Hughes, P. L. Johnson, M. Legan, J. Lockler, L. L. Reed, and A. J. Streckok, "National Energy Software Center," ANL-7411 Supplement 2, Argonne National Lab. (May 1986).
17. "RSIC Computer Code and Data Collections: A Capsule Review of the

Computer Code Collection (CCC), Peripheral Shielding ROUTINES (PSR), and Data Library Collection (DLC) Packaged by the Radiation Shielding Information Center," Oak Ridge National Lab. (1988).
18. R. W. Hardie and W. W. Little, Jr., "1DX, A One-Dimensional Diffusion Code for Generating Effective Nuclear Cross Sections," BNWL-954, Battelle Northwest Lab. (1969).
19. W. M. Stacey, Jr., B. J. Toppel, H. Henryson II, B. A. Zolotar, R. N. Hwang, and C. G. Stenberg, "A New Space-Dependent Fast Neutron Multigroup Cross Section Preparation Capability," *Trans. Am. Nucl. Soc.*, **15,** 292 (1972).
20. W. J. Davis, M. B. Yarbrough, and A. B. Bortz, "SPHINX, A One-Dimensional Diffusion and Transport Nuclear Cross Section Processing Code," WARD-XS-3045-17, PSR-129 (1977).
21. W. W. Little, Jr. and R. W. Hardie, "2DB: A Two-Dimensional Diffusion Burnup Code for Fast Reactor Analysis," BNWL-640, Pacific Northwest Lab. (Jan. 1968); also see W. W. Little, Jr. and R. W. Hardie, "2DB User's Manual," BNWL-831, Pacific Northwest Lab. (July 1968).
22. R. W. Hardie and W. W. Little, Jr., "3DB: A Three-Dimensional Diffusion Theory Burnup Code," BNWL-1264, Pacific Northwest Lab. (1970).
23. D. R. Vondy, T. B. Fowler, G. W. Cunningham III, "VENTURE: A Code Block for Solving Multigroup Neutronics Problems Applying the Finite Difference Diffusion Theory Approximation to Neutron Transport, Version II," ORNL-5062/R1, Oak Ridge National Lab. (1977).
24. K. L. Derstine, "DIF3D: A Code to Solve One-, Two-, and Three Dimensional Finite-Difference Diffusion Theory Problems," ANL-82-64, Argonne National Lab. (1982).
25. K. D. Lathrop "DTF-IV, A FORTRAN-IV Program for Solving the Multigroup Transport Equation with Anisotropic Scattering," LA-3373, Los Alamos Scientific Lab. (1965).
26. W. W. Engle, Jr., "ANISN-ORNL: A One-Dimensional Discrete Ordinates Transport Code with Anisotropic Scattering," CCC-254, Oak Ridge National Lab. (1975).
27. N. M. Greene and L. M. Petrie, "XSDRNPM-S: A One-Dimensional Discrete-Ordinates Code for Transport Analysis," NUREG/CR-0200, Vol. 2, Sec. F3, Oak Ridge National Lab. (1983).
28. R. D. O'Dell, F. W. Brinkley, Jr., and D. R. Marr, "User's Manual for ONEDANT: A Code Package for One-Dimensional Diffusion Accelerated Neutral-Particle Transport," LA-9184-M, Los Alamos National Lab. (1982).
29. R. E. Alcouffe, F. W. Brinkley, D. R. Marr, and R. D. O'Dell, "User's Guide for TWODANT: A Code Package for Two-Dimensional, Diffusion-Accelerated Neutral-Particle Transport," LA-10049-M, Los Alamos National Lab. (1984).
30. W. A. Rhoades and R. L. Childs, "The DORT Two-Dimensional Discrete Ordinates Transport Code," *Nucl. Sci. Eng.*, **99,** 88 (1988).
31. W. F. Walters, F. W. Brinkley, and D. R. Marr, "User's Guide for TWOHEX: A Code Package for Two-Dimensional, Neutral Particle Transport in Equilateral Triangular Meshes," LA-10258-M, Los Alamos National Lab. (1984).

32. K. D. Lathrop, "THREETRAN: A Program To Solve the Multigroup Discrete Ordinates Transport Equation in (x,y,z) Geometry," LA-6333-MS, Los Alamos National Lab. (1976).
33. Los Alamos Radiation Transport Group X-6, "MCNP—A Generalized Monte Carlo Code for Neutron and Photon Transport," LA-7396-M, Los Alamos National Lab. (1981).
34. A. Baur, L. Bourdet, G. Dejonghe, J. Gonnord, A. Monnier, J. C. Nimal, and T. Vergnaud, "Programme de Monte Carlo polycinetique a trois dimensions, TRIPOLI-2," Centre d'Etude Nucleaires de Saclay (1980).
35. M. B. Emmett, "The MORSE Monte Carlo Radiation Transport Code System," ORNL-4972, Oak Ridge National Lab. (1975).
36. H. C. Honeck, "ENDF/B: Specifications for an Evaluated Nuclear Data File for Reactor Application," BNL-50066, Brookhaven National Lab. (1966).
37. H. C. Honeck and D. R. Finch, "FLANGE II, A Code To Process Thermal Neutron Data from an ENDF/B Tape," DP-1278 (ENDF-152), Savannah River Lab. (1971).
38. D. E. Kusner, R. A. Dannels, and S. Kellman, "ETOG-1, A FORTRAN IV Program To Process Data from the ENDF/B File to the MUFT, GAM, and ANISN Formats," WCAP-3845-1, Westinghouse Electric Corp. (1969).
39. N. M. Greene, J. L. Lucius, J. E. White, R. Q. Wright, C. W. Craven, Jr., and M. L. Tobias, "XLACS: A Program To Produce Weighted Multigroup Neutron Cross Sections from ENDF/B," ORNL/TM-3646, Oak Ridge National Lab. (1972).
40. R. F. Barry, "LEOPARD—A Spectrum Dependent Non-Spatial Depletion Code for the IBM-7094," WCAP-3269-26, Westinghouse Electric Corp. (1963).
41. C. G. Poncelet, "LASER—A Depletion Program for Lattice Calculations Based on MUFT and THERMOS," WCAP-6073, Westinghouse Electric Corp. (1966).
42. W. R. Cobb, W. J. Eich, and D. E. Tively, "EPRI-CELL Code Description," Advanced Recycle Methodology Program System Documentation Part I, Chapter 5, Electric Power Research Institute, Palo Alto, California (1977).
43. H. Bohl, Jr., E. M. Gelbard, and G. H. Ryan, "MUFT-4: Fast Neutron Spectrum Code for the IBM-704," WAPD-TM-72, Westinghouse Electric Corp. (1957).
44. G. D. Joanou, E. J. Leshan, and J. S. Dudek, "GAM-I: A Consistent P_1 Multigroup Code for the Calculation of the Fast Neutron Spectrum and Multigroup Constants," GA-1850, General Atomic (1961).
45. H. H. Honeck, "THERMOS: A Thermalization Transport Theory Code for Reactor Lattice Calculations," BNL-5826, Brookhaven National Lab. (1961).
46. H. J. Amster and R. Suarez, "The Calculation of Thermal Constants Averaged Over a Wigner-Wilkins Flux Spectrum: Description of the SOFOCATE Code," USAEC Report WAPD-TM-39, Westinghouse Electric Corp. (1957).
47. M. Edenius and A. Ahlin, "CASMO-3, A Fuel Assembly Burnup Program, User's Manual," Studsvik/NFA-86/7 (1986).
48. T. Q. Nguyen, Y. S. Liu, C. Durston, and A. L. Casadei, "Benchmarking of the PHOENIX-P/ANC Advanced Nuclear Design System," presented at In-

ternational Reactor Physics Conference, Jackson Hole, Wyoming, September 18–22, 1988.
49. M. Nelkin, "Scattering of Slow Neutrons by Water," *Phys. Rev.*, **119,** 741 (1960).
50. I. I. Bondarenko, "Group Constants for Nuclear Reactor Calculations," Translation-Consultants Bureau Enterprises, Inc., New York (1964).
51. C. R. Weisbin, P. D. Soran, R. B. MacFarlane, D. R. Morris, R. J. Lebauve, J. S. Hendricks, J. E. White, and R. B. Kidman, "MINX: A Multigroup Interpretation of Nuclear Cross Sections from ENDF/B, LA-6486-MS (ENDF-237), Los Alamos National Lab. (1976).
52. R. E. MacFarlane, D. Muir, and R. M. Boicourt, "The NJOY Nuclear Data Processing System: User's Manual," LA-9303-MS (ENDF-324), Los Alamos National Lab. (1985).
53. R. E. Schenter, J. L. Baker, and R. B. Kidman, "ETOX, A Code to Calculate Group Constants for Nuclear Reactor Calculations," BNWL-1002, Battelle Northwest Lab. (1969).
54. B. A. Hutchins, C. L. Cowan, M. D. Kelly, and J. B. Turner, "ENDRUN-II, A Computer Code To Generate a Generalized Multigroup Data File from ENDF/B," GEAP-13704, General Electric Co. (1971).
55. N. M. Greene, W. E. Ford III, J. L. Lucius, J. E. White, L. M. Petrie, and R. Q. Wright, "AMPX-II: A Modular Code System for Generating Coupled Multigroup Neutron-Gamma Libraries from ENDF/B," PSR-63, Oak Ridge National Lab. (1984).
56. H. Henryson II, B. J. Toppel, and C. G. Stenberg, "MC^2–2: A Code To Calculate Fast Neutron Spectra and Multigroup Cross Sections," ANL-8144 (ENDF 239), Argonne National Lab. (1976).
57. J. Lewins and M. Becker, eds., "Sensitivity and Uncertainty Analysis of Reactor Performance Parameters," Vol. 14 of *Advances in Nuclear Science and Technology,* Plenum Press, New York (1982).
58. D. L. Delp, D. L. Fischer, J. R. Harriman, and M. J. Stedwel, "FLARE—A Three-Dimensional Boiling Water Reactor Simulator," GEAP-4598, General Electric Co. (1964).
59. L. Goldstein, F. Nakache, and A. Veras, "Calculation of Fuel-Cycle Burnup and Power Distribution of Dresden-I Reactor with the TRILUX Fuel Management Program," *Trans. Am. Nucl. Soc.,* **10,** 300 (1967).
60. D. M. Ver Planck, "Manual for the Reactor Analysis Program SIMULATE," YAEC-1158, Electric Power Research Institute (Aug. 1978).
61. S. Borresen, "A Simplified, Coarse-Mesh, Three-Dimensional Diffusion Scheme for Calculating the Gross Power Distribution in a Boiling Water Reactor," *Nucl. Sci. Eng.,* **44,** 37 (1971).
62. H. Graves, Jr., *Nuclear Fuel Management,* John Wiley & Sons, New York (1979).
63. K. Smith, "SIMULATE-3 Pin Power Reconstruction: Benchmarking Against the B&W Critical Experiments," *Trans. Am. Nucl. Soc.,* **56** (1988).

APPENDIX A

Volume Elements, Differential Operators, and the Solid Angle

Coordinate Systems

The standard coordinate systems employed in nuclear reactor theory are the Cartesian, cylindrical, and spherical (polar) coordinate systems in one, two, and three spatial dimensions. In Fig. A-1, the general space vector $\boldsymbol{r}$ is depicted in these systems in three dimensions. The Cartesian coordinate system is also shown as a basis for the definition of the polar angle, ϑ, and the azimuthal angle, α.

Volume Elements

Cartesian:

$$dV = dx\, dy\, dz \quad . \tag{A.1}$$

Cylindrical:

$$dV = r dr\, d\alpha\, dz \quad . \tag{A.2}$$

Spherical:

$$dV = r^2\, dr\, \sin\vartheta d\vartheta\, d\alpha \quad . \tag{A.3}$$

Differential Operators

Gradients

Cartesian:

$$\nabla = \begin{pmatrix} \dfrac{\partial}{\partial x} \\ \dfrac{\partial}{\partial y} \\ \dfrac{\partial}{\partial z} \end{pmatrix} \quad . \tag{A.4}$$

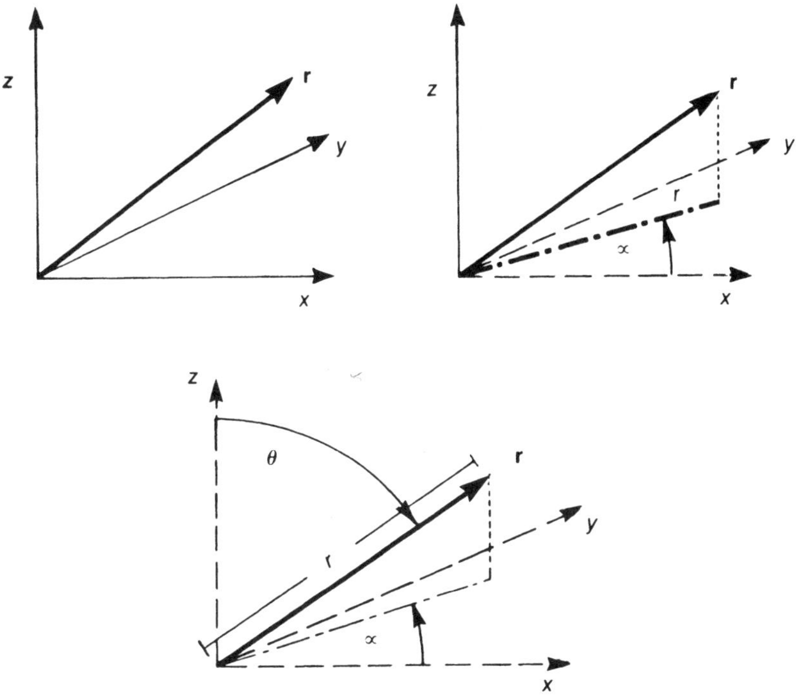

Fig. A-1. Representation of the space vector, ***r*** (the heavy solid arrow), in Cartesian, cylindrical, and spherical coordinate systems. The dotted lines represent projections; the dashed-dotted lines represent projections of ***r*** in the *x*-*y* plane; α represents the azimuthal angle; and ϑ represents the polar angle.

Cylindrical:

$$\nabla = \begin{pmatrix} \dfrac{\partial}{\partial r} \\ \dfrac{1}{r}\dfrac{\partial}{\partial \alpha} \\ \dfrac{\partial}{\partial z} \end{pmatrix} . \tag{A.5}$$

Spherical:

$$\nabla = \begin{pmatrix} \dfrac{\partial}{\partial r} \\ \dfrac{1}{r}\dfrac{\partial}{\partial \vartheta} \\ \dfrac{1}{r \sin\vartheta}\dfrac{\partial}{\partial \alpha} \end{pmatrix} . \tag{A.6}$$

Laplace Operators

Cartesian:

$$\nabla^2 = \frac{\partial^2}{\partial x^2} + \frac{\partial^2}{\partial y^2} + \frac{\partial^2}{\partial z^2} \quad . \tag{A.7}$$

Cylindrical:

$$\nabla^2 = \frac{1}{r}\frac{\partial}{\partial r}\left(r\frac{\partial}{\partial r}\right) + \frac{1}{r^2}\frac{\partial^2}{\partial \alpha^2} + \frac{\partial^2}{\partial^2 z} \quad . \tag{A.8}$$

Spherical:

$$\nabla^2 = \frac{1}{r^2}\frac{\partial}{\partial r}\left(r^2\frac{\partial}{\partial r}\right) + \frac{1}{r^2 \sin\vartheta}\frac{\partial}{\partial\vartheta}\left(\sin\vartheta\frac{\partial}{\partial\vartheta}\right) + \frac{1}{r^2 \sin^2\vartheta}\frac{\partial^2}{\partial\alpha^2} \quad . \tag{A.9}$$

The Solid Angle Vector $\mathbf{\Omega}$

The solid angle vector $\mathbf{\Omega}$ is normally in polar coordinates. Since the length of $\mathbf{\Omega}$ is normalized to unity, the three components of $\mathbf{\Omega}$ contain only two independent variables, the two angular coordinates ϑ_s and α_s:

$$\mathbf{\Omega} = \begin{pmatrix} \Omega x \\ \Omega y \\ \Omega z \end{pmatrix} = \begin{pmatrix} \sin\vartheta \cos\alpha \\ \sin\vartheta \sin\alpha \\ \cos\vartheta \end{pmatrix} = \begin{pmatrix} \sqrt{1-\mu^2}\cos\alpha \\ \sqrt{1-\mu^2}\sin\alpha \\ \mu \end{pmatrix} \tag{A.10}$$

with

$$\mu = \cos\vartheta \quad . \tag{A.11}$$

Normalization:

$$\mathbf{\Omega}^2 = \mathbf{\Omega}\cdot\mathbf{\Omega} = \Omega_x^2 + \Omega_y^2 + \Omega_z^2 = 1 \tag{A.12}$$

and

$$d\Omega = d\mu \, d\alpha \quad . \tag{A.13}$$

The most important Ω integrals are given in Sec. 6-4A.

APPENDIX B

Neutron Wave Terminology

Following Newton's example of decomposing light by a prism, physicists studied the variation of light intensity as a function of its wave length, the so-called "spectrum." Use of a collimating slot led to the discovery of the "line spectrum," which is different for each chemical element. The "lines" are the optical image of the collimating slot. The early investigations emphasized the study of the simplest spectrum, the one of hydrogen. Apparently, some of the lines had similar features, and grouping them together produced the so-called "spectral series." There were four dominant series named the "sharp," "principle," "diffuse," and "fundamental" series.

It was not before the discovery of quantum mechanics in 1925 that the differences in these series were correctly related to the angular momenta of the respective atomic states. These series correspond to states with $l = 0$, 1, 2, or 3, respectively, with l being the (integer) angular momentum quantum number. This suggested the following terminology, which was first introduced by Rydberg:

s-state, for $l = 0$
p-state, for $l = 1$
d-state, for $l = 2$
f-state, for $l = 3$, then continued in alphabetical order ($g,h, \ldots$).

The same terminology is used for the denotation of scattering states in their decomposition into quantized angular momentum components. One distinguishes *s*-waves, *p*-waves, etc., of neutrons where the corresponding l values are the angular momentum quantum numbers of the neutron relative to the nucleus. If the nucleus and neutron were mathematical points, the angular momentum of a neutron with a momentum mv would be rmv, with r being the (closest) distance of the neutron trajectory from the "point" nucleus.

In this idealized case, a collision with zero angular momentum would require a direct "hit," i.e., $r = 0$. However, due to Heisenberg's uncertainty principle, the positions of the neutron and nucleus are not precisely defined; in fact, the position of the neutron is defined only within its wave length, $\lambda = h/mv$. Furthermore, the angular momentum varies only in quantum steps. Zero angular momentum ($l = 0$) holds predominantly for the neutrons in the inner cylindrical segment of the approaching wave.

For small neutron energies, (say, ≤ 1 keV) the wave length is considerably larger than the diameter of the nucleus. Thus, the $l = 0$ segment of the neutron wave has a diameter larger than the nucleus. Then, the preponderance of low-energy collisions is with *s*-wave neutrons.

For neutron energies of several kiloelectron volts and more, *p*-wave neutrons react with nucleii in increasing numbers. For even larger energies, interaction with *d*-wave neutrons also becomes significant. It should be noted, however, that these energy limits (between *s*- and *p*- and *d*-waves) are not sharp. They have only probabilistic significance. A small fraction of low-energy collisions occurs with *p*-wave neutrons. A typical example of a low-energy *p*-wave reaction is a resonance in ^{238}U cross sections near 10 eV.

APPENDIX C

The δ Function

Dirac introduced the δ function in the mid 1920s for application in quantum mechanics. It turned out to be a very powerful concept, now widely applied in theoretical and applied science fields. The δ function is not a function in the mathematical definition of this concept, it is merely defined by the properties of its integral, i.e., it is a "generalized function," a "distribution." Nevertheless, the historical name, δ function, is preferred to the correct name "δ distribution."

The standard definition of Dirac's δ function as it appears in mathematical textbooks is:

$$\delta(x-x') = 0 \quad \text{for } x' \neq x \tag{C.1}$$

and

$$\int_a^b f(x')\delta(x-x')\,dx' = \begin{cases} f(x) & \text{for } x \in (a,b) \\ 0 & \text{for } x \notin [a,b] \end{cases} \quad , \tag{C.2}$$

where (a,b) is the open interval

$$a<x<b$$

and $[a,b]$ includes the boundaries

$$a \leq x \leq b \quad .$$

The function $f(x)$ must be continuous in (a,b).

For $f(x) = 1$ in particular, the following relationship holds:

$$\int_a^b \delta(x-x')\,dx' = \begin{cases} 1 & \text{for } x \in (a,b) \\ 0 & \text{for } x \notin [a,b] \end{cases} \quad . \tag{C.3}$$

The δ function singles out the value of a function upon which it acts at any particular point x. This is sometimes called the "sifting property" of the δ function.

The δ function is an *even* "generalized function," i.e., a change of the sign in its argument does not change the results; thus

$$\int_a^b f(x')\delta(x'-x)\,dx' = \int_a^b f(x')\delta(x-x')\,dx' \quad . \tag{C.4}$$

In the same fashion, derivatives of the δ function are defined; e.g., $\delta'(x-x')$:

$$\int_a^b f(x')\delta'(x-x')\,dx' = f'(x) \quad , \tag{C.5}$$

with $f'(x)$ being the derivative of f at a point x within the same interval as above. Then $f(x)$ must be differentiable in the interval.

APPENDIX D

Mathematical Formulas Useful in Reactor Theory

Some additional mathematical formulas useful in reactor theory are provided here. They are presented without proof; for details of their development, standard calculus references should be consulted, e.g., Ref. 1.

Derivative of a Definite Integral

The derivative of a definite integral takes the form:

$$\frac{d}{dt}\int_{a(t)}^{b(t)} f(x,t)\, dx = f[b(t),t]b'(t) - f[a(t),t]a'(t) + \int_{a(t)}^{b(t)} \frac{\partial f(x,t)}{\partial t}\, dx \quad , \tag{D.1}$$

where $a'(t)$ and $b'(t)$ are the derivatives of $a(t)$ and $b(t)$.

First-Order Linear Differential Equation Solution

A first-order linear differential equation takes the form:

$$\frac{df(x)}{dx} = a(x)f(x) + g(x) \quad . \tag{D.2}$$

The solution is obtained by use of the integrating factor, i.e.,

$$\exp\left[\int^{x} a(x')\, dx'\right] \quad ,$$

and is:

$$f(x) = C \exp\left[\int^{x} a(x')\, dx'\right] + \int^{x} g(x') \cdot \exp\left[\int_{x'}^{x} a(x'')\, dx''\right] dx' \quad . \tag{D.3}$$

The value is determined by the boundary condition that then determines the unique solution of Eq. (D.2). Typical examples are boundary conditions at $x = 0$ and $x = \infty$:

$$f(0) = f_0$$

and

$$f(\infty) = f_\infty, \quad \text{which often equals zero.}$$

The corresponding solutions are:

$$f(x) = f_0 \exp\left[\int_0^x a(x')\, dx'\right] + \int_0^x g(x') \exp\left[\int_{x'}^x a(x'')\, dx''\right] dx' \tag{D.4}$$

and

$$f(E) = \int_\infty^E g(E') \exp\left[\int_{E'}^E a(E'')\, dE''\right] dE', \tag{D.5}$$

if $x = E$ and $f_\infty = 0$.

Taylor Series

The Taylor series expansion of a function, $f(x)$, around the point, x_0, is given by the following expression:

$$f(x) = f(x_0) + \frac{f'(x_0)}{1!}(x - x_0) + \frac{f''(x_0)}{2!}(x - x_0)^2 \quad , \ldots , \tag{D.6}$$

where the prime denotes the derivative of f with respect to x.

REFERENCE

1. W. Kaplan, *Advanced Calculus*, Addison-Wesley Publishing Co., Reading, Massachusetts (1952).

APPENDIX E

The Factorial and Related Integrals

The most important definite integral in nuclear reactor theory is

$$\int_0^\infty x^a e^{-x}\, dx = a! \quad . \tag{E.1}$$

For the most part, a is an integer. Then the first three integrals are:

$$\int_0^\infty e^{-x}\, dx = 0! = 1 \quad ,$$

$$\int_0^\infty x e^{-x}\, dx = 1! = 1 \quad ,$$

$$\int_0^\infty x^2 e^{-x}\, dx = 2! = 2 \quad , \quad \text{etc.} \tag{E.2}$$

But there appear also the integrals for $a = \pm 0.5$:

$$\int_0^\infty \sqrt{x} e^{-x}\, dx = \left(\frac{1}{2}\right)! = \frac{1}{2}\sqrt{\pi}$$

and

$$\int_0^\infty \frac{1}{\sqrt{x}} e^{-x}\, dx = \left(-\frac{1}{2}\right)! = \sqrt{\pi} \quad . \tag{E.3}$$

An important factorial to know is $\left(\frac{1}{2}\right)! = \frac{1}{2}\sqrt{\pi}$. The recursion formula that defines the factorial

$$a! = a(a-1)! \tag{E.4}$$

allows us to obtain readily the other factorials for $a = n + \frac{1}{2}$ with n being an integer. For example, for $a = \frac{1}{2}$, one obtains from Eq. (E.4):

$$\left(\frac{1}{2}\right)! = \frac{1}{2}\left(-\frac{1}{2}\right)!, \quad \text{thus} \quad \left(-\frac{1}{2}\right)! = 2\left(\frac{1}{2}\right)! \quad . \tag{E.5}$$

For $a = \frac{3}{2}$ follows:

$$\left(\frac{3}{2}\right)! = \frac{3}{2}\left(\frac{1}{2}\right)! \quad , \quad \text{etc.} \tag{E.6}$$

INDEX

ACKNOWLEDGMENTS

Figure 6-6 is from *The Physical Theory of Neutron Chain Reactors* by Alvin M. Weinberg and Eugene P. Wigner. Copyright © 1958 by The University of Chicago Press. Reprinted by permission of The University of Chicago Press.

Figure 7-4 is from *Nuclear Reactor Theory* by George I. Bell and Samuel Glasstone. Copyright © 1970 by Litton Education Publishing, Inc. Reprinted by permission of Van Nostrand Reinhold Company.